Relativity, Gravitation and Cosmology

T0177365

This textbook provides an introduction to general relativity and its applications in astrophysics and cosmology that is suitable for advanced undergraduates with a background in maths, physics and astronomy. The book begins with two chapters devoted to special relativity that introduce spacetime and discuss its implications for the laws of mechanics and electromagnetism. These are followed by chapters that survey the mathematical and physical concepts that underlie general relativity, including the idea of a tensor, the principle of equivalence and the Einstein field equations. Subsequent chapters describe the Schwarzschild solution of the field equations, its use in the study of black holes, and a range of tests of general relativity. The book concludes with a chapter on relativistic cosmology that stresses fundamental principles, Robertson-Walker spacetime and the use of the Friedmann equations in the analysis of cosmic evolution. Produced by academics drawing on decades of Open University experience in supported open learning, the book is completely self-contained with numerous worked examples and exercises (with full solutions provided), and illustrated in full-colour throughout. Designed to be worked through sequentially by a self-guided student, it also includes clearly identified key facts and equations as well as informative chapter summaries and an Appendix of useful data.

Robert J. A. Lambourne is Professor of Educational Physics in the Department of Physical Sciences, The Open University. In 2002 he was awarded the Bragg medal and prize of the Institute of Physics for his contributions to physics education, and was made a National Teaching Fellow in 2006.

Relativity, Gravitation and Cosmology

Author:

Robert J. A. Lambourne

Consultant authors

Chapters 1 and 2: Jim Hague

Chapters 3, 4 and 7: Derek Capper

Chapters 5, 6 and 8: Aiden Droogan

The Open University

CAMBRIDGE
UNIVERSITY PRESS

CAMBRIDGE UNIVERSITY PRESS

Cambridge, New York, Melbourne, Madrid, Cape Town, Singapore, São Paulo, Delhi, Dubai, Tokyo

Cambridge University Press
The Edinburgh Building, Cambridge CB2 8RU, UK

In association with THE OPEN UNIVERSITY

The Open University, Walton Hall, Milton Keynes MK7 6AA, UK

Published in the United States of America by Cambridge University Press, New York.

www.cambridge.org
Information on this title: www.cambridge.org/9780521131384

First published 2010.

Copyright © The Open University 2010.

Edited and designed by The Open University.

Typeset by The Open University.

Printed and bound in the United Kingdom by Page Bros, Norwich.

This book forms part of an Open University course S383 *The Relativistic Universe*. Details of this and other Open University courses can be obtained from the Student Registration and Enquiry Service, The Open University, PO Box 197, Milton Keynes MK7 6BJ, United Kingdom: tel. +44 (0)845 300 60 90, email general-enquiries@open.ac.uk

http://www.open.ac.uk

British Library Cataloguing in Publication Data available on request.

Library of Congress Cataloguing in Publication Data available on request.

ISBN   978-0-521-76119-2 Hardback
ISBN   978-0-521-13138-4 Paperback

Additional resources for this publication at www.cambridge.org/9780521131384

1.2

The Open University has had Woodland Carbon Code Pending Issuance Units assigned from Doddington North forest creation project (IHS ID103/26819) that will, as the trees grow, compensate for the greenhouse gas emissions from the manufacture of the paper in S383 Book 1. More information can be found at https://www.woodlandcarboncode.org.uk/

RELATIVITY, GRAVITATION AND COSMOLOGY

Contents

Contents

Introduction

On the cosmic scale, gravitation dominates the universe. Nuclear and electromagnetic forces account for the detailed processes that allow stars to shine and astronomers to see them. But it is gravitation that shapes the universe, determining the geometry of space and time and thus the large-scale distribution of galaxies. Providing insight into gravitation – its effects, its nature and its causes – is therefore rightly seen as one of the most important goals of physics and astronomy.

Through more than a thousand years of human history the common explanation of gravitation was based on the Aristotelian belief that objects had a natural place in an Earth-centred universe that they would seek out if free to do so. For about two and a half centuries the Newtonian idea of gravity as a force held sway. Then, in the twentieth century, came Einstein's conception of gravity as a manifestation of spacetime curvature. It is this latter view that is the main concern of this book.

The story of Einsteinian gravitation begins with a failure. Einstein's theory of special relativity, published in 1905 while he was working as a clerk in the Swiss Patent Office in Bern, marked an enormous step forward in theoretical physics and soon brought him academic recognition and personal fame. However, it also showed that the Newtonian idea of a gravitational force was inconsistent with the relativistic approach and that a new theory of gravitation was required. Ten years later, Einstein's general theory of relativity met that need, highlighting the important role of geometry in accounting for gravitational phenomena and leading on to concepts such as black holes and gravitational waves. Within a year and a half of its completion, the new theory was providing the basis for a novel approach to cosmology – the science of the universe – that would soon have to take account of the astronomy of galaxies and the physics of cosmic expansion. The change in thinking demanded by relativity was radical and profound. Its mastery is one of the great challenges and greatest delights of any serious study of physical science.

This book begins with two chapters devoted to special relativity. These are followed by a mainly mathematical chapter that provides the background in geometry that is needed to appreciate Einstein's subsequent development of the theory. Chapter 4 examines the basic principles and assumptions of general relativity – Einstein's theory of gravity – while Chapters 5 and 6 apply the theory to an isolated spherical body and then extend that analysis to non-rotating and rotating black holes. Chapter 7 concerns the testing of general relativity, including the use of astronomical observations and gravitational waves. Finally, Chapter 8 examines modern relativistic cosmology, setting the scene for further and ongoing studies of observational cosmology.

The text before you is the result of a collaborative effort involving a team of authors and editors working as part of the broader effort to produce the Open University course S383 *The Relativistic Universe*. Details of the team's membership and responsibilities are listed elsewhere but it is appropriate to acknowledge here the particular contributions of Jim Hague regarding Chapters 1 and 2, Derek Capper concerning Chapters 3, 4 and 7, and Aiden Droogan in relation to Chapters 5, 6 and 8. Robert Lambourne was responsible for planning and producing the final unified text which benefited greatly from the input of the S383 Course Team Chair, Andrew Norton, and the attention of production editor

Figure 1 Albert Einstein (1879–1955) depicted during the time that he worked at the Patent Office in Bern. While there, he published a series of papers relating to special relativity, quantum physics and statistical mechanics. He was awarded the Nobel Prize for Physics in 1921, mainly for his work on the photoelectric effect.

Peter Twomey. The whole team drew heavily on the work and wisdom of an earlier Open University Course Team that was responsible for the production of the course S357 *Space, Time and Cosmology*.

A major aim for this book is to allow upper-level undergraduate students to develop the skills and confidence needed to pursue the independent study of the many more comprehensive texts that are now available to students of relativity, gravitation and cosmology. To facilitate this the current text has largely adopted the notation used in the outstanding book by Hobson et al.

General Relativity : *An Introduction for Physicists*, M. P. Hobson, G. Efstathiou and A. N. Lasenby, Cambridge University Press, 2006.

Other books that provide valuable further reading are (roughly in order of increasing mathematical demand):

An Introduction to Modern Cosmology, A. Liddle, Wiley, 1999.
Relativity, Gravitation and Cosmology : *A Basic Introduction*, T-P. Cheng, Oxford University Press: 2005.
Introducing Einstein's Relativity, R. d'Inverno, Oxford University Press, 1992.
Relativity : *Special, General and Cosmological*, W. Rindler, Oxford University Press, 2001.
Cosmology, S. Weinberg, Cambridge University Press, 2008.

Two useful sources of reprints of original papers of historical significance are:

The Principle of Relativity, A. Einstein et al., Dover, New York, 1952.
Cosmological Constants, edited by J. Bernstein and G. Feinberg, Columbia University Press, 1986.

Those wishing to undertake background reading in astronomy, physics and mathematics to support their study of this book or of any of the others listed above might find the following particularly helpful:

An Introduction to Galaxies and Cosmology, edited by M. H. Jones and R. J. A. Lambourne, Cambridge University Press, 2003.
The seven volumes in the series
The Physical World, edited by R. J. A. Lambourne, A. J. Norton et al., Institute of Physics Publishing, 2000.
(Go to www.physicalworld.org for further details.)
The paired volumes
Basic Mathematics for the Physical Sciences, edited by R. J. A. Lambourne and M. H. Tinker, Wiley, 2000.
Further Mathematics for the Physical Sciences, edited by M. H. Tinker and R. J. A. Lambourne, Wiley, 2000.

Numbered exercises appear throughout this book. Complete solutions to these exercises can be found at the back of the book. There are a number of lengthy worked examples; these are highlighted by a blue background. There are also several shorter in-text questions that are immediately followed by their answers. These may be treated as questions or examples. The questions are indented and indicated by a filled circle; their answers are also indented and shown by an open circle.

Chapter 1 Special relativity and spacetime

Introduction

In two seminal papers in 1861 and 1864, and in his treatise of 1873, James Clerk Maxwell (Figure 1.1), Scottish physicist and genius, wrote down his revolutionary unified theory of electricity and magnetism, a theory that is now summarized in the equations that bear his name. One of the deep results of the theory introduced by Maxwell was the prediction that wave-like excitations of combined electric and magnetic fields would travel through a vacuum with the same speed as light. It was soon widely accepted that light itself was an electromagnetic disturbance propagating through space, thus unifying electricity and magnetism with optics.

The fundamental work of Maxwell opened the way for an understanding of the universe at a much deeper level. Maxwell himself, in common with many scientists of the nineteenth century, believed in an all-pervading medium called the **ether**, through which electromagnetic disturbances travelled, just as ocean waves travelled through water. Maxwell's theory predicted that light travels with the same speed in all directions, so it was generally assumed that the theory predicted the results of measurements made using equipment that was at rest with respect to the ether. Since the Earth was expected to move through the ether as it orbited the Sun, measurements made in terrestrial laboratories were expected to show that light actually travelled with different speeds in different directions, allowing the speed of the Earth's movement through the ether to be determined. However, experiments, most notably by A. A. Michelson and E. W. Morley in 1887, failed to detect any variations in the measured speed of light. This led some to suspect that measurements of the speed of light in a vacuum would always yield the same result irrespective of the motion of the measuring equipment. Explaining how this could be the case was a major challenge that prompted ingenious proposals from mathematicians and physicists such as Henri Poincaré, George Fitzgerald and Hendrik Lorentz. However, it was the young Albert Einstein who first put forward a coherent and comprehensive solution in his 1905 paper 'On the electrodynamics of moving bodies', which introduced the **special theory of relativity**. With the benefit of hindsight, we now realize that Maxwell had unintentionally formulated the first major theory that was consistent with special relativity, a revolutionary new way of thinking about space and time.

This chapter reviews the implications of special relativity theory for the understanding of space and time. The narrative covers the fundamentals of the theory, concentrating on some of the major differences between our intuition about space and time and the predictions of special relativity. By the end of this chapter, you should have a broad conceptual understanding of special relativity, and be able to derive its basic equations, *the Lorentz transformations*, from the postulates of special relativity. You will understand how to use *events* and *intervals* to describe properties of space and time far from gravitating bodies. You will also have been introduced to Minkowski spacetime, a four-dimensional fusion of space and time that provides the natural setting for discussions of special relativity.

Figure 1.1 James Clerk Maxwell (1831–1879) developed a theory of electromagnetism that was already compatible with special relativity theory several decades before Einstein and others developed the theory. He is also famous for major contributions to statistical physics and the invention of colour photography.

1.1   Basic concepts of special relativity

1.1.1   Events, frames of reference and observers

When dealing with special relativity it is important to use language very precisely in order to avoid confusion and error. Fundamental to the precise description of physical phenomena is the concept of an *event*, the spacetime analogue of a point in space or an instant in time.

> **Events**
>
> An **event** is an instantaneous occurrence at a specific point in space.

An exploding firecracker or a small light that flashes once are good approximations to events, since each happens at a definite time and at a definite position.

To know when and where an event happened, we need to assign some coordinates to it: a time coordinate t and an ordered set of spatial coordinates such as the *Cartesian coordinates* (x, y, z), though we might equally well use *spherical coordinates* (r, θ, ϕ) or any other suitable set. The important point is that we should be able to assign a unique set of clearly defined coordinates to any event. This leads us to our second important concept, a *frame of reference*.

> **Frames of reference**
>
> A **frame of reference** is a system for assigning coordinates to events. It consists of a system of synchronized clocks that allows a unique value of the time to be assigned to any event, and a system of spatial coordinates that allows a unique position to be assigned to any event.

In much of what follows we shall make use of a Cartesian coordinate system with axes labelled x, y and z. The precise specification of such a system involves selecting an origin and specifying the orientation of the three orthogonal axes that meet at the origin. As far as the system of clocks is concerned, you can imagine that space is filled with identical synchronized clocks all ticking together (we shall need to say more about how this might be achieved later). When using a particular frame of reference, the time assigned to an event is the time shown on the clock at the site of the event when the event happens. It is particularly important to note that the time of an event is not the time at which the event is seen at some far off point — it is the time at the event itself that matters.

Reference frames are often represented by the letter S. Figure 1.2 provides what we hope is a memorable illustration of the basic idea, in this case with just two spatial dimensions. This might be called the frame S_{gnome}.

Among all the frames of reference that might be imagined, there is a class of frames that is particularly important in special relativity. This is the class of *inertial frames*. An inertial frame of reference is one in which a body that is not subject to any net force maintains a constant velocity. This is *Newton's first law of motion*, so we can say the following.

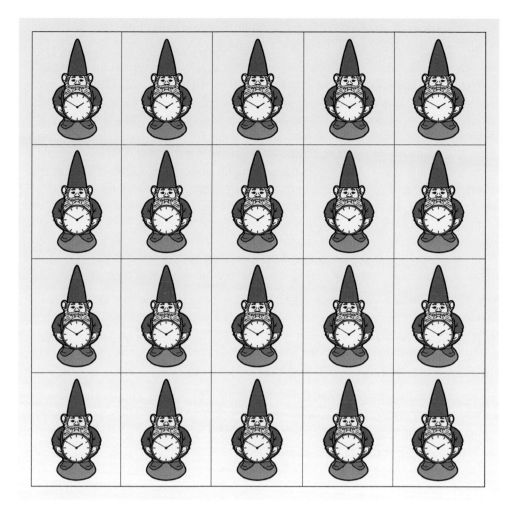

Figure 1.2 A jocular representation of a frame of reference in two spatial dimensions. Gnomes pervade all of space and time. Each gnome has a perfectly reliable clock. When an event occurs, the gnome nearest to the event communicates the time t and location (x, y) of the event to the observer.

Inertial frames of reference

An **inertial frame of reference** is a frame of reference in which Newton's first law of motion holds true.

Any frame that moves with constant velocity relative to an inertial frame will also be an inertial frame. So, if you can identify or establish one inertial frame, then you can find an infinite number of such frames, each having a constant velocity relative to any of the others. Any frame that accelerates relative to an inertial frame cannot be an inertial frame. Since rotation involves changing velocity, any frame that rotates relative to an inertial frame is also disqualified from being inertial.

One other concept is needed to complete the basic vocabulary of special relativity. This is the idea of an *observer*.

Observers

An **observer** is an individual dedicated to using a particular frame of reference for recording events.

We might speak of an observer O using frame S, or a different observer O′ (read as 'O-prime') using frame S′ (read as 'S-prime').

Though you may think of an observer as a person, just like you or me, at rest in their chosen frame of reference, it is important to realize that an observer's location is of no importance for reporting the coordinates of events in special relativity. The position that an observer assigns to an event is the place where it happened. The time that an observer assigns is the time that would be shown on a clock at the site of the event when the event actually happened, and where the clock concerned is part of the network of synchronized clocks always used in that observer's frame of reference. An observer might see the explosion of a distant star tonight, but would report the time of the explosion as the time long ago when the explosion actually occurred, not the time at which the light from the explosion reached the observer's location. To this extent, 'seeing' and 'observing' are very different processes. It is best to avoid phrases such as 'an observer sees ...' unless that is what you really mean. An observer measures and observes.

Any observer who uses an inertial frame of reference is said to be an **inertial observer**. Einstein's special theory of relativity is mainly concerned with observations made by inertial observers. That's why it's called *special* relativity — the term 'special' is used in the sense of 'restricted' or 'limited'. We shall not really get away from this limitation until we turn to general relativity in Chapter 4.

Exercise 1.1   For many purposes, a frame of reference fixed in a laboratory on the Earth provides a good approximation to an inertial frame. However, such a frame is not really an inertial frame. How might its true, non-inertial, nature be revealed experimentally, at least in principle?   ■

1.1.2   The postulates of special relativity

Physicists generally treat the laws of physics as though they hold true everywhere and at all times. There is some evidence to support such an assumption, though it is recognized as a hypothesis that might fail under extreme conditions. To the extent that the assumption is true, it does not matter where or when observations are made to test the laws of physics since the time and place of a test of fundamental laws should not have any influence on its outcome.

Where and when laws are tested might not influence the outcome, but what about motion? We know that inertial and non-inertial observers will not agree about Newton's first law. But what about different inertial observers in uniform relative motion where one observer moves at constant velocity with respect to the other? A pair of inertial observers *would* agree about Newton's first law; might they also agree about other laws of physics?

It has long been thought that they would at least agree about the laws of mechanics. Even before Newton's laws were formulated, the great Italian physicist Galileo Galilei (1564–1642) pointed out that a traveller on a smoothly moving boat had exactly the same experiences as someone standing on the shore. A ball game could be played on a uniformly moving ship just as well as it could be played on shore. To the early investigators, uniform motion alone appeared to have no detectable consequences as far as the laws of mechanics were concerned. An observer shut up in a sealed box that prevented any observation of the outside world would be unable to perform any mechanics experiment that would reveal the uniform velocity of the box, even though any acceleration could be easily detected. (We are all familiar with the feeling of being pressed back in our seats when a train or car accelerates forward.) These notions provided the basis for the first theory of relativity, which is now known as **Galilean relativity** in honour of Galileo's original insight. This theory of relativity assumes that all inertial observers will agree about the laws of Newtonian mechanics.

Einstein believed that inertial observers would agree about the laws of physics quite generally, not just in mechanics. But he was not convinced that Galilean relativity was correct, which brought Newtonian mechanics into question. The only statement that he wanted to presume as a law of physics was that all inertial observers agreed about the speed of light in a vacuum. Starting from this minimal assumption, Einstein was led to a new theory of relativity that was markedly different from Galilean relativity. The new theory, the special theory of relativity, supported Maxwell's laws of electromagnetism but caused the laws of mechanics to be substantially rewritten. It also provided extraordinary new insights into space and time that will occupy us for the rest of this chapter.

Einstein based the special theory of relativity on two *postulates*, that is, two statements that he believed to be true on the basis of the physics that he knew. The first postulate is often referred to as the **principle of relativity**.

The first postulate of special relativity

The laws of physics can be written in the same form in all inertial frames.

This is a bold extension of the earlier belief that observers would agree about the laws of mechanics, but it is not at first sight exceptionally outrageous. It will, however, have profound consequences.

The second postulate is the one that gives primacy to the behaviour of light, a subject that was already known as a source of difficulty. This postulate is sometimes referred to as the **principle of the constancy of the speed of light**.

The second postulate of special relativity

The speed of light in a vacuum has the same constant value, $c = 3 \times 10^8 \, \mathrm{m \, s^{-1}}$, in all inertial frames.

This postulate certainly accounts for Michelson and Morley's failure to detect any variations in the speed of light, but at first sight it still seems crazy. Our experience with everyday objects moving at speeds that are small compared with the speed of light tells us that if someone in a car that is travelling forward at speed v throws something forward at speed w relative to the car, then, according to an observer standing on the roadside, the thrown object will move forward with speed $v + w$. But the second postulate tells us that if the traveller in the car turns on a torch, effectively throwing forward some light moving at speed c relative to the car, then the roadside observer will find that the light travels at speed c, not the $v + c$ that might have been expected. Einstein realized that for this to be true, space and time must behave in previously unexpected ways.

The second postulate has another important consequence. Since all observers agree about the speed of light, it is possible to use light signals (or any other electromagnetic signal that travels at the speed of light) to ensure that the network of clocks we imagine each observer to be using is properly synchronized. We shall not go into the details of how this is done, but it is worth pointing out that if an observer sent a radar signal (which travels at the speed of light) so that it arrived at an event just as the event was happening and was immediately reflected back, then the time of the event would be midway between the times of transmission and reception of the radar signal. Similarly, the distance to the event would be given by half the round trip travel time of the signal, multiplied by the speed of light.

1.2   Coordinate transformations

A **theory of relativity** concerns the relationship between observations made by observers in relative motion. In the case of special relativity, the observers will be inertial observers in uniform relative motion, and their most fundamental observations will be the time and space coordinates of events.

For the sake of definiteness and simplicity, we shall consider two inertial observers O and O′ whose respective frames of reference, S and S′, are arranged in the following **standard configuration** (see Figure 1.3):

1. The origin of frame S′ moves along the x-axis of frame S, in the direction of increasing values of x, with constant velocity V as measured in S.

2. The x-, y- and z-axes of frame S are always parallel to the corresponding x'-, y'- and z'-axes of frame S′.

3. The event at which the origins of S and S′ coincide occurs at time $t = 0$ in frame S and at time $t' = 0$ in frame S′.

We shall make extensive use of 'standard configuration' in what follows. The arrangement does not entail any real loss of generality since any pair of inertial frames in uniform relative motion can be placed in standard configuration by choosing to reorientate the coordinate axes in an appropriate way, shifting the origin, and resetting the clocks appropriately.

In general, the observers using the frames S and S′ will not agree about the coordinates of an event, but since each observer is using a well-defined frame of reference, there must exist a set of equations relating the coordinates (t, x, y, z) assigned to a particular event by observer O, to the coordinates (t', x', y', z') assigned to the same event by observer O′. The set of equations that performs the task of relating the two sets of coordinates is called a **coordinate transformation**. This section considers first the *Galilean transformations* that provide the basis of Galilean relativity, and then the *Lorentz transformations* on which Einstein's special relativity is based.

1.2.1   The Galilean transformations

Before the introduction of special relativity, most physicists would have said that the coordinate transformation between S and S′ was 'obvious', and they would have written down the following **Galilean transformations**:

$$t' = t, \tag{1.1}$$
$$x' = x - Vt, \tag{1.2}$$
$$y' = y, \tag{1.3}$$
$$z' = z, \tag{1.4}$$

where $V = |\boldsymbol{V}|$ is the relative speed of S′ with respect to S.

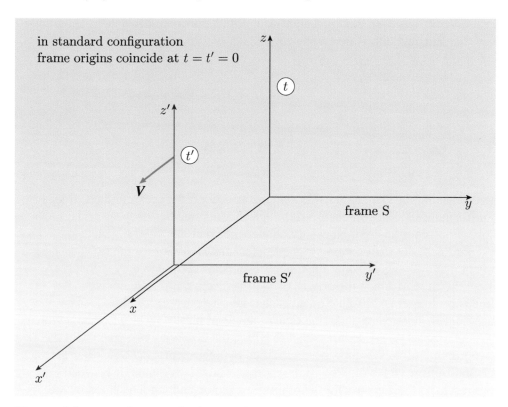

in standard configuration
frame origins coincide at $t = t' = 0$

Figure 1.3 Two frames of reference in *standard configuration*. Note that the speed V is measured in frame S.

To justify this result, it might have been argued that since the observers agree about the time of the event at which the origins coincide (see point 3 in the definition of standard configuration), they must also agree about the times of all other events. Further, since at time t the origin of S′ will have travelled a distance Vt along the x-axis of frame S, it must be the case that any event that occurs at time t with position coordinate x in frame S must occur at $x' = x - Vt$ in frame S′, while the values of y and z will be unaffected by the motion. However, as Einstein realized, such an argument contains many assumptions about the behaviour of time and space, and those assumptions might not be correct. For example, Equation 1.1 implies that time is in some sense *absolute*, by which we mean that the time interval between any two events is the same for all observers. Newton certainly believed this to be the case, but without supporting evidence it was really nothing more than a plausible assumption. It was intuitively appealing, but it was fundamentally untested.

1.2.2   The Lorentz transformations

Rather than rely on intuition and run the risk of making unjustified assumptions, Einstein chose to set out his two postulates and use them to deduce the appropriate coordinate transformation between S and S′. A derivation will be given later, but before that let's examine the result that Einstein found. The equations that he derived had already been obtained by the Dutch physicist Hendrik Lorentz (Figure 1.4) in the course of his own investigations into light and electromagnetism. For that reason, they are known as the **Lorentz transformations** even though Lorentz did not interpret or utilize them in the same way that Einstein did. Here are the equations:

$$t' = \frac{t - Vx/c^2}{\sqrt{1 - V^2/c^2}},$$
$$x' = \frac{x - Vt}{\sqrt{1 - V^2/c^2}},$$
$$y' = y,$$
$$z' = z.$$

Figure 1.4   Hendrik Lorentz (1853–1928) wrote down the Lorentz transformations in 1904. He won the 1902 Nobel Prize for Physics for work on electromagnetism, and was greatly respected by Einstein.

It is clear that the Lorentz transformations are very different from the Galilean transformations. They indicate a thorough mixing together of space and time, since the t'-coordinate of an event now depends on both t and x, just as the x'-coordinate does. According to the Lorentz transformations, the two observers do not generally agree about the time of events, even though they still agree about the time at which the origins of their respective frames coincided. So, time is no longer an absolute quantity that all observers agree about. To be meaningful, statements about the time of an event must now be associated with a particular observer. Also, the extent to which the observers disagree about the x-coordinate of an event has been modified by a factor of $1/\sqrt{1 - V^2/c^2}$. In fact, this multiplicative factor is so common in special relativity that it is usually referred to as the **Lorentz factor** or **gamma factor** and is represented by the symbol $\gamma(V)$, emphasizing that its value depends on the relative speed V of the two frames. Using this factor, the Lorentz transformations can be written in the following compact form.

The Lorentz transformations

$$t' = \gamma(V)(t - Vx/c^2), \tag{1.5}$$
$$x' = \gamma(V)(x - Vt), \tag{1.6}$$
$$y' = y, \tag{1.7}$$
$$z' = z, \tag{1.8}$$

where

$$\gamma(V) = \frac{1}{\sqrt{1 - V^2/c^2}}. \tag{1.9}$$

Figure 1.5 shows how the Lorentz factor grows as the relative speed V of the two frames increases. For speeds that are small compared with the speed of light, $\gamma(V) \approx 1$, and the Lorentz transformations approximate the Galilean transformations provided that x is not too large. As the relative speed of the two frames approaches the speed of light, however, the Lorentz factor grows rapidly and so do the discrepancies between the Galilean and Lorentz transformations.

Exercise 1.2   Compute the Lorentz factor $\gamma(V)$ when the relative speed V is (a) 10% of the speed of light, and (b) 90% of the speed of light.   ∎

The Lorentz transformations are so important in special relativity that you will see them written in many different ways. They are often presented in matrix form, as

$$\begin{pmatrix} ct' \\ x' \\ y' \\ z' \end{pmatrix} = \begin{pmatrix} \gamma(V) & -\gamma(V)V/c & 0 & 0 \\ -\gamma(V)V/c & \gamma(V) & 0 & 0 \\ 0 & 0 & 1 & 0 \\ 0 & 0 & 0 & 1 \end{pmatrix} \begin{pmatrix} ct \\ x \\ y \\ z \end{pmatrix}. \tag{1.10}$$

You should convince yourself that this matrix multiplication gives equations equivalent to the Lorentz transformations. (The equation for transforming the time coordinate is multiplied by c.) We can also represent this relationship by the equation

$$[x'^{\mu}] = [\Lambda^{\mu}{}_{\nu}][x^{\nu}], \tag{1.11}$$

where we use the symbol $[x^{\mu}]$ to represent the column vector with components $(x^0, x^1, x^2, x^3) = (ct, x, y, z)$, and the symbol $[\Lambda^{\mu}{}_{\nu}]$ to represent the **Lorentz transformation matrix**

$$[\Lambda^{\mu}{}_{\nu}] \equiv \begin{pmatrix} \Lambda^0{}_0 & \Lambda^0{}_1 & \Lambda^0{}_2 & \Lambda^0{}_3 \\ \Lambda^1{}_0 & \Lambda^1{}_1 & \Lambda^1{}_2 & \Lambda^1{}_3 \\ \Lambda^2{}_0 & \Lambda^2{}_1 & \Lambda^2{}_2 & \Lambda^2{}_3 \\ \Lambda^3{}_0 & \Lambda^3{}_1 & \Lambda^3{}_2 & \Lambda^3{}_3 \end{pmatrix}$$

$$= \begin{pmatrix} \gamma(V) & -\gamma(V)V/c & 0 & 0 \\ -\gamma(V)V/c & \gamma(V) & 0 & 0 \\ 0 & 0 & 1 & 0 \\ 0 & 0 & 0 & 1 \end{pmatrix}. \tag{1.12}$$

At this stage, when dealing with an individual matrix element $\Lambda^{\mu}{}_{\nu}$, you can simply regard the first index as indicating the row to which it belongs and the second index as indicating the column. It then makes sense that each of the elements x^{μ} in the column vector $[x^{\mu}]$ should have a raised index. However, as you will see later, in the context of relativity the positioning of these indices actually has a much greater significance.

The quantity $[x^{\mu}]$ is sometimes called the **four-position** since its four components (ct, x, y, z) describe the position of the event in time and space. Note that by using ct to convey the time information, rather than just t, all four components of the four-position are measured in units of distance. Also note that the Greek indices μ and ν take the values 0 to 3. It is conventional in special and general relativity to start the indexing of the vectors and matrices from zero, where $x^0 = ct$. This is because the time coordinate has special properties.

Using the individual components of the four-position, another way of writing the Lorentz transformation is in terms of summations:

$$x'^{\mu} = \sum_{\nu=0}^{3} \Lambda^{\mu}{}_{\nu} x^{\nu} \quad (\mu = 0, 1, 2, 3). \tag{1.13}$$

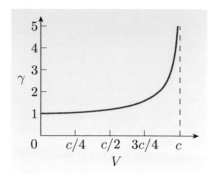

Figure 1.5   Plot of the Lorentz factor, $\gamma(V) = 1/\sqrt{1 - V^2/c^2}$. The factor is close to 1 for speeds much smaller than the speed of light, but increases rapidly as V approaches c. Note that $\gamma > 1$ for all values of V.

This one line really represents four different equations, one for each value of μ. When an index is used in this way, it is said to be a **free index**, since we are free to give it any value between 0 and 3, and whatever choice we make indicates a different equation. The index ν that appears in the summation is not free, since whatever value we choose for μ, we are required to sum over all possible values of ν to obtain the final equation. This means that we could replace all appearances of ν by some other index, α say, without actually changing anything. An index that is summed over in this way is said to be a **dummy index**.

Familiarity with the summation form of the Lorentz transformations is particularly useful when beginning the discussion of general relativity; you will meet many such sums. Before moving on, you should convince yourself that you can easily switch between the use of separate equations, matrices (including the use of four-positions) and summations when representing Lorentz transformations.

Given the coordinates of an event in frame S, the Lorentz transformations tell us the coordinates of that same event as observed in frame S$'$. It is equally important that there is some way to transform coordinates of an event in frame S$'$ back into the coordinates in frame S. The transformations that perform this task are known as the **inverse Lorentz transformations**.

The inverse Lorentz transformations

$$t = \gamma(V)(t' + Vx'/c^2), \tag{1.14}$$
$$x = \gamma(V)(x' + Vt'), \tag{1.15}$$
$$y = y', \tag{1.16}$$
$$z = z'. \tag{1.17}$$

Note that the only difference between the Lorentz transformations and their inverses is that all the primed and unprimed quantities have been interchanged, and the relative speed of the two frames, V, has been replaced by the quantity $-V$. (This changes the transformations but not the value of the Lorentz factor, which depends only on V^2, so can still be written as $\gamma(V)$.) This relationship between the transformations is expected, since frame S$'$ is moving with speed V in the positive x-direction as measured in frame S, while frame S is moving with speed V in the negative x'-direction as measured in frame S$'$. You should confirm that performing a Lorentz transformation and its inverse transformation in succession really does lead back to the original coordinates, i.e. $(ct, x, y, z) \rightarrow (ct', x', y', z') \rightarrow (ct, x, y, z)$.

● An event occurs at coordinates $(ct = 3\,\text{m},\ x = 4\,\text{m},\ y = 0,\ z = 0)$ in frame S according to an observer O. What are the coordinates of the same event in frame S$'$ according to an observer O$'$, moving with speed $V = 3c/4$ in the positive x-direction, as measured in S?

○ First, the Lorentz factor $\gamma(V)$ should be computed:

$$\gamma(3c/4) = 1/\sqrt{1 - 3^2/4^2} = 4/\sqrt{7}.$$

The new coordinates are then given by the Lorentz transformations:

$$ct' = c\gamma(3c/4)(t - 3x/4c) = (4/\sqrt{7})(3\,\text{m} - 3c \times 4\,\text{m}/4c) = 0\,\text{m},$$

$$x' = \gamma(3c/4)(x - 3tc/4) = (4/\sqrt{7})(4\,\text{m} - 3 \times 3\,\text{m}/4) = \sqrt{7}\,\text{m},$$

$$y' = y = 0\,\text{m},$$

$$z' = z = 0\,\text{m}.$$

Exercise 1.3   The matrix equation

$$\begin{pmatrix} ct' \\ x' \end{pmatrix} = \begin{pmatrix} \gamma(V) & -\gamma(V)V/c \\ -\gamma(V)V/c & \gamma(V) \end{pmatrix} \begin{pmatrix} ct \\ x \end{pmatrix}$$

can be inverted to determine the coordinates (ct, x) in terms of (ct', x'). Show that inverting the 2×2 matrix leads to the inverse Lorentz transformations in Equations 1.14 and 1.15. ■

1.2.3   A derivation of the Lorentz transformations

This subsection presents a derivation of the Lorentz transformations that relates the coordinates of an event in two inertial frames, S and S', that are in standard configuration. It mainly ignores the y- and z-coordinates and just considers the transformation of the t- and x-coordinates of an event. A general transformation relating the coordinates (t', x') of an event in frame S' to the coordinates (t, x) of the same event in frame S may be written as

$$t' = a_0 + a_1 t + a_2 x + a_3 t^2 + a_4 x^2 + \cdots, \tag{1.18}$$

$$x' = b_0 + b_1 x + b_2 t + b_3 x^2 + b_4 t^2 + \cdots, \tag{1.19}$$

where the dots represent additional terms involving higher powers of x or t.

Now, we know from the definition of standard configuration that the event marking the coincidence of the origins of frames S and S' has the coordinates $(t, x) = (0, 0)$ in S and $(t', x') = (0, 0)$ in S'. It follows from Equations 1.18 and 1.19 that the constants a_0 and b_0 are zero.

The transformations in Equations 1.18 and 1.19 can be further simplified by the requirement that the observers are using inertial frames of reference. Since Newton's first law must hold in all inertial frames of reference, it is necessary that an object not accelerating in one set of coordinates is also not accelerating in the other set of coordinates. If the higher-order terms in x and t were not zero, then an object observed to have no acceleration in S (such as a spaceship with its thrusters off moving on the line $x = vt$, shown in the upper part of Figure 1.6) would be observed to accelerate in terms of x' and t' (i.e. $x' \neq v't'$, as indicated in the lower part of Figure 1.6). Observer O would report no force on the spaceship, while observer O' would report some unknown force acting on it. In this way, the two observers would register different laws of physics, violating the first postulate of special relativity. The higher-order terms are therefore inconsistent with the required physics and must be removed, leaving only a linear transformation.

So we expect the special relativistic coordinate transformation between two frames in standard configuration to be represented by linear equations of the form

$$t' = a_1 t + a_2 x, \tag{1.20}$$

$$x' = b_1 x + b_2 t. \tag{1.21}$$

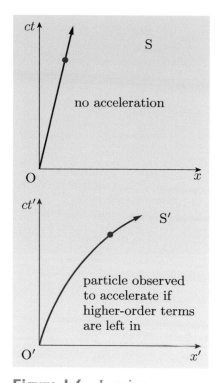

Figure 1.6   Leaving higher-order terms in the coordinate transformations would cause uniform motion in one inertial frame S to be observed as accelerated motion in the other inertial frame S'. These diagrams, in which the vertical axis represents time multiplied by the speed of light, show that if the t^2 terms were left in the transformations, then motion with no acceleration in frame S would be transformed into motion with non-zero acceleration in frame S'. This would imply change in velocity without force in S', in conflict with Newton's first law.

The remaining task is to determine the coefficients a_1, a_2, b_1 and b_2.

To do this, use is made of known relations between coordinates in both frames of reference. The first step is to use the fact that at any time t, the origin of S' (which is always at $x' = 0$ in S') will be at $x = Vt$ in S. It follows from Equation 1.21 that

$$0 = b_1 Vt + b_2 t,$$

from which we see that

$$b_2 = -b_1 V. \tag{1.22}$$

Dividing Equation 1.21 by Equation 1.20, and using Equation 1.22 to replace b_2 by $-b_1 V$, leads to

$$\frac{x'}{t'} = \frac{b_1 x - b_1 Vt}{a_1 t + a_2 x}. \tag{1.23}$$

Now, as a second step we can use the fact that at any time t', the origin of frame S (which is always at $x = 0$ in S) will be at $x' = -Vt'$ in S'. Substituting these values for x and x' into Equation 1.23 gives

$$\frac{-Vt'}{t'} = \frac{-b_1 Vt}{a_1 t}, \tag{1.24}$$

from which it follows that

$$b_1 = a_1.$$

If we now substitute $a_1 = b_1$ into Equation 1.23 and divide the numerator and denominator on the right-hand side by t, then

$$\frac{x'}{t'} = \frac{b_1 (x/t) - Vb_1}{b_1 + a_2 (x/t)}. \tag{1.25}$$

As a third step, the coefficient a_2 can be found using the principle of the constancy of the speed of light. A pulse of light emitted in the positive x-direction from $(ct = 0,\ x = 0)$ has speed $c = x'/t'$ and also $c = x/t$. Substituting these values into Equation 1.25 gives

$$c = \frac{b_1 c - Vb_1}{b_1 + a_2 c},$$

which can be rearranged to give

$$a_2 = -Vb_1/c^2 = -Va_1/c^2. \tag{1.26}$$

Now that a_2, b_1 and b_2 are known in terms of a_1, the coordinate transformations between the two frames can be written as

$$t' = a_1(t - Vx/c^2), \tag{1.27}$$
$$x' = a_1(x - Vt). \tag{1.28}$$

All that remains for the fourth step is to find an expression for a_1. To do this, we first write down the inverse transformations to Equations 1.27 and 1.28, which are found by exchanging primes and replacing V by $-V$. (We are implicitly assuming that a_1 depends only on some even power of V.) This gives

$$t = a_1(t' + Vx'/c^2), \tag{1.29}$$
$$x = a_1(x' + Vt'). \tag{1.30}$$

Substituting Equations 1.29 and 1.30 into Equation 1.28 gives

$$x' = a_1 \left(a_1(x' + Vt') - Va_1 \left(t' + \frac{V}{c^2}x' \right) \right).$$

The second and third terms involving $a_1 Vt'$ cancel in this expression, leaving an expression in which the x' cancels on both sides:

$$x' = a_1^2 \left(1 - \frac{V^2}{c^2} \right) x'.$$

By rearranging this equation and taking the positive square root, the coefficient a_1 is determined to be

$$a_1 = \frac{1}{\sqrt{1 - V^2/c^2}}. \tag{1.31}$$

Thus a_1 is seen to be the Lorentz factor $\gamma(V)$, which completes the derivation.

Some further arguments allow the Lorentz transformations to be extended to one time and three space dimensions. There can be no y and z contributions to the transformations for t' and x' since the y- and z-axes could be oriented in any of the perpendicular directions without affecting the events on the x-axis. Similarly, there can be no contributions to the transformations for y' and z' from any other coordinates, as space would become distorted in a non-symmetric manner.

1.2.4 Intervals and their transformation rules

Knowing how the coordinates of an event transform from one frame to another, it is relatively simple to determine how the coordinate intervals that separate pairs of events transform. As you will see in the next section, the rules for transforming intervals are often very useful.

Intervals

An **interval** between two events, measured along a specified axis in a given frame of reference, is the difference in the corresponding coordinates of the two events.

To develop transformation rules for intervals, consider the Lorentz transformations for the coordinates of two events labelled 1 and 2:

$$t_1' = \gamma(V)(t_1 - Vx_1/c^2), \qquad x_1' = \gamma(V)(x_1 - Vt_1),$$
$$y_1' = y_1, \qquad\qquad\qquad z_1' = z_1$$
$$t_2' = \gamma(V)(t_2 - Vx_2/c^2), \qquad x_2' = \gamma(V)(x_2 - Vt_2),$$
$$y_2' = y_2, \qquad\qquad\qquad z_2' = z_2.$$

Subtracting the transformation equation for t_1' from that for t_2', and subtracting the transformation equation for x_1' from that for x_2', and so on, gives the following **transformation rules for intervals**:

$$\Delta t' = \gamma(V)(\Delta t - V\Delta x/c^2), \tag{1.32}$$
$$\Delta x' = \gamma(V)(\Delta x - V\Delta t), \tag{1.33}$$
$$\Delta y' = \Delta y, \tag{1.34}$$
$$\Delta z' = \Delta z, \tag{1.35}$$

where $\Delta t = t_2 - t_1$, $\Delta x = x_2 - x_1$, $\Delta y = y_2 - y_1$ and $\Delta z = z_2 - z_1$ denote the various time and space intervals between the events. The inverse transformations for intervals have the same form, with V replaced by $-V$:

$$\Delta t = \gamma(V)(\Delta t' + V\Delta x'/c^2), \tag{1.36}$$
$$\Delta x = \gamma(V)(\Delta x' + V\Delta t'), \tag{1.37}$$
$$\Delta y = \Delta y', \tag{1.38}$$
$$\Delta z = \Delta z'. \tag{1.39}$$

The transformation rules for intervals are useful because they depend only on coordinate differences and not on the specific locations of events in time or space.

1.3 Consequences of the Lorentz transformations

In this section, some of the extraordinary consequences of the Lorentz transformations will be examined. In particular, we shall consider the findings of different observers regarding the rate at which a clock ticks, the length of a rod and the simultaneity of a pair of events. In each case, the trick for determining how the relevant property transforms between frames of reference is to carefully specify how intuitive concepts such as length or duration should be defined consistently in different frames of reference. This is most easily done by identifying each concept with an appropriate *interval* between two events: 1 and 2. Once this has been achieved, we can determine which intervals are known and then use the interval transformation rules (Equations 1.32–1.35 and 1.36–1.39) to find relationships between them. The rest of this section will give examples of this process.

1.3.1 Time dilation

One of the most celebrated consequences of special relativity is the finding that 'moving clocks run slow'. More precisely, any inertial observer must observe that the clocks used by another inertial observer, in uniform relative motion, will run slow. Since clocks are merely indicators of the passage of time, this is really the assertion that any inertial observer will find that time passes more slowly for any other inertial observer who is in relative motion. Thus, according to special relativity, if you and I are inertial observers, and we are in uniform relative motion, then I can perform measurements that will show that time is passing more slowly for you and, simultaneously, you can perform measurements that will show that time is passing more slowly for me. Both of us will be right because time is a relative quantity, not an absolute one. To show how this effect follows from the Lorentz transformations, it is essential to introduce clear, unambiguous definitions of the time intervals that are to be related.

Rather than deal with ticking clocks, our discussion here will refer to short-lived sub-nuclear particles of the sort routinely studied at CERN and other particle physics laboratories. For the purpose of the discussion, a short-lived particle is considered to be a point-like object that is created at some event, labelled 1, and subsequently decays at some other event, labelled 2. The time interval between these two events, as measured in any particular inertial frame, is the **lifetime** of the particle in that frame. This interval is analogous to the time between successive ticks of a clock.

We shall consider the lifetime of a particular particle as observed by two different inertial observers O and O′. Observer O uses a frame S that is fixed in the laboratory, in which the particle travels with constant speed V in the positive x-direction. We shall call this the **laboratory frame**. Observer O′ uses a frame S′ that moves with the particle. Such a frame is called the **rest frame** of the particle since the particle is always at rest in that frame. (You can think of the observer O′ as riding on the particle if you wish.)

According to observer O′, the birth and decay of the (stationary) particle happen at the same place, so if event 1 occurs at (t_1', x'), then event 2 occurs at (t_2', x'), and the lifetime of the particle will be $\Delta t' = t_2' - t_1'$. In special relativity, the time between two events measured in a frame in which the events happen at the same position is called the **proper time** between the events and is usually denoted by the symbol $\Delta\tau$. So, in this case, we can say that in frame S′ the intervals of time and space that separate the two events are $\Delta t' = \Delta\tau = t_2' - t_1'$ and $\Delta x' = 0$.

According to observer O in the laboratory frame S, event 1 occurs at (t_1, x_1) and event 2 at (t_2, x_2), and the lifetime of the particle is $\Delta t = t_2 - t_1$, which we shall call ΔT. Thus in frame S the intervals of time and space that separate the two events are $\Delta t = \Delta T = t_2 - t_1$ and $\Delta x = x_2 - x_1$.

These events and intervals are represented in Figure 1.7, and everything we know about them is listed in Table 1.1. Such a table is helpful in establishing which of the interval transformations will be useful.

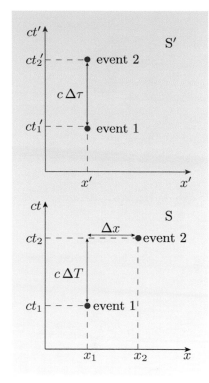

Figure 1.7 Events and intervals for establishing the relation between the lifetime of a particle in its rest frame (S′) and in a laboratory frame (S). Note that we show the coordinate on the vertical axis as 'ct' rather than 't' to ensure that both axes have the dimension of length. To convert time intervals such as $\Delta\tau$ and ΔT to this coordinate, simply multiply them by the constant c.

Table 1.1   A tabular approach to time dilation. The coordinates of the events are listed and the intervals between them worked out, taking account of any known values. The last row is used to show which of the intervals relates to a named quantity (such as the lifetimes ΔT and $\Delta\tau$) or has a known value (such as $\Delta x' = 0$). Any interval that is neither known nor related to a named quantity is shown as a question mark.

Event	S (laboratory)	S′ (rest frame)
2	(t_2, x_2)	(t_2', x')
1	(t_1, x_1)	(t_1', x')
Intervals	$(t_2 - t_1, x_2 - x_1)$ $\equiv (\Delta t, \Delta x)$	$(t_2' - t_1', 0)$ $\equiv (\Delta t', \Delta x')$
Relation to known intervals	$(\Delta T, ?)$	$(\Delta\tau, 0)$

Four of the interval transformation rules that were introduced in the previous section involve three intervals. But only Equation 1.36 involves the three known intervals. Substituting the known intervals into that equation gives

Figure 1.8   Henri Poincaré (1854–1912).

$\Delta T = \gamma(V)(\Delta\tau + 0)$. Therefore the particle lifetimes measured in S and S′ are related by

$$\Delta T = \gamma(V)\,\Delta\tau. \tag{1.40}$$

Since $\gamma(V) > 1$, this result tells us that the particle is observed to live longer in the laboratory frame than it does in its own rest frame. This is an example of the effect known as **time dilation**. A process that occupies a (proper) time $\Delta\tau$ in its own rest frame has a longer duration ΔT when observed from some other frame that moves relative to the rest frame. If the process is the ticking of a clock, then a consequence is that moving clocks will be observed to run slow.

The time dilation effect has been demonstrated experimentally many times. It provides one of the most common pieces of evidence supporting Einstein's theory of special relativity. If it did not exist, many experiments involving short-lived particles, such as *muons*, would be impossible, whereas they are actually quite routine.

It is interesting to note that the French mathematician Henri Poincaré (Figure 1.8) proposed an effect similar to time dilation shortly before Einstein formulated special relativity.

Exercise 1.4   A particular muon lives for $\Delta\tau = 2.2\,\mu s$ in its own rest frame. If that muon is travelling with speed $V = 3c/5$ relative to an observer on Earth, what is its lifetime as measured by that observer?   ∎

1.3.2   Length contraction

There is another curious relativistic effect that relates to the length of an object observed from different frames of reference. For the sake of simplicity, the object that we shall consider is a rod, and we shall start our discussion with a definition of the rod's length that applies whether or not the rod is moving.

In any inertial frame of reference, the **length** of a rod is the distance between its end-points at a single time as measured in that frame.

Thus, in an inertial frame S in which the rod is oriented along the x-axis and moves along that axis with constant speed V, the length L of the rod can be related to two events, 1 and 2, that happen at the ends of the rod at the same time t. If event 1 is at (t, x_1) and event 2 is at (t, x_2), then the length of the rod, as measured in S at time t, is given by $L = \Delta x = x_2 - x_1$.

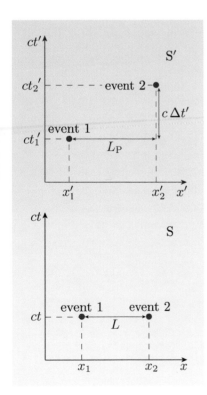

Figure 1.9   Events and intervals for establishing the relation between the length of a rod in its rest frame (S′) and in a laboratory frame (S).

Now consider these same two events as observed in an inertial frame S′ in which the rod is oriented along the x'-axis but is always at rest. In this case we still know that event 1 and event 2 occur at the end-points of the rod, but we have no reason to suppose that they will occur at the same time, so we shall describe them by the coordinates (t'_1, x'_1) and (t'_2, x'_2). Although these events may not be simultaneous, we know that in frame S′ the rod is not moving, so its end-points are always at x'_1 and x'_2. Consequently, we can say that the length of the rod in its own rest frame — a quantity sometimes referred to as the **proper length** of the rod and denoted L_P — is given by $L_P = \Delta x' = x'_2 - x'_1$.

These events and intervals are represented in Figure 1.9, and everything we know about them is listed in Table 1.2.

Table 1.2 Events and intervals for length contraction.

Event	S (laboratory)	S′ (rest frame)
2	(t, x_2)	(t'_2, x'_2)
1	(t, x_1)	(t'_1, x'_1)
Intervals	$(0, x_2 - x_1)$	$(t'_2 - t'_1, x'_2 - x'_1)$
	$\equiv (\Delta t, \Delta x)$	$\equiv (\Delta t', \Delta x')$
Relation to known intervals	$(0, L)$	$(?, L_P)$

On this occasion, the one unknown interval is $\Delta t'$, so the interval transformation rule that relates the three known intervals is Equation 1.33. Substituting the known intervals into that equation gives $L_P = \gamma(V)(L - 0)$. So the lengths measured in S and S′ are related by

$$L = L_P / \gamma(V). \tag{1.41}$$

Since $\gamma(V) > 1$, this result tells us that the rod is observed to be shorter in the laboratory frame than in its own rest frame. In short, moving rods contract. This is an example of the effect known as **length contraction**. The effect is not limited to rods. Any moving body will be observed to contract along its direction of motion, though it is particularly important in this case to remember that this does not mean that it will necessarily be *seen* to contract. There is a substantial body of literature relating to the visual appearance of rapidly moving bodies, which generally involves factors apart from the observed length of the body.

Length contraction is sometimes known as *Lorentz–Fitzgerald contraction* after the physicists (Figure 1.4 and Figure 1.10) who first suggested such a phenomenon, though their interpretation was rather different from that of Einstein.

Exercise 1.5 There is an alternative way of defining length in frame S based on two events, 1 and 2, that happen at *different* times in that frame. Suppose that event 1 occurs at $x = 0$ as the front end of the rod passes that point, and event 2 also occurs at $x = 0$ but at the later time when the rear end passes. Thus event 1 is at $(t_1, 0)$ and event 2 is at $(t_2, 0)$. Since the rod moves with uniform speed V in frame S, we can define the length of the rod, as measured in S, by the relation $L = V(t_2 - t_1)$. Use this alternative definition of length in frame S to establish that the length of a moving rod is less than its proper length. (The events are represented in Figure 1.11.) ■

1.3.3 The relativity of simultaneity

It was noted in the discussion of length contraction that two events that occur at the same time in one frame do not necessarily occur at the same time in another frame. Indeed, looking again at Figure 1.9 and Table 1.2 but now calling on the interval transformation rule of Equation 1.32, it is clear that if the events 1 and 2 are observed to occur at the same time in frame S (so $\Delta t = 0$) but are separated by a distance L along the x-axis, then in frame S′ they will be separated by the time

$$\Delta t' = -\gamma(V)VL/c^2.$$

Figure 1.10 George Fitzgerald (1851–1901) was an Irish physicist interested in electromagnetism. He was influential in understanding that length contracts.

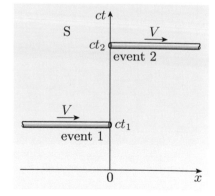

Figure 1.11 An alternative set of events that can be used to determine the length of a uniformly moving rod.

Two events that occur at the same time in some frame are said to be **simultaneous** in that frame. The above result shows that the condition of being simultaneous is a relative one not an absolute one; two events that are simultaneous in one frame are not necessarily simultaneous in every other frame. This consequence of the Lorentz transformations is referred to as the **relativity of simultaneity**.

1.3.4   The Doppler effect

A physical phenomenon that was well known long before the advent of special relativity is the **Doppler effect**. This accounts for the difference between the emitted and received frequencies (or wavelengths) of radiation arising from the relative motion of the emitter and the receiver. You will have heard an example of the Doppler effect if you have listened to the siren of a passing ambulance: the frequency of the siren is higher when the ambulance is approaching (i.e. travelling towards the receiver) than when it is receding (i.e. travelling away from the receiver).

Astronomers routinely use the Doppler effect to determine the speed of approach or recession of distant stars. They do this by measuring the received wavelengths of narrow lines in the star's spectrum, and comparing their results with the proper wavelengths of those lines that are well known from laboratory measurements and represent the wavelengths that would have been emitted in the star's rest frame.

Despite the long history of the Doppler effect, one of the consequences of special relativity was the recognition that the formula that had traditionally been used to describe it was wrong. We shall now obtain the correct formula.

Consider a lamp at rest at the origin of an inertial frame S emitting electromagnetic waves of proper frequency f_{em} as measured in S. Now suppose that the lamp is observed from another inertial frame S′ that is in standard configuration with S, moving away at constant speed V (see Figure 1.12). A detector fixed at the origin of S′ will show that the radiation from the receding lamp is received with frequency f_{rec} as measured in S′. Our aim is to find the relationship between f_{rec} and f_{em}.

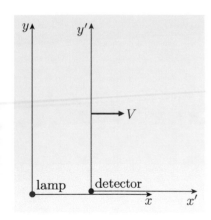

Figure 1.12   The Doppler effect arises from the relative motion of the emitter and receiver of radiation.

The emitted waves have regularly positioned *nodes* (points of zero disturbance) that are separated by a proper wavelength $\lambda_{em} = c/f_{em}$ as measured in S. In that frame the time interval between the emission of one node and the next, Δt, represents the proper period of the wave, T_{em}, so we can write $\Delta t = T_{em} = 1/f_{em}$.

Due to the phenomenon of time dilation, an observer in frame S′ will find that the time separating the emission of successive equivalent nodes is $\Delta t' = \gamma(V)\,\Delta t$. However, this is not the time that separates the arrival of those nodes at the detector because the detector is moving away from the emitter at a constant rate. In fact, during the interval $\Delta t'$ the detector will increase its distance from the emitter by $V\Delta t'$ as measured in S′, and this will cause the reception of the two nodes to be separated by a total time $\Delta t' + V\Delta t'/c$ as measured in S′. This represents the received period of the wave and is therefore the reciprocal of the received frequency, so we can write

$$\frac{1}{f_{rec}} = \Delta t' + \frac{V\Delta t'}{c} = \gamma(V)\,\Delta t\left(1 + \frac{V}{c}\right).$$

We can now identify Δt with the reciprocal of the emitted frequency and use the identity $\gamma(V) = 1/\sqrt{(1 - V/c)(1 + V/c)}$ to write

$$\frac{1}{f_{\text{rec}}} = \frac{1}{f_{\text{em}}} \frac{1}{\sqrt{(1 - V/c)(1 + V/c)}} \left(1 + \frac{V}{c}\right),$$

which can be rearranged to give

$$f_{\text{rec}} = f_{\text{em}} \sqrt{\frac{c - V}{c + V}}. \tag{1.42}$$

This shows that the radiation received from a receding source will have a frequency that is less than the proper frequency with which the radiation was emitted. It follows that the received wavelength $\lambda_{\text{rec}} = c/f_{\text{rec}}$ will be greater than the proper wavelength λ_{em}. Consequently, the spectral lines seen in the light of receding stars will be shifted towards the red end of the spectrum; a phenomenon known as **redshift** (see Figure 1.13). In a similar way, the spectra of approaching stars will be subject to a **blueshift** described by an equation similar to Equation 1.42 but with V replaced by $-V$ throughout. The correct interpretation of these **Doppler shifts** is of great importance.

Exercise 1.6 Some astronomers are studying an unusual phenomenon, close to the centre of our galaxy, involving a jet of material containing sodium. The jet is moving almost directly along the line between the Earth and the galactic centre. In a laboratory, a stationary sample of sodium vapour absorbs light of wavelength $\lambda = 5850 \times 10^{-10}$ m. Spectroscopic studies show that the wavelength of the sodium absorption line in the jet's spectrum is $\lambda' = 4483 \times 10^{-10}$ m. Is the jet approaching or receding? What is the speed of the jet relative to Earth? (Note that the main challenge in this question is the mathematical one of using Equation 1.42 to obtain an expression for V in terms of λ/λ'.) ■

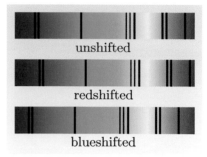

Figure 1.13 Spectral lines are redshifted (that is, reduced in frequency) when the source is receding, and blueshifted (increased in frequency) when the source is approaching.

1.3.5 The velocity transformation

Suppose that an object is observed to be moving with velocity $\boldsymbol{v} = (v_x, v_y, v_z)$ in an inertial frame S. What will its velocity be in a frame S$'$ that is in standard configuration with S, travelling with uniform speed V in the positive x-direction? The Galilean transformation would lead us to expect $\boldsymbol{v}' = (v_x - V, v_y, v_z)$, but we know that is not consistent with the observed behaviour of light. Once again we shall use the interval transformation rules that follow directly from the Lorentz transformations to find the velocity transformation rule according to special relativity.

We know from Equations 1.32 and 1.33 that the time and space intervals between two events 1 and 2 that occur on the x-axis in frame S, transform according to

$$\Delta t' = \gamma(V)(\Delta t - V\Delta x/c^2),$$
$$\Delta x' = \gamma(V)(\Delta x - V\Delta t).$$

Dividing the second of these equations by the first gives

$$\frac{\Delta x'}{\Delta t'} = \frac{\gamma(V)(\Delta x - V\Delta t)}{\gamma(V)(\Delta t - V\Delta x/c^2)}.$$

Dividing the upper and lower expressions on the right-hand side of this equation by Δt, and cancelling the Lorentz factors, gives

$$\frac{\Delta x'}{\Delta t'} = \frac{(\Delta x/\Delta t - V)}{(1 - (\Delta x/\Delta t)V/c^2)}.$$

Now, if we suppose that the two events that we are considering are very close together — indeed, if we consider the limit as Δt and Δx go to zero — then the quantities $\Delta x/\Delta t$ and $\Delta x'/\Delta t'$ will become the instantaneous velocity components v_x and v_x' of a moving object that passes through the events 1 and 2. Extending these arguments to three dimensions by considering events that are not confined to the x-axis leads to the following **velocity transformation** rules:

$$v_x' = \frac{v_x - V}{1 - v_x V/c^2}, \tag{1.43}$$

$$v_y' = \frac{v_y}{\gamma(V)(1 - v_x V/c^2)}, \tag{1.44}$$

$$v_z' = \frac{v_z}{\gamma(V)(1 - v_x V/c^2)}. \tag{1.45}$$

These equations may look rather odd at first sight but they make good sense in the context of special relativity. When v_x and V are small compared to the speed of light c, the term $v_x V/c^2$ is very small and the denominator is approximately 1. In such cases, the Galilean velocity transformation rule, $v_x' = v_x - V$, is recovered as a low-speed approximation to the special relativistic result. At high speeds the situation is even more interesting, as the following question will show.

● An observer has established that two objects are receding in opposite directions. Object 1 has speed c, and object 2 has speed V. Using the velocity transformation, compute the velocity with which object 1 recedes as measured by an observer travelling on object 2.

○ Let the line along which the objects are travelling be the x-axis of the original observer's frame, S. We can then suppose that a frame of reference S′ that has its origin on object 2 is in standard configuration with frame S, and apply the velocity transformation to the velocity components of object 1 with $\boldsymbol{v} = (-c, 0, 0)$ (see Figure 1.14). The velocity transformation predicts that as observed in S′, the velocity of object 1 is $\boldsymbol{v}' = (v_x', 0, 0)$, where

$$v_x' = \frac{v_x - V}{1 - v_x V/c^2} = \frac{-c - V}{1 - (-c)V/c^2} = -c.$$

So, as observed from object 2, object 1 is travelling in the $-x'$-direction at the speed of light, c. This was inevitable, since the second postulate of special relativity (which was used in the derivation of the Lorentz transformations) tells us that all observers agree about the speed of light. It is nonetheless pleasing to see how the velocity transformation delivers the required result in this case. It is worth noting that this result does not depend on the value of V.

Exercise 1.7   According to an observer on a spacestation, two spacecraft are moving away, travelling in the same direction at different speeds. The nearer spacecraft is moving at speed $c/2$, the further at speed $3c/4$. What is the speed of one of the spacecraft as observed from the other? ∎

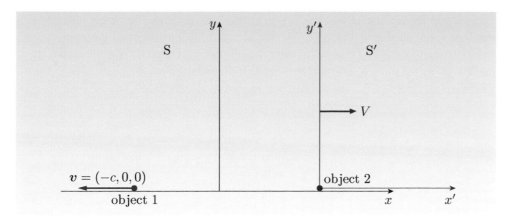

Figure 1.14 Two objects move in opposite directions along the x-axis of frame S. Object 1 travels with speed c; object 2 travels with speed V and is the origin of a second frame of reference S'.

1.4 Minkowski spacetime

In 1908 Einstein's former mathematics teacher, Hermann Minkowski (Figure 1.15), gave a lecture in which he introduced the idea of **spacetime**. He said in the lecture: 'Henceforth space by itself, and time by itself are doomed to fade away into mere shadows, and only a kind of union of the two will preserve an independent reality'. This section concerns that four-dimensional union of space and time, the set of all possible events, which is now called **Minkowski spacetime**.

1.4.1 Spacetime diagrams, lightcones and causality

We have already seen how the Lorentz transformations lead to some very counter-intuitive consequences. This subsection introduces a graphical tool known as a **spacetime diagram** or a **Minkowski diagram** that will help you to visualize events in Minkowski spacetime and thereby develop a better intuitive understanding of relativistic effects. The spacetime diagram for a frame of reference S is usually presented as a plot of ct against x, and each point on the diagram represents a possible event as observed in frame S. The y- and z-coordinates are usually ignored.

Given two inertial frames, S and S', in standard configuration, it is instructive to plot the ct'- and x'-axes of frame S' on the spacetime diagram for frame S. The x'-axis of frame S' is defined by the set of events for which $ct' = 0$, and the ct'-axis is defined by the set of events for which $x' = 0$. The coordinates of these events in S are related to their coordinates in S' by the following Lorentz transformations. (Note that the time transformation of Equation 1.5 has been multiplied by c so that each coordinate can be measured in units of length.)

$$ct' = \gamma(V)(ct - Vx/c),$$
$$x' = \gamma(V)(x - Vt).$$

Setting $ct' = 0$ in the first of these equations gives $0 = \gamma(V)(ct - Vx/c)$. This shows that in the spacetime diagram for frame S, the x'-axis of frame S' is

Figure 1.15 Hermann Minkowski (1864–1909) was one of Einstein's mathematics teachers at the Swiss Federal Polytechnic in Zurich. In 1907 he moved to the University of Göttingen, and while there he introduced the idea of spacetime. Einstein was initially unimpressed but later acknowledged his indebtedness to Minkowski for easing the transition from special to general relativity.

represented by the line $ct = (V/c)x$, a straight line through the origin with gradient V/c. Similarly, setting $x' = 0$ in the second transformation equation gives $0 = \gamma(V)(x - Vt)$, showing that the ct'-axis of frame S′ is represented by the line $ct = (c/V)x$, a straight line through the origin with gradient c/V in the spacetime diagram of S. These lines are shown in Figure 1.16.

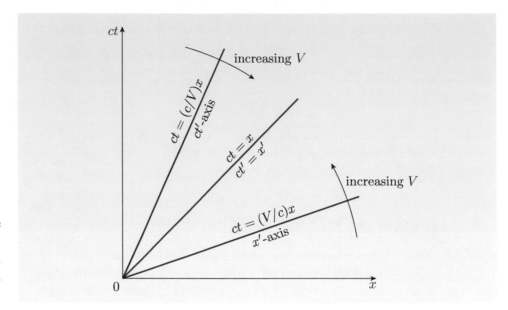

Figure 1.16 The spacetime diagram of frame S, showing the events that make up the ct'- and x'-axes of frame S′, and the path of a light ray that passes through the origin.

There is another feature of interest in the diagram. The straight line through the origin of gradient 1 links all the events where $x = ct$ and thus shows the path of a light ray that passes through $x = 0$ at time $t = 0$. Using the inverse Lorentz transformations shows that this line also passes through all the events where $\gamma(V)(x' + Vt') = \gamma(V)(ct' + Vx'/c)$, that is (after some cancelling and rearranging), where $x' = ct'$. So the line of gradient 1 passing through the origin also represents the path of a light ray that passes through the origin of frame S′ at $t' = 0$. In fact, any line with gradient 1 on a spacetime diagram must always represent the possible path of a light ray, and thanks to the second postulate of special relativity, we can be sure that all observers will agree about that.

As the relative speed V of the frames S and S′ increases, the lines representing the x'- and ct'-axes of S′ close in on the line of gradient 1 from either side, rather like the closing of a clapper board. This behaviour reflects the fact that Lorentz transformations will generally alter the coordinates of events but will not change the behaviour of light on which all observers must agree.

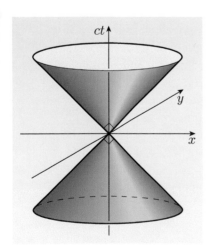

Figure 1.17 In three dimensions (one time and two space) it becomes clear that a line of gradient 1 in a spacetime diagram is part of a lightcone.

In the somewhat unusual case when we include a second spatial axis (the y-axis, say) in the spacetime diagram, the original line of gradient 1 is seen to be part of a cone, as indicated in Figure 1.17. This cone, which connects the event at the origin to all those events, past and future, that might be linked to it by a signal travelling at the speed of light, is an example of a **lightcone**. A horizontal slice (at $ct =$ constant) through the (pseudo) three-dimensional diagram at any particular time shows a circle, but in a fully four-dimensional diagram with all three spatial axes included, such a fixed-time slice would be a sphere, and would represent a spherical shell of light surrounding the origin. At times earlier than $t = 0$, the shell would represent incoming light signals closing in on the origin. At times later than $t = 0$, the shell would represent outgoing light signals travelling away from the origin. Although observers O and O′, using frames S and S′, would not generally agree about the coordinates of events, they would agree about which events were on the lightcone, which were inside the lightcone and which were outside. This agreement between observers makes lightcones very useful in discussions about which events might cause, or be caused by, other events.

Going back to an ordinary two-dimensional spacetime diagram of the kind shown in Figure 1.18, it is straightforward to read off the coordinates of an event in frame S or in frame S′. The event 1 in the diagram clearly has coordinates (ct_1, x_1) in frame S. In frame S′, it has a different set of coordinates. These can be determined by drawing construction lines *parallel* to the lines representing the primed axes. Where a construction line parallel to one primed axis intersects the other primed axis, the coordinate can be found. By doing this on both axes, both coordinates are found. In the case of Figure 1.18, the dashed construction lines show that, as observed in frame S′, event 1 occurs at the same time as event 2, and at the same position as event 3.

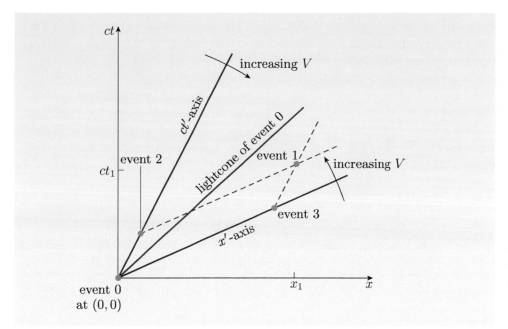

Figure 1.18 A spacetime diagram for frame S with four events, 0, 1, 2 and 3. Event coordinates in S′ can be found by drawing construction lines parallel to the appropriate axes.

Another lesson that can be drawn from Figure 1.18 concerns the order of events. Starting from the bottom of the ct-axis and working upwards, it is clear that in frame S, the four events occur in the order 0, 2, 3 and 1. But it is equally clear from the dashed construction lines that in frame S′, event 3 happens at the same time as event 0 (they are simultaneous in S′), and both happen at an earlier time than event 2 and event 1, which are also simultaneous in S′. This illustrates the relativity of simultaneity, but more importantly it also shows that the order of events 2 and 3 will be different for observers O and O′.

At first sight it is quite shocking to learn that the relative motion of two observers can reverse the order in which they observe events to happen. This has the potential to overthrow our normal notion of **causality**, the principle that all observers must agree that any effect is *preceded* by its cause. It is easy to imagine observing the pressing of a plunger and then observing the explosion that it causes. It would be very shocking if some other observer, simply by moving sufficiently fast in the right direction, was able to observe the explosion first and then the pressing of the plunger that caused it. (It is important to remember that we are discussing observing, not seeing.)

Fortunately, such an overthrow of causality is not permitted by special relativity, *provided that we do not allow signals to travel at speeds greater than c*. Although observers will disagree about the order of some events, they will not disagree about the order of any two events that might be linked by a light signal or any signal that travels at less than the speed of light. Such events are said to be **causally related**.

To see how the order of causally related events is preserved, look again at Figure 1.18, noting that all the events that are causally related to event 0 are contained within its lightcone, and that includes event 2. Events that are not causally related to event 0, such as event 1 and event 3, are outside the lightcone of event 0 and could only be linked to that event by signals that travel faster than light. Now, remember that as the relative speed V of the observers O and O' increases, the line representing the ct'-axis closes in on the lightcone. As a result, there will not be any value of V that allows the causally related events 0 and 2 to change their order. Event 2 will always be at a higher value of ct' than event 0. However, when you examine the corresponding behaviour of events 0 and 3, which are not causally related, the conclusion is quite different. Figure 1.18 shows the condition in which event 0 and event 3 occur at the same time $t' = 0$, according to O'. When O and O' have a lower relative speed, event 3 occurs after event 0, but as V increases and the line representing the x'-axis (where all events occur at $ct' = 0$) closes in on the lightcone, we see that there will be a value of V above which the order of event 0 and event 3 is reversed.

So, if event 0 represents the pressing of a plunger and event 2 and event 3 represent explosions, all observers will agree that event 0 might have caused event 2, which happened later. However, those same observers will not agree about the order of event 0 and event 3, though they will agree that event 0 could not have *caused* event 3 unless bodies or signals can travel faster than light. It is the desire to preserve causal relationships that is the basis for the requirement that no material body or signal of any kind should be able to travel faster than light.

● Is event 1 in Figure 1.18 causally related to event 0? Is event 1 causally related to event 3? Justify your answers.

○ Event 1 is outside the lightcone of event 0, so the two cannot be causally related. The diagram does not show the lightcone of event 3, but if you imagine a line of gradient 1, parallel to the shown lightcone, passing through event 3, it is clear that event 1 is inside the lightcone of event 3, so those two events are causally related. The earlier event may have caused the later one, and all observers will agree about that.

An important lesson to learn from this question is the significance of drawing lightcones for events other than those at the origin. Every event has a lightcone, and that lightcone is of great value in determining causal relationships.

1.4.2 Spacetime separation and the Minkowski metric

In three-dimensional space, the separation between two points (x_1, y_1, z_1) and (x_2, y_2, z_2) can be conveniently described by the square of the distance Δl between them:

$$(\Delta l)^2 = (\Delta x)^2 + (\Delta y)^2 + (\Delta z)^2, \tag{1.46}$$

where $\Delta x = x_2 - x_1$, $\Delta y = y_2 - y_1$ and $\Delta z = z_2 - z_1$. This quantity has the useful property of being unchanged by rotations of the coordinate system. So, if we choose to describe the points using a new coordinate system with axes x', y' and z', obtained by rotating the old system about one or more of its axes, then the spatial separation of the two points would still be described by an expression of the form

$$(\Delta l')^2 = (\Delta x')^2 + (\Delta y')^2 + (\Delta z')^2, \tag{1.47}$$

and we would find in addition that

$$(\Delta l)^2 = (\Delta l')^2. \tag{1.48}$$

We describe this situation by saying that the spatial separation of two points is **invariant** under rotations of the coordinate system used to describe the positions of the two points.

These ideas can be extended to four-dimensional Minkowski spacetime, where the most useful expression for the **spacetime separation** of two events is the following.

Spacetime separation

$$(\Delta s)^2 = (c\,\Delta t)^2 - (\Delta x)^2 - (\Delta y)^2 - (\Delta z)^2. \tag{1.49}$$

The reason why this particular form is chosen is that it turns out to be invariant under Lorenz transformations. So, if O and O′ are inertial observers using frames S and S′, they will generally not agree about the coordinates that describe two events 1 and 2, or about the distance or the time that separates them, but they will agree that the two events have an invariant spacetime separation

$$(\Delta s)^2 = (c\,\Delta t)^2 - (\Delta l)^2 = (c\,\Delta t')^2 - (\Delta l')^2 = (\Delta s')^2. \tag{1.50}$$

Exercise 1.8   Two events occur at $(ct_1, x_1, y_1, z_1) = (3, 7, 0, 0)$ m and $(ct_2, x_2, y_2, z_2) = (5, 5, 0, 0)$ m. What is their spacetime separation?

Exercise 1.9   In the case that $\Delta y = 0$ and $\Delta z = 0$, use the interval transformation rules to show that the spacetime separation given by Equation 1.49 really is invariant under Lorentz transformations.   ■

A convenient way of writing the spacetime separation is as a summation:

$$(\Delta s)^2 = \sum_{\mu, \nu = 0}^{3} \eta_{\mu\nu}\,\Delta x^\mu\,\Delta x^\nu, \tag{1.51}$$

where the four quantities Δx^0, Δx^1, Δx^2 and Δx^3 are the components of $[\Delta x^\mu] = (c\,\Delta t, \Delta x, \Delta y, \Delta z)$, and the new quantities $\eta_{\mu\nu}$ that have been introduced are the sixteen components of an entity called the **Minkowski metric**, which can be represented as

$$[\eta_{\mu\nu}] \equiv \begin{pmatrix} \eta_{00} & \eta_{01} & \eta_{02} & \eta_{03} \\ \eta_{10} & \eta_{11} & \eta_{12} & \eta_{13} \\ \eta_{20} & \eta_{21} & \eta_{22} & \eta_{23} \\ \eta_{30} & \eta_{31} & \eta_{32} & \eta_{33} \end{pmatrix} = \begin{pmatrix} 1 & 0 & 0 & 0 \\ 0 & -1 & 0 & 0 \\ 0 & 0 & -1 & 0 \\ 0 & 0 & 0 & -1 \end{pmatrix}. \tag{1.52}$$

It's worth noting that the Minkowski metric has been shown as a matrix only for convenience; Equation 1.51 is not a matrix equation, though it is a well-defined sum. The important point is that the quantity $[\eta_{\mu\nu}]$ has sixteen components, and from Equation 1.52 you can uniquely identify each of them. The metric provides a valuable reminder of how the spacetime separation is related to the coordinate intervals. Metrics will have a crucial role to play in the rest of this book. The Minkowski metric is just the first of many that you will meet.

The spacetime separation of two events is an important quantity for several reasons. Its sign alone tells us about the possible causal relationship between the events. In fact, we can identify three classes of relationship, corresponding to the cases $(\Delta s)^2 > 0$, $(\Delta s)^2 = 0$ and $(\Delta s)^2 < 0$.

Time-like, light-like and space-like separations

Events with a positive spacetime separation, $(\Delta s)^2 > 0$, are said to be **time-like** separated. Such events are causally related, and there will exist a frame in which the two events happen at the same place but at different times.

Events with a zero (or null) spacetime separation, $(\Delta s)^2 = 0$, are said to be **light-like** separated. Such events are causally related, and all observers will agree that they could be linked by a light signal.

Events with a negative spacetime separation, $(\Delta s)^2 < 0$, are said to be **space-like** separated. Such events are not causally related, and there will exist a frame in which the two events happen at the same time but at different places.

These different kinds of spacetime separation correspond to different regions of spacetime defined by the lightcone of an event. Figure 1.19 shows the lightcone of event 0. All the events that have a time-like separation from event 0 are within the future or past lightcone of event 0; all the events that are light-like separated from event 0 are on its lightcone; and all the events that are space-like separated from event 0 are outside its lightcone. This emphasizes the role that lightcones play in revealing the causal structure of Minkowski spacetime.

Another reason why spacetime separation is important relates to *proper time*. You will recall that in the earlier discussion of time dilation, it was said that the proper time between two events was the time separating those events as measured in a frame where the events happen at the same position. In such a frame, the spacetime separation of the events is $(\Delta s)^2 = c^2(\Delta t)^2 = c^2(\Delta \tau)^2$. However,

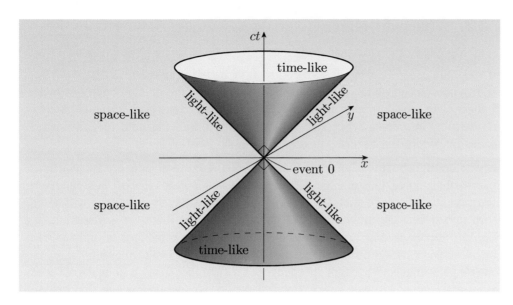

Figure 1.19 Events that are time-like separated from event 0 are found inside its lightcone. Events that are light-like separated are found on the lightcone, and events that are space-like separated from event 0 are outside the lightcone.

since the spacetime separation of events is an invariant quantity, we can use it to determine the proper time between two time-like separated events, irrespective of the frame in which the events are described. For two time-like separated events with positive spacetime separation $(\Delta s)^2$, the proper time $\Delta\tau$ between those two events is given by the following.

Proper time related to spacetime separation

$$(\Delta\tau)^2 = (\Delta s)^2/c^2. \tag{1.53}$$

The relation between proper time and the invariant spacetime separation is extremely useful in special relativity. The reason for this relates to the length of a particle's pathway through four-dimensional Minkowski spacetime. Such a pathway, with all its twists and turns, records the whole history of the particle and is sometimes called its **world-line**. (One well-known relativist called his autobiography *My worldline*.) By adding together the spacetime separations between successive events along a particle's world-line, and dividing the sum by c^2, we can determine the total time that has passed according to a clock carried by the particle. This simple principle will be used to help to explain a troublesome relativistic effect in the next subsection.

In this book, a positive sign will always be associated with the square of the time interval in the spacetime separation, and a negative sign with the spatial intervals. This choice of sign is just a convention, and the opposite set of signs could have been used. The convention used here ensures that the spacetime separation of events on the world-line of an object moving slower than light is positive. Nonetheless, you will find that many authors adopt the opposite convention, so when consulting other works, always pay attention to the sign convention that they are using.

Exercise 1.10 Given two time-like separated events, show that the proper time between those events is the least amount of time that any inertial observer will measure between them. ∎

1.4.3   The twin effect

We end this chapter with a discussion of a well-known relativistic effect, the **twin effect**. This caused a great deal of controversy early in the theory's history. It is usually presented as a thought experiment concerning the phenomenon of time dilation. The thought experiment involves two twins, Astra and Terra. The twins are identical in every way, except that Astra likes to travel around very fast in her spaceship, while Terra prefers to stay at home on Earth.

As was demonstrated earlier in this chapter, fast-moving objects are subject to observable time dilation effects. This indicates that if Astra jets off in some fixed direction at close to the speed of light, then, as measured by Terra, she will age more slowly because 'moving clocks run slow'. This is fine — it is just what relativity theory predicts, and agrees with the observed behaviour of high-speed particles. But now suppose that Astra somehow manages to turn around and return to Earth at equally high speed. It seems clear that Terra will again observe that Astra's clock will run slow and will therefore not be surprised to find that on her return, Astra has aged less than her stay-at-home twin Terra.

The supposed problem arises when this process is examined from Astra's point of view. Would it not be the case, some argued, that Astra would observe the same events apart from a reversal of velocities, so that Terra would be the travelling twin and it would be Terra's clock that would be running slow during both parts of the journey? Consequently, shouldn't Astra expect Terra to be the younger when they were reunited? Clearly, it's not possible for each twin to be younger than the other when they meet at the same place, so if the arguments are equally sound, it was said, there must be something wrong with special relativity.

In fact, the arguments are not equally sound. The basic problem is that the presumed symmetry between Terra's view and Astra's view is illusory. It is Astra who would be the younger at the reunion, as will now be explained with the aid of a spacetime diagram and a proper use of spacetime separations in Minkowski space.

The first point to make clear is that although velocity is a purely relative quantity, acceleration is not. According to the first postulate of special relativity, the laws of physics do not distinguish one inertial frame from another, so a traveller in a closed box cannot determine his or her speed by performing a physics experiment. However, such a traveller would certainly be able to feel the effect of any acceleration, as we all know from everyday experience. In order to leave the Solar System, jet around the galaxy and return, Astra must have undergone a change in velocity, and that would involve a detectable acceleration. To a first approximation, Terra does not accelerate (her velocity changes due to the rotation and revolution of the Earth are very small compared with Astra's accelerations). A single inertial frame of reference is sufficient to represent Terra's view of events, but no single *inertial* frame can adequately represent Astra's view. There is no symmetry between these two observers; only Terra is an (approximately) inertial observer.

In order to be clear about what's going on and to avoid the use of non-inertial frames, it is convenient to use three inertial frames when discussing the twin effect. The first is Terra's frame, which we can treat as fixed on a non-rotating, non-revolving Earth. The second, which we shall call Astra's frame, moves at a high but constant speed V relative to Terra's frame. You can think of this as the frame of Astra's spaceship, and you can think of Astra as simply jumping aboard her passing ship at the departure, event 0, when she leaves Terra to begin the outward leg of her journey. The third inertial frame, called Stella's frame, belongs to another space traveller who happens to be approaching Earth at speed V along the same line that Astra leaves along. At some point, Stella's ship will pass Astra's, and at that point we can imagine that Astra jumps from her ship to Stella's ship to make the return leg of her journey. Of course, this is unrealistic since the 'jump' would kill Astra, so you may prefer to imagine that Astra is actually a conscious robot or even that she can somehow 'teleport' from one ship to another. In any case, the important point is that the transfer is abrupt and has no effect on Astra's age.

The event at which Astra makes the transfer to Stella's ship we shall call event 1, and the event at which Astra and Terra are eventually reunited we shall call event 2. Astra's quick transfer from one ship to the other allows us to discuss the essential features of the twin effect without getting bogged down in details about the nature of the acceleration that Astra experiences. It is vital that Astra is accelerated, but exactly how that happens is unimportant. Note that we may treat each of these frames as being in standard configuration with either of the others. We can set up the frames in such a way that the origins of Terra's frame and Astra's frame coincide at event 0, the origins of Astra's frame and Stella's frame coincide at event 1, and the origins of Stella's frame and Terra's frame coincide at event 2.

Figure 1.20 is a spacetime diagram for Terra's frame, showing all these events and making clear the coordinates that Terra assigns to them.

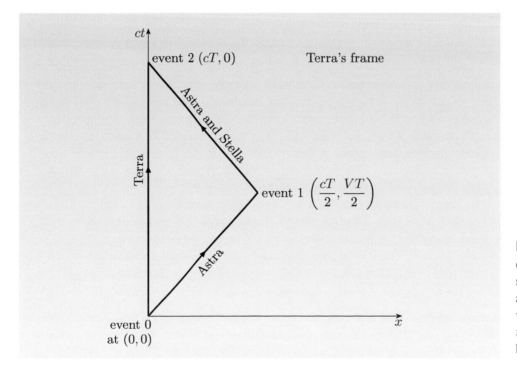

Figure 1.20 A spacetime diagram for Terra's frame, showing the departure, transfer and reunion events together with their coordinates. The t-coordinate has been multiplied by c, as usual.

It is clear from the figure that the proper time between departure and reunion (both of which happen at Terra's location) is T. A little calculation using the relation $(\Delta\tau)^2 = (\Delta s)^2/c^2$ makes it equally clear that the proper time between event 0 and event 1 is given by

$$(\Delta\tau_{0,1})^2 = \frac{(\Delta s_{0,1})^2}{c^2} = \frac{1}{c^2}\left[\left(\frac{cT}{2}\right)^2 - \left(\frac{VT}{2}\right)^2\right]$$

$$= \frac{T^2}{4}\left(1 - \frac{V^2}{c^2}\right) = \left(\frac{T}{2\gamma}\right)^2. \qquad (1.54)$$

So

$$\Delta\tau_{0,1} = \frac{T}{2\gamma}. \qquad (1.55)$$

Although we have arrived at this result using the coordinates assigned by Terra, it is important to note that proper time is an invariant, so all inertial observers will agree on the proper time between two events no matter how it is calculated.

A similar calculation for the proper time separating event 1 and event 2 shows that

$$\Delta\tau_{1,2} = \frac{T}{2\gamma}. \qquad (1.56)$$

So the total proper time that elapses along the world-line followed by Astra is $\Delta\tau_{0,1} + \Delta\tau_{1,2} = T/\gamma$. As expected, this shows that Astra will be the younger twin at the time of the reunion.

How is it possible for Terra and Astra to disagree about the time between events 0 and 2? The answer to this question is that when the whole trip is considered, Astra is *not* an inertial observer; she undergoes an acceleration that Terra does not. Given two time-like separated events, the time that elapses between those events, as measured by an observer present at both events, will depend on the observer's world-line. The total time between the events, measured along the world-line of a non-inertial observer, is generally *less* than the proper time between these events as measured along the world-line of an inertial observer.

The analysis that we have just completed is really sufficient to settle any questions about the twin effect. However, it is still instructive to examine the same events from Astra's frame (which she leaves at event 1). The spacetime diagram for Astra's frame is shown in Figure 1.21. The coordinates of the events have been worked out from those given in Terra's frame using the Lorentz transformations.

● Confirm the coordinate assignments shown in Figure 1.21.

○ In Terra's frame, event 0 is at $(ct, x) = (0, 0)$, event 1 at $(cT/2, VT/2)$, and event 2 at $(cT, 0)$. Treating Terra's frame as frame S and Astra's frame as S', and using the Lorentz transformations $t' = \gamma(t - Vx/c^2)$ and $x' = \gamma(x - Vt)$, it follows immediately that in Astra's frame, event 0 is at $(ct', x') = (0, 0)$, event 1 is at $(ct', x') = (cT/2\gamma, 0)$ (remember that $\gamma(V) = 1/\sqrt{1 - V^2/c^2}$), and event 2 is at $(ct', x') = (c\gamma T, -\gamma VT)$.

Note that again there is a kink in Astra's world-line due to the acceleration that she undergoes. There is no such kink in Terra's world-line since she is an inertial observer. Once again we can work out the proper time that Astra experiences while passing between the three events: this represents the time that would have elapsed according to a clock that Astra carries between each of the events. The proper time between event 0 and event 1 is simply $\Delta\tau_{0,1} = T/2\gamma$, since those events happen at the same place in Astra's frame. The proper time between event 1 and event 2 is given by

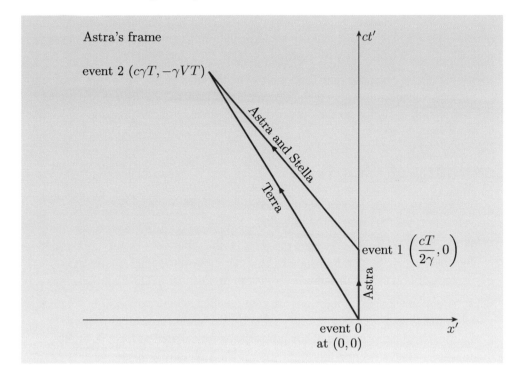

Figure 1.21 A spacetime diagram for Astra's frame, showing the departure, transfer and reunion events with their coordinates. Note that Astra leaves this frame at event 1.

$$(\Delta\tau_{1,2})^2 = \frac{(\Delta s_{1,2})^2}{c^2} = \frac{1}{c^2}\left[\left(c\gamma T - \frac{cT}{2\gamma}\right)^2 - (-\gamma VT)^2\right]$$

$$= T^2\left[\left(\gamma - \frac{1}{2\gamma}\right)^2 - \left(\gamma\frac{V}{c}\right)^2\right]$$

$$= T^2\left[\gamma^2 - 1 + \frac{1}{4\gamma^2} - \frac{\gamma^2 V^2}{c^2}\right]$$

$$= T^2\left[\gamma^2\left(1 - \frac{V^2}{c^2}\right) - 1 + \frac{1}{4\gamma^2}\right].$$

Since $\gamma^2(1 - V^2/c^2) = 1$, the above expression simplifies to give

$$\Delta\tau_{1,2} = \frac{T}{2\gamma}.$$

So once again the theory predicts that the time for the round trip recorded by Astra is $\Delta\tau_{0,1} + \Delta\tau_{1,2} = T/\gamma$.

There is one other point to notice using Astra's frame. Time dilation tells us that, as measured in Astra's frame, Terra's clock will be running slow. From Astra's frame, a 1-second tick of Terra's clock will be observed to last γ seconds. But in Astra's frame, it is also the case that the time of the reunion is γT, which is greater than the time of the reunion as observed in Terra's frame. According to an observer who uses Astra's frame, this longer journey time compensates for the slower ticking of Terra's clock, with the result that such an observer will fully expect Terra to have aged by T while Astra herself has aged by only T/γ. Using the coordinates of event 0 and event 2 in Astra's frame, it is easy to confirm that the proper time between them is T, which is another way of stating the same result.

Exercise 1.11   Using the velocity transformation, show that Astra observes the speed of approach of Stella's spaceship to be $2V/(1 + V^2/c^2)$.

Exercise 1.12   Suppose that Terra sends regular time signals towards Astra and Stella at one-second intervals. Write down expressions for the frequency at which Astra receives the signals on the outward and return legs of her journey.   ■

Summary of Chapter 1

1. Basic terms in the vocabulary of relativity include: event, frame of reference, inertial frame and observer.

2. A theory of relativity concerns the relationships between observations made by observers in a specified state of relative motion. Special relativity is essentially restricted to inertial observers in uniform relative motion.

3. Einstein based special relativity on two postulates: the principle of relativity (that the laws of physics can be written in the same form in all inertial frames) and the principle of the constancy of the speed of light (that all inertial observers agree that light travels through empty space with the same fixed speed, c, in all directions).

4. Given two inertial frames S and S′ in standard configuration, the coordinates of an event observed in frame S are related to the coordinates of the same event observed in frame S′ by the Lorentz transformations

$$t' = \gamma(V)(t - Vx/c^2), \tag{Eqn 1.5}$$
$$x' = \gamma(V)(x - Vt), \tag{Eqn 1.6}$$
$$y' = y, \tag{Eqn 1.7}$$
$$z' = z, \tag{Eqn 1.8}$$

where the Lorentz factor is

$$\gamma(V) = \frac{1}{\sqrt{1 - V^2/c^2}}. \tag{Eqn 1.9}$$

These transformations may also be represented by matrices,

$$\begin{pmatrix} ct' \\ x' \\ y' \\ z' \end{pmatrix} = \begin{pmatrix} \gamma(V) & -\gamma(V)V/c & 0 & 0 \\ -\gamma(V)V/c & \gamma(V) & 0 & 0 \\ 0 & 0 & 1 & 0 \\ 0 & 0 & 0 & 1 \end{pmatrix} \begin{pmatrix} ct \\ x \\ y \\ z \end{pmatrix}, \tag{Eqn 1.10}$$

or as a set of summations

$$x'^\mu = \sum_{\nu=0}^{3} \Lambda^\mu{}_\nu \, x^\nu \quad (\mu = 0, 1, 2, 3). \tag{Eqn 1.13}$$

5. The inverse Lorentz transformations may be written as

$$t = \gamma(V)(t' + Vx'/c^2),$$ (Eqn 1.14)

$$x = \gamma(V)(x' + Vt'),$$ (Eqn 1.15)

$$y = y',$$ (Eqn 1.16)

$$z = z'.$$ (Eqn 1.17)

6. Similar equations describe the transformation of intervals, Δt, Δx, etc., between the two frames.

7. The consequences of special relativity, deduced by considering the transformation of events and intervals, include the following.

   (a) Time dilation:

$$\Delta T = \gamma(V)\,\Delta\tau.$$ (Eqn 1.40)

   (b) Length contraction:

$$L = L_{\mathrm{P}}/\gamma(V).$$ (Eqn 1.41)

   (c) The relativity of simultaneity.

   (d) The relativistic Doppler effect (Eqn 1.42):

$$f_{\mathrm{rec}} = f_{\mathrm{em}}\sqrt{(c+V)/(c-V)} \quad \text{(for an approaching source)},$$

$$f_{\mathrm{rec}} = f_{\mathrm{em}}\sqrt{(c-V)/(c+V)} \quad \text{(for a receding source)}.$$

   (e) The velocity transformation:

$$v'_x = \frac{v_x - V}{1 - v_x V/c^2},$$ (Eqn 1.43)

$$v'_y = \frac{v_y}{\gamma(V)(1 - v_x V/c^2)},$$ (Eqn 1.44)

$$v'_z = \frac{v_z}{\gamma(V)(1 - v_x V/c^2)}.$$ (Eqn 1.45)

8. Four-dimensional Minkowski spacetime contains all possible events.

9. Spacetime diagrams showing events as observed by a particular observer are a valuable tool that can provide pictorial insights into relativistic effects and the structure of Minkowski spacetime.

10. Lightcones are particularly useful for understanding causal relationships between events in Minkowski spacetime.

11. The invariant spacetime separation between two events has the form

$$(\Delta s)^2 = (c\,\Delta t)^2 - (\Delta x)^2 - (\Delta y)^2 - (\Delta z)^2,$$ (Eqn 1.49)

and may be positive (time-like), zero (light-like) or negative (space-like).

12. The spacetime separation may be conveniently written as

$$(\Delta s)^2 = \sum_{\mu,\nu=0}^{3} \eta_{\mu\nu}\,\Delta x^\mu\,\Delta x^\nu,$$ (Eqn 1.51)

where the $\eta_{\mu\nu}$ are the components of the Minkowski metric

$$[\eta_{\mu\nu}] \equiv \begin{pmatrix} \eta_{00} & \eta_{01} & \eta_{02} & \eta_{03} \\ \eta_{10} & \eta_{11} & \eta_{12} & \eta_{13} \\ \eta_{20} & \eta_{21} & \eta_{22} & \eta_{23} \\ \eta_{30} & \eta_{31} & \eta_{32} & \eta_{33} \end{pmatrix} = \begin{pmatrix} 1 & 0 & 0 & 0 \\ 0 & -1 & 0 & 0 \\ 0 & 0 & -1 & 0 \\ 0 & 0 & 0 & -1 \end{pmatrix}. \quad \text{(Eqn 1.52)}$$

13. The proper time $\Delta\tau$ between two time-like separated events is given by

$$(\Delta\tau)^2 = (\Delta s)^2/c^2. \quad \text{(Eqn 1.53)}$$

This is the time that would be recorded on a clock that moves uniformly between the two events.

14. The proper time between two events is an invariant under Lorentz transformations.

15. The time between two time-like separated events, as measured by an observer present at both events, depends on the world-line of the observer. The time between the events, measured by a non-inertial observer, is generally *less* than the proper time between these events as measured by an inertial observer. This is the basis of the twin effect.

Chapter 2   Special relativity and physical laws

Introduction

Physical laws are usually expressed mathematically, as equations. They are used by physicists to summarize their findings regarding the basic principles that govern the Universe. From the late 1600s to the mid-1800s, Newton's laws and the Galilean relativity underpinning them were believed to be the fundamental rules. The precision engineering of the nineteenth century and the clock-like regularity of the Solar System all seemed to be consistent with this view.

However, as we have already seen, the investigations of electricity and magnetism, and their unification with optics through Maxwell's demonstration of the electromagnetic nature of light, exposed a new conflict between fundamental laws. Lorentz and others worked on this problem but it was Einstein who recognized most clearly and completely that its essence was in a conflict between the invariance of the speed of light in a vacuum and the requirement of Galilean relativity that observers in relative motion should disagree about the speed of light. Einstein's response was to extend the principle of relativity from the laws of mechanics to all the laws of physics, including specifically the constancy of the speed of light, and to accept as a consequence the need for a new theory of relativity based on the Lorentz transformations rather than Galilean transformations.

The requirement of special relativity, that physical laws should take the same form in all inertial frames, is highly restrictive. It prevents many candidates from being accepted as genuine physical laws. The principle of relativity cannot tell us which proposed laws are correct — that must be done by experiment — but it can show up those that are not acceptable in principle. When the coordinates used in two different frames are related by the Lorentz transformations, it is soon seen that the laws of Newtonian mechanics do not take the same form in all inertial frames. So an immediate implication of special relativity is the need for an extensive rewriting of the laws of mechanics. The new laws must be consistent with the well-established successes of Newtonian mechanics, but they must also show the invariance under Lorentz transformations required by the principle of relativity. In this chapter we shall consider those new laws of mechanics and see the extent to which Newtonian concepts had to be modified or replaced. We shall then go on to see what special relativity has to say about the laws of electricity and magnetism.

The discussion of physical laws in this chapter will introduce some important mathematical entities that may be new to you. These entities, called *four-vectors* and *four-tensors*, are of particular relevance to special relativity but they set the scene for the introduction of more general tensors in the later chapters that deal with general relativity. Pay special attention to these four-vectors and four-tensors. Appreciating their role in the formulation of physical laws that are consistent with special relativity is at least as important as learning about any specific feature of those laws.

2.1   Invariants and physical laws

2.1.1   The invariance of physical quantities

Central to the formulation of physical laws in special relativity are invariant quantities or *invariants* for short. You have already met a number of these invariants: most obviously, the speed of light in a vacuum, but also the spacetime separation between events $(\Delta s)^2 = (c\,\Delta t)^2 - (\Delta x)^2 - (\Delta y)^2 - (\Delta z)^2$ and, in the case of time-like separated events, the closely related *proper time* interval $\Delta \tau$ given by $(\Delta \tau)^2 = (\Delta s)^2/c^2$.

An alternative way of defining the proper time between two events is as the time between those events measured in a frame in which the two events occur at the same spatial position. (The fact that the events are time-like separated guarantees that such a frame exists.) This is an interesting definition since it uses a measurement made in one inertial frame to define a quantity that can then be used in all inertial frames. This approach to defining invariants is quite common. For example, we can and will say that the electric charge of a particle is the charge that it has when measured in the frame in which the particle is at rest. The charge is then defined in an invariant way, even though the prescription for measuring it involves a particular frame — the rest frame of the particle.

A similar approach can be used to provide an invariant value for the mass of a particle. In keeping with the common practice of particle physicists, we shall say that the mass of a particle is the mass that would be measured in a frame in which the particle is at rest. This provides a mass that all observers can agree about. Some authors refer to this quantity as the *rest mass* of the particle, but we have no need to do so here since this is the only sense in which we shall use the term mass in this chapter. Incidentally, if you have studied relativity before, you may have encountered the idea of a *relativistic mass* that increases with the speed of the particle. This is based on a quite different definition of mass that will not be used in this book. The masses that we shall refer to are defined invariantly and will never depend on speed. Other invariant quantities — some of them very important — will be introduced later, but for the moment here is a summary of what we have said about invariants.

Invariants

An **invariant** is a quantity that has the same value in all inertial frames.

Invariant quantities include:

- the speed of light in a vacuum, c
- the spacetime separation $(\Delta s)^2 = (c\,\Delta t)^2 - (\Delta x)^2 - (\Delta y)^2 - (\Delta z)^2$
- the proper time $(\Delta \tau)^2 = (\Delta s)^2/c^2$ between time-like separated events
- the charge of a particle, q
- the mass of a particle, m.

2.1.2 The invariance of physical laws

The requirement that the laws of physics should take the same form in all inertial frames involves extending the idea of invariance from invariance of a quantity to invariance of the form of an equation. The easiest way to appreciate this is by means of an example. So, although it is mainly of historical interest, we shall now demonstrate the **form invariance** under the Galilean coordinate transformation of Newton's laws of motion.

Newton's laws of motion can be stated as follows.

1. A body maintains a constant velocity unless acted upon by an unbalanced external force.

2. A body acted upon by an unbalanced force accelerates in the direction of that force at a rate that is proportional to the force and inversely proportional to the body's mass.

3. When body A exerts a force on body B, body B exerts a force on body A that has the same magnitude but acts in the opposite direction. (This law is often stated as: to every action there is an equal and opposite reaction.)

The first law is really telling us that in order to use the other laws, we should make sure that we observe from an inertial frame of reference. So we don't need to give any further thought to this law as long as we restrict ourselves to inertial frames. The third law also presents no difficulty. Provided that oppositely directed forces of equal magnitude transform in the same way in Galilean relativity, there will not be any problem about agreeing on the form of the third law. This is true even for forces that act at a distance, such as the gravitational force acting on a person due to the Earth and the reaction to that force that acts simultaneously at the Earth's centre of mass.

The real challenge comes with Newton's second law of motion. Let's start by writing the second law as an equation

$$\boldsymbol{f} = m\boldsymbol{a}, \tag{2.1}$$

where $\boldsymbol{f}$ is the applied force, m is the mass of the body, and $\boldsymbol{a}$ is its acceleration. If we take this to be the form of Newton's second law in some particular inertial frame S with Cartesian coordinate axes x, y and z, we can relate the acceleration to the coordinates of the body in frame S by writing

$$\boldsymbol{f} = m\left(\frac{\mathrm{d}^2x}{\mathrm{d}t^2}, \frac{\mathrm{d}^2y}{\mathrm{d}t^2}, \frac{\mathrm{d}^2z}{\mathrm{d}t^2}\right). \tag{2.2}$$

Now suppose that we have a second frame of reference S′ in standard configuration with S, so that the coordinates in the two frames are related by the Galilean transformations

$$t' = t, \tag{2.3}$$
$$x' = x - Vt, \tag{2.4}$$
$$y' = y, \tag{2.5}$$
$$z' = z. \tag{2.6}$$

Differentiating the expressions for the position coordinates twice with respect to t', and noting that this is equivalent to differentiating with respect to t (since

$t' = t$), we see that

$$\frac{\mathrm{d}^2 x'}{\mathrm{d}t'^2} = \frac{\mathrm{d}^2 x}{\mathrm{d}t^2}, \quad \frac{\mathrm{d}^2 y'}{\mathrm{d}t'^2} = \frac{\mathrm{d}^2 y}{\mathrm{d}t^2}, \quad \frac{\mathrm{d}^2 z'}{\mathrm{d}t'^2} = \frac{\mathrm{d}^2 z}{\mathrm{d}t^2}.$$

Mass is certainly an invariant in Galilean relativity, so, under a Galilean transformation from frame S to frame S', the right-hand side of Equation 2.2 becomes

$$m\left(\frac{\mathrm{d}^2 x'}{\mathrm{d}t'^2}, \frac{\mathrm{d}^2 y'}{\mathrm{d}t'^2}, \frac{\mathrm{d}^2 z'}{\mathrm{d}t'^2}\right) \equiv m\boldsymbol{a}', \tag{2.7}$$

where the quantity $\boldsymbol{a}'$ has been introduced to emphasize the form-invariance of the right-hand side of Newton's second law under a Galilean transformation. This is a promising start, but what about the left-hand side: how does the force $\boldsymbol{f}$ transform under a Galilean transformation? To answer that question, we need to know how the force depends on the coordinates.

For the sake of definiteness, let's consider the case in which a body of mass m at position $\boldsymbol{r} = (x, y, z)$ is acted upon by a gravitational force due to a body of mass M at position $\boldsymbol{R} = (X, Y, Z)$ (see Figure 2.1). According to Newton's law of universal gravitation, in frame S the force will be

$$\boldsymbol{f} = -G\,\frac{mM}{d^2}\,\widehat{\boldsymbol{d}},$$

where G is Newton's gravitational constant (an invariant constant with the value $6.673 \times 10^{-11}\,\mathrm{N\,m^2\,kg^{-2}}$), the distance d is the magnitude of the displacement vector $\boldsymbol{d} = \boldsymbol{r} - \boldsymbol{R}$ from the body of mass M to the body of mass m, and $\widehat{\boldsymbol{d}}$ is a unit vector in the direction of $\boldsymbol{d}$.

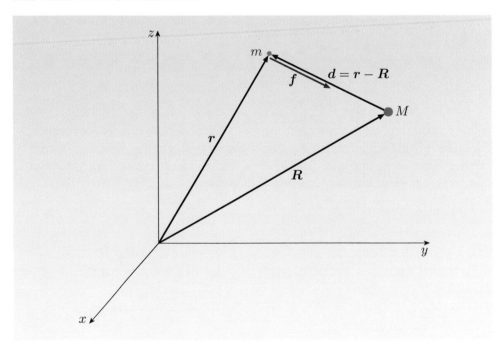

Figure 2.1   The gravitational force $\boldsymbol{f}$ on a body of mass m at position $\boldsymbol{r}$ due to a body of mass M at position $\boldsymbol{R}$.

Under a Galilean transformation from frame S to frame S', the position vectors of the two bodies will change, becoming $\boldsymbol{r}' \equiv (x', y', z') = (x - Vt, y, z)$

and $\boldsymbol{R}' \equiv (X', Y', Z') = (X - Vt, Y, Z)$, but the displacement between the bodies will be $\boldsymbol{d}' = \boldsymbol{r}' - \boldsymbol{R}' = (x - X, y - Y, z - Z)$, which is identical to the displacement $\boldsymbol{d}$ in frame S. It follows that the magnitude of the displacement d' and the unit vector $\widehat{\boldsymbol{d}'}$ in the direction of the displacement will also retain their old values. Since masses are invariant in Galilean relativity, we thus see that Newton's law of universal gravitation takes the same form in S and S'. Consequently, we can conclude that, at least in the case of gravitational forces, Newton's second law of motion, $\boldsymbol{f} = m\boldsymbol{a}$, also takes the same form in frames S and S', and by implication in all inertial frames. All we have to do to find the form of the law in frame S' is to add primes to all the old quantities, remembering that in the case of invariants the primes will be irrelevant since the primed quantities will have the same values as the unprimed quantities.

An equation that is form-invariant under a given coordinate transformation is sometimes said to be **covariant** under that transformation. In the particular case that we have been considering, not only have we shown that Newton's second law is covariant under the Galilean transformation, we have also concluded that the forces, masses and accelerations will have the same values in all inertial frames. So in this case, in addition to establishing the *covariance* of the equations, we have also shown the *invariance* of the quantities involved. Later in this chapter you will meet examples of physical laws that are covariant under a transformation but where the quantities involved are certainly not invariant.

The argument that we have already applied to Newton's second law in the case of gravitational forces can be extended to any force that depends only on a combination of displacements and invariants. Such an extension would include Hooke's law (for the force produced by the stretching of a spring) and even Coulomb's law of electrostatic forces. However, the argument cannot be extended to all conceivable forces. It does not, for example, work for electromagnetic forces that depend on the velocity of a charged particle. Of course, this failure is not a great concern to us since we have already seen that it was problems arising from electromagnetism and light that persuaded Einstein to reject Galilean relativity in favour of special relativity, even at the price of having to accept new laws of mechanics.

So, now that the idea of covariance or form-invariance has been introduced in the relatively simple context of Galilean relativity, let us return to special relativity and go in search of laws of mechanics that are covariant under the Lorentz transformations.

2.2 The laws of mechanics

2.2.1 Relativistic momentum

The best place to start the reformulation of mechanics is with the concept of momentum. This quantity plays a crucial role in the analysis of high-speed collisions between fundamental particles, one of the main areas where relativistic mechanics (i.e. Lorentz-covariant mechanics) is routinely used. Relativistic mechanics will be essential to the analysis of the high-energy proton–proton collisions in the Large Hadron Collider (Figure 2.2 overleaf) at CERN, near Geneva.

In Newtonian mechanics, the momentum of a particle of mass m travelling with velocity $\boldsymbol{v}$ is given by $\boldsymbol{p}_{\text{Newtonian}} = m\boldsymbol{v}$. The importance of momentum comes mainly from the observation that, provided that no external forces act on a system,

Figure 2.2   The Large Hadron Collider at CERN: a proton–proton collider, based on a 27 km-circle of bending magnets, accelerating cavities and gigantic detectors.

the total momentum of that system is conserved (i.e. constant). This means, as indicated in Figure 2.3, that if a particle of mass m_A travelling with some initial velocity $\boldsymbol{u}_A$ collides with a particle of mass m_B travelling with initial velocity $\boldsymbol{u}_B$, then after the collision the final velocities $\boldsymbol{v}_A$ and $\boldsymbol{v}_B$ of those two particles will be related by

$$m_A\boldsymbol{u}_A + m_B\boldsymbol{u}_B = m_A\boldsymbol{v}_A + m_B\boldsymbol{v}_B. \tag{2.8}$$

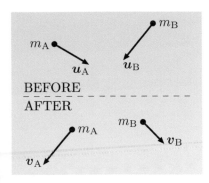

Figure 2.3   Two particles before and after a collision. The particles have velocities $\boldsymbol{u}_A$ and $\boldsymbol{u}_B$ before the collision, and $\boldsymbol{v}_A$ and $\boldsymbol{v}_B$ after the collision.

In special relativity, as you saw in Chapter 1, the rule for transforming velocities is rather complicated. This implies that momentum defined in the Newtonian way will obey an equally complicated transformation rule, and raises doubts about the covariance of Equation 2.8 under Lorentz transformations. Detailed calculations show that these doubts are justified. Even if Newtonian momentum is conserved in one inertial frame, the Lorentz velocity transformation shows that it cannot be conserved in all inertial frames. All this suggests that we should seek a new definition of momentum, sometimes called *relativistic momentum*, that will transform simply under Lorentz transformations and will provide a conservation law that is Lorentz-covariant. Of course, in formulating a new definition of momentum, we should not forget that physicists spent many years believing that experiments supported the conservation of Newtonian momentum — we should also aim to account for that.

Consider a particle of mass m travelling with uniform velocity $\boldsymbol{v}$ between two events, labelled 1 and 2, separated by the coordinate intervals $(\Delta t, \Delta x, \Delta y, \Delta z)$. What makes the Newtonian momentum of such a particle transform in a complicated way is its direct relationship to the particle's velocity:

$$\boldsymbol{v} \equiv (v_x, v_y, v_z) = \left(\frac{\Delta x}{\Delta t}, \frac{\Delta y}{\Delta t}, \frac{\Delta z}{\Delta t} \right). \tag{2.9}$$

This involves ratios such as $\Delta x / \Delta t$ where both Δx and Δt transform in moderately complicated ways. Momentum would transform far more simply if all references to the time between the two events, Δt, were replaced by references to the proper time between the events, $\Delta \tau$, which is an invariant and therefore transforms very simply. This suggests that a simple definition of the relativistic momentum of the particle would be

$$\boldsymbol{p} \equiv (p_x, p_y, p_z) = m \left(\frac{\Delta x}{\Delta \tau}, \frac{\Delta y}{\Delta \tau}, \frac{\Delta z}{\Delta \tau} \right). \tag{2.10}$$

Since the particle mass m and the proper time interval $\Delta \tau$ are both invariants, relativistic momentum defined like this will transform in the same way as the displacement vector $(\Delta x, \Delta y, \Delta z)$. Moreover, it follows from our discussion of proper time in Chapter 1 that since a particle travelling with speed v is present at both event 1 and event 2, the time between those events, Δt, is related to the proper time between them by $\Delta t = \gamma(v) \Delta \tau$, so we can rewrite the definition of relativistic momentum as

$$\boldsymbol{p} \equiv (p_x, p_y, p_z) = m\gamma(v) \left(\frac{\Delta x}{\Delta t}, \frac{\Delta y}{\Delta t}, \frac{\Delta z}{\Delta t} \right) = m\gamma(v)\boldsymbol{v}. \tag{2.11}$$

We now have a clear definition of relativistic momentum that is guaranteed to transform simply between different inertial frames. However, several issues remain to be resolved before we can accept it. First, does it lead to a Lorentz-covariant conservation law, so that the observed conservation of momentum in one inertial frame implies the conservation of momentum in all inertial frames? Second, is such a conservation law correct: is momentum defined in this new way really conserved in any inertial frame? (Remember, covariance establishes the acceptability of a law in principle, but only experiment can establish its truth in practice.) Third, how does this newly defined relativistic momentum relate to Newtonian momentum? Let's deal with the last of these questions first.

The relativistic momentum $\boldsymbol{p} = m\gamma(v)\boldsymbol{v}$ differs from Newtonian momentum only by a Lorentz factor $\gamma(v)$. This means that at speeds that are small compared with the speed of light, where $\gamma(v) \approx 1$, the two will be almost indistinguishable and all the apparent successes of Newtonian momentum conservation can be recovered.

As far as the covariance of relativistic momentum conservation is concerned, the question is this: if in some frame S

$$m_A \gamma(u_A)\boldsymbol{u}_A + m_B \gamma(u_B)\boldsymbol{u}_B = m_A \gamma(v_A)\boldsymbol{v}_A + m_B \gamma(v_B)\boldsymbol{v}_B, \tag{2.12}$$

will the velocity transformations also show that in some other inertial frame S$'$

$$m_A \gamma(u'_A)\boldsymbol{u}'_A + m_B \gamma(u'_B)\boldsymbol{u}'_B = m_A \gamma(v'_A)\boldsymbol{v}'_A + m_B \gamma(v'_B)\boldsymbol{v}'_B \,? \tag{2.13}$$

Note that there are no primes on any of the masses in this last equation — that's because they are invariant.

We could perform a detailed calculation to show that the law of relativistic momentum conservation is covariant under Lorentz transformations, but it's really not necessary. There is a much neater way of reaching the same conclusion based on the fact that the relativistic momentum (p_x, p_y, p_z) transforms in the same way as the displacement vector $(\Delta x, \Delta y, \Delta z)$. Suppose that we let the initial momenta in frame S be $\boldsymbol{p}_A$ and $\boldsymbol{p}_B$, and let the final momenta be $\overline{\boldsymbol{p}}_A$ and $\overline{\boldsymbol{p}}_B$. Then relativistic momentum conservation implies that $\boldsymbol{p}_A + \boldsymbol{p}_B = \overline{\boldsymbol{p}}_A + \overline{\boldsymbol{p}}_B$, or, after a slight rearrangement,

$$\boldsymbol{p}_A + \boldsymbol{p}_B + (-\overline{\boldsymbol{p}}_A) + (-\overline{\boldsymbol{p}}_B) = \boldsymbol{0}. \tag{2.14}$$

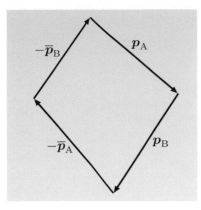

Figure 2.4   If the final momenta are reversed in direction, the conservation of momentum can be represented graphically by a closed figure in which arrows representing the particle momenta join head to tail.

Now, this equation can be represented geometrically as in Figure 2.4. With the arrows corresponding to the final momenta reversed in direction, the four arrows representing the individual momenta in frame S form a closed figure when drawn head to tail. Under a Lorentz transformation to some other frame, all of these momenta may change, but since they transform like displacement vectors, it will still be the case, even after transformation, that they will form a closed figure. Hence we can be sure that the transformed momenta will obey

$$\boldsymbol{p}'_A + \boldsymbol{p}'_B + (-\overline{\boldsymbol{p}}'_A) + (-\overline{\boldsymbol{p}}'_B) = \boldsymbol{0}, \tag{2.15}$$

and consequently $\boldsymbol{p}'_A + \boldsymbol{p}'_B = \overline{\boldsymbol{p}}'_A + \overline{\boldsymbol{p}}'_B$. Thus we see, in this case at least, that if relativistic momentum is conserved in one inertial frame, then it will be conserved in all inertial frames. This geometric argument can be extended to as many colliding particles as we want, so the argument shows that relativistic momentum conservation is a Lorentz-covariant result.

● Why can't this same geometric argument be used to show that Newtonian momentum, if conserved in some inertial frame S, will also be conserved in all other inertial frames, even under Lorentz transformations?

○ This is because Newtonian momentum does not transform in the same way as a displacement vector under a Lorentz transformation. Even if the Newtonian momentum vectors formed a closed figure in frame S, the complicated transformation law of Newtonian momentum would ensure that they did not form a closed figure in all other inertial frames.

Now the only remaining question is: 'does nature really make use of this possibility?' Here experiment is the arbiter, and the analysis of an enormous number of high-speed particle collisions clearly indicates that nature does so. It is relativistic momentum that is found to be conserved in nature. So we can conclude the following.

Relativistic momentum

In Lorentz-covariant mechanics, the relativistic momentum of a particle of mass m moving with velocity $\boldsymbol{v}$ is defined as

$$\boldsymbol{p} = \gamma(v)m\boldsymbol{v} = \frac{m\boldsymbol{v}}{\sqrt{1 - v^2/c^2}}. \tag{2.16}$$

The total relativistic momentum of a system is conserved in the absence of external forces.

Exercise 2.1   An electron of mass $m = 9.11 \times 10^{-31}$ kg has speed $4c/5$. What is the magnitude of its (relativistic) momentum?   ∎

2.2.2   Relativistic kinetic energy

Another quantity of importance in mechanics is kinetic energy. As in the case of momentum, special relativity demands that we modify the definition of kinetic energy before it can take its proper place in a Lorentz-covariant formulation of mechanics.

In Newtonian mechanics, the kinetic energy of a particle travelling with speed v can be found from the work W done in accelerating that particle from rest to its final speed v. If we consider the case of a particle with speed u accelerated along the x-axis by a force of magnitude f, we can write the kinetic energy as

$$E_K = W = \int_{u=0}^{u=v} f \, dx. \tag{2.17}$$

In Newtonian mechanics, the applied force is the same as the rate of change of momentum, $f = ma = m \, dv/dt = dp/dt$, so

$$E_K = \int_{u=0}^{u=v} \frac{dp}{dt} \, dx. \tag{2.18}$$

The integral can be rewritten in a much more useful form by changing integration variables and using the chain rule:

$$\int \frac{dp}{dt} \, dx = \int \frac{dp}{dt} \frac{dx}{dp} \, dp = \int \frac{dx}{dt} \, dp = \int u \, dp.$$

In this way a Newtonian expression for kinetic energy that initially involved distance and force can be re-expressed in terms of speed u and momentum magnitude p. This latter expression can be taken over to special relativity, where we already know the relationship between speed and the magnitude of momentum.

So, in special relativity, a reasonable starting point from which to define the relativistic kinetic energy of a particle of mass m moving with speed v is

$$E_K = \int_0^v u \, d\left(\frac{mu}{\sqrt{1 - (u/c)^2}} \right).$$

This integral can be evaluated using the technique of integration by parts:

$$E_K = \left[\frac{mu^2}{\sqrt{1 - (u/c)^2}} \right]_0^v - \int_0^v \frac{mu}{\sqrt{1 - (u/c)^2}} \, du.$$

The remaining integral can be performed by inspection, giving

$$E_K = \left[\frac{mu^2}{\sqrt{1 - (u/c)^2}} + mc^2 \sqrt{1 - (u/c)^2} \right]_0^v.$$

A compact final result can be found by putting both terms over a common denominator:

$$E_K = \left[\frac{mc^2}{\sqrt{1 - (u/c)^2}} \right]_0^v = mc^2 \left(\frac{1}{\sqrt{1 - (v/c)^2}} - 1 \right).$$

Thus the suggested expression for the **relativistic kinetic energy** of a particle of mass m moving with speed v is

$$E_K = (\gamma(v) - 1)mc^2. \tag{2.19}$$

There is no general principle of conservation of kinetic energy for us to consider in this case, but in Newtonian physics, kinetic energy is conserved in **elastic collisions**. In an elastic collision, the particles do not change their number, state or nature, so what goes in is also what comes out. As far as covariance under

Lorentz transformations is concerned, it is possible to show that in the case of
elastic collisions, the proposed expression for relativistic kinetic energy does
ensure that an elastic collision in one inertial frame will also be elastic in all other
inertial frames.

How does the relativistic kinetic energy relate to the Newtonian kinetic energy?
At first sight the relationship is not at all obvious, but it soon becomes clear if we
use the following mathematical expansion of the Lorentz factor, $\gamma(v)$, obtained
via Taylor's theorem or the binomial expansion:

$$\gamma(v) = \frac{1}{\sqrt{1 - v^2/c^2}} = 1 + \frac{1}{2}\frac{v^2}{c^2} + \frac{3}{8}\left(\frac{v^2}{c^2}\right)^2 + \cdots . \tag{2.20}$$

The expansion continues with higher orders of v^2/c^2. In Newtonian physics, the
speed v will generally be small compared with the speed of light c, so these
higher-order terms can be ignored. Substituting the truncated expression for $\gamma(v)$
into Equation 2.19 gives

$$E_K = \left[1 + \frac{1}{2}\frac{v^2}{c^2} + \frac{3}{8}\left(\frac{v^2}{c^2}\right)^2 - 1\right]mc^2$$

$$\approx \tfrac{1}{2}mv^2 + \text{terms of order } \frac{v^4}{c^2} + \cdots . \tag{2.21}$$

So the Newtonian expression for kinetic energy emerges as a low-speed
approximation to the relativistic expression. All the low-speed experiments that
support the Newtonian expression will also support the more general expression
of relativistic mechanics.

Henceforth we shall adopt the proposed definition, so we can say the following.

Relativistic kinetic energy

In Lorentz-covariant mechanics, the relativistic kinetic energy of a particle
of mass m moving with speed v is

$$E_K = (\gamma(v) - 1)mc^2 = \frac{mc^2}{\sqrt{1 - v^2/c^2}} - mc^2. \tag{2.22}$$

Exercise 2.2   Compute the kinetic energy of a muon (mass
$m_\mu = 1.88 \times 10^{-28}$ kg) travelling with speed $9c/10$.   ∎

2.2.3   Total relativistic energy and mass energy

In the 1905 paper in which Einstein introduced the special theory of relativity, he
considered the acceleration of an electron and arrived at the expressions for
momentum and kinetic energy that have been introduced in this chapter. However,
our next topic is one that was not considered in that first paper. It concerns the
best known result of special relativity, $E = mc^2$, and the 'equivalence' between
mass and energy that it is usually said to indicate. It was first discussed in a
three-page paper ('Does the inertia of a body depend upon its energy content?')

published a few months after the first paper, and then more fully developed in later publications.

The crucial result is already suggested by the expression for relativistic kinetic energy $E_K = (\gamma(v) - 1)mc^2$, which can be rewritten as

$$\gamma(v)mc^2 = E_K + mc^2. \tag{2.23}$$

This is now interpreted as showing that in an inertial frame S, where a particle of mass m has speed v, that particle will have a **total relativistic energy** $E = \gamma(v)mc^2$ that is the sum of a relativistic kinetic energy E_K and a **mass energy** $E_0 = mc^2$. As a mere rearrangement and renaming of terms, this is a harmless exercise. The revolutionary step is in the proposal that when no external forces act, in relativistic mechanics generally, and high-speed particle collisions in particular, *it is the total relativistic energy that is conserved.* In high-speed collisions, neither kinetic energy nor mass energy will necessarily be conserved, but their sum, represented by the total relativistic energy E, will be.

The startling possibility opened up by this suggestion is that in high-speed collisions, particles with mass may be created at the expense of relativistic kinetic energy. It is also possible for some or all of the particles involved in a collision to be annihilated, releasing mass energy that may emerge from the collision either as the mass energy of particles created in the collision or as a contribution to the kinetic energy of all the particles that emerge, or both. This takes relativistic mechanics into an important domain that was completely unexplored by Newtonian mechanics.

It's worth noting that the relationship between mass and energy represented by the formula $E_0 = mc^2$ is not limited to high-speed particle collisions. The initial arguments in favour of such a relationship were based on considerations of the emission of radiation from a body, and it has often been stressed that in the case of a composite body, such as a piece of metal, the simple act of heating it so as to raise its temperature will increase its internal energy and thereby increase its mass. Note that this has nothing to do with the speed of the body; it is a change in the invariant mass that we are discussing.

When Einstein first proposed the equivalence of mass and energy, he suggested that it might account for the energy associated with radioactive decay. This is now known to be the case. $E_0 = mc^2$ plays a vital role in explaining many nuclear phenomena, and particle creation (see Figure 2.5) is the basis of much of the work carried out in particle physics laboratories. Ironically, Einstein's famous relation has also become indissolubly linked with the awesome energy release of nuclear weapons (Figure 2.6) despite Einstein's many pronouncements on the need for world peace.

Figure 2.5 Tracks of particles produced in a high-energy collision between two elementary particles.

Figure 2.6 An atomic explosion — a horrifying reminder of mass–energy equivalence.

Total relativistic energy and mass energy

In Lorentz-covariant mechanics, the total relativistic energy E and the mass energy E_0 (sometimes called the rest energy) of a particle of mass m with speed v are given by

$$E = \gamma(v)mc^2 = \frac{mc^2}{\sqrt{1 - v^2/c^2}}, \tag{2.24}$$

$$E_0 = mc^2. \tag{2.25}$$

Exercise 2.3   The proton has mass $m_{\mathrm{p}} = 1.67 \times 10^{-27}$ kg. Compute the total (relativistic) energy of a proton moving with speed $v = 3c/5$.

Exercise 2.4   At what speed is the total energy of a particle twice the mass energy?

Exercise 2.5   (a) In a nuclear fission of uranium-235 caused by the absorption of a neutron, nuclei of krypton and barium are produced, and three neutrons are emitted. The difference in the total mass of the particles present at the start of the process and those present at the end is 3.08×10^{-28} kg. What is the energy released in this process, in both joules and electronvolts ($1 \, \mathrm{eV} = 1.60 \times 10^{-19}$ J)?

(b) Given that the binding energy of a hydrogen atom is 13.6 eV, what is the difference between the mass of a hydrogen atom and the masses of its constituent electron and proton?   ■

2.2.4   Four-momentum

In Chapter 1 you were briefly introduced to the four-position, that is, the four-component object

$$[x^{\mu}] \equiv (x^0, x^1, x^2, x^3) = (ct, x, y, z), \tag{2.26}$$

which usefully combined space and time coordinates while using ct rather than t to ensure that they could all be expressed in units of distance. Often when writing $[x^{\mu}]$, it is convenient to write $\boldsymbol{r}$ instead of the three space components x, y and z, so we can write

$$[x^{\mu}] = (ct, \boldsymbol{r}).$$

Now suppose that the four-position $[x^{\mu}]$ describes the events on the spacetime pathway (i.e. the world-line) of a particle of mass m. We can imagine that the particle carries a clock with it that records the proper time τ between successive events as it moves along its world-line. We can then regard each component x^{μ} of the particle's four-position $[x^{\mu}]$ as a function of proper time τ. Differentiating each of those components x^{μ} with respect to τ gives us four so-called *proper derivatives* $\mathrm{d}x^{\mu}/\mathrm{d}\tau$ that we can use as the components of another four-component entity that we shall denote $[U^{\mu}]$. Thus

$$[U^{\mu}] = \left[\frac{\mathrm{d}x^{\mu}}{\mathrm{d}\tau}\right] = \left(c\frac{\mathrm{d}t}{\mathrm{d}\tau}, \frac{\mathrm{d}x}{\mathrm{d}\tau}, \frac{\mathrm{d}y}{\mathrm{d}\tau}, \frac{\mathrm{d}z}{\mathrm{d}\tau}\right). \tag{2.27}$$

The derivatives $\mathrm{d}x/\mathrm{d}\tau$, $\mathrm{d}y/\mathrm{d}\tau$ and $\mathrm{d}z/\mathrm{d}\tau$ that appear on the right can be regarded as infinitesimal limits of the ratios $\Delta x/\Delta\tau$, $\Delta y/\Delta\tau$ and $\Delta z/\Delta\tau$ that we considered earlier when introducing relativistic momentum. On that earlier occasion, the relation $\Delta t = \gamma(v)\,\Delta\tau$ was used to relate those ratios to the scaled velocity components $\gamma(v)v_x$, $\gamma(v)v_y$ and $\gamma(v)v_z$. Doing the same here, and also noting that $\mathrm{d}t/\mathrm{d}\tau$ is the limit of $\Delta t/\Delta\tau = \gamma(v)$, we can write

$$[U^{\mu}] \equiv (U^0, U^1, U^2, U^3) = (c\gamma(v), \gamma(v)\boldsymbol{v}). \tag{2.28}$$

This quantity is called the **four-velocity** of the particle. Since $[U^{\mu}]$ is the derivative of $[x^{\mu}]$ with respect to the invariant τ, the four-velocity $[U^{\mu}]$ behaves just as the four-position does under Lorentz transformations.

The four-velocity has an interesting property that becomes apparent when $[U^\mu]$ is combined with the Minkowski metric $[\eta_{\mu\nu}]$ that was introduced in Chapter 1. In that earlier case, we met the invariant $(\Delta s)^2 = \sum_{\mu,\nu=0}^{3} \eta_{\mu\nu} \, \Delta x^\mu \, \Delta x^\nu$. Now we can see that

$$\sum_{\mu,\nu=0}^{3} \eta_{\mu\nu} \, U^\mu \, U^\nu = \gamma(v)^2 c^2 - \gamma(v)^2 [(v_x)^2 + (v_y)^2 + (v_z)^2]$$
$$= \gamma(v)^2 [c^2 - v^2]. \tag{2.29}$$

But since

$$\gamma(v)^2 = \frac{c^2}{c^2 - v^2}, \tag{2.30}$$

it is clear that the original sum has the invariant value

$$\sum_{\mu,\nu=0}^{3} \eta_{\mu\nu} \, U^\mu \, U^\nu = c^2. \tag{2.31}$$

Multiplying the four-velocity $[U^\mu]$ by the invariant mass m gives a related four-component entity called the **four-momentum**:

$$[P^\mu] = m[U^\mu] = (\gamma(v)mc, \gamma(v)mv_x, \gamma(v)mv_y, \gamma(v)mv_z). \tag{2.32}$$

All the terms on the right should already be familiar. The first is the total relativistic energy divided by the speed of light, E/c; the other three are the components of the relativistic momentum $\boldsymbol{p}$, so we can write

$$[P^\mu] \equiv (P^0, P^1, P^2, P^3) = (E/c, \boldsymbol{p}). \tag{2.33}$$

It is clear that the four-momentum contains all the information about the relativistic energy and relativistic momentum of any particle.

The crucial point about all this is that under a Lorentz transformation from one inertial frame to another (S to S$'$, say), the four-momentum $[P^\mu]$ transforms in exactly the same way as the four-position.

- ● Why must the four-momentum transform in the same way as the four-position?

- ○ Because τ is an invariant, the four-velocity $[U^\mu] = [\mathrm{d}x^\mu/\mathrm{d}\tau]$ will transform in the same way as the four-position $[x^\mu]$. Since m is also invariant, it follows that the four-momentum $[P^\mu] = [mU^\mu]$ must also transform like $[x^\mu]$.

As a result of the simple behaviour of $[P^\mu]$ under a Lorentz transformation, we can say that if the frames S and S$'$ are in standard configuration, then a particle of mass m with velocity $\boldsymbol{v}$, that has relativistic energy E and relativistic momentum $\boldsymbol{p}$ in inertial frame S, will be found to have energy E' and momentum $\boldsymbol{p}'$ in S$'$, where

$$E' = \gamma(V)(E - Vp_x), \tag{2.34}$$
$$p'_x = \gamma(V)(p_x - VE/c^2), \tag{2.35}$$
$$p'_y = p_y, \tag{2.36}$$
$$p'_z = p_z. \tag{2.37}$$

Note that the particle speeds v and v' that help to determine the energy and momentum in S and S$'$ are quite distinct from V, which represents the speed of frame S′ as measured in frame S. For a particle travelling along the x-axis, so that $v = (v, 0, 0)$, the relation between v, v' and V follows from the velocity transformation of Chapter 1, and is given by $v' = (v - V)/(1 - vV/c^2)$.

As was the case with the four-position, the transformation rule for four-momentum can be written in a number of equivalent ways using the Lorentz transformation matrix $[\Lambda^\mu{}_\nu]$. In terms of matrices,

$$
\begin{pmatrix} E'(v')/c \\ p'_x(v') \\ p'_y(v') \\ p'_z(v') \end{pmatrix} = \begin{pmatrix} \gamma(V) & -\gamma(V)V/c & 0 & 0 \\ -\gamma(V)V/c & \gamma(V) & 0 & 0 \\ 0 & 0 & 1 & 0 \\ 0 & 0 & 0 & 1 \end{pmatrix} \begin{pmatrix} E(v)/c \\ p_x(v) \\ p_y(v) \\ p_z(v) \end{pmatrix}, \quad (2.38)
$$

which may be represented more compactly as

$$
[P'^\mu] = [\Lambda^\mu{}_\nu][P^\nu]. \tag{2.39}
$$

Alternatively, we can represent the transformation using components and summations:

$$
P'^\mu = \sum_{\nu=0}^{3} \Lambda^\mu{}_\nu P^\nu \quad (\mu = 0, 1, 2, 3). \tag{2.40}
$$

The fact that the four-momentum transforms in exactly the same way as the four-position says something quite profound about energy and momentum. Under Lorentz transformation, the energy and momentum components intertwine, and can be thought of as aspects of a single quantity, just as space and time are unified into spacetime. That which is energy to one observer is a mix of energy and momentum to another.

Exercise 2.6   In frame of reference S, an electron moving along the x-axis has energy $3m_e c^2$ and momentum magnitude $\sqrt{8}m_e c$. Use the transformations of energy and momentum to find the energy and momentum magnitude observed in frame S′ moving with speed $4c/5$ relative to S in the positive x-direction.   ■

2.2.5   The energy–momentum relation

It was shown in Equation 2.31 that

$$
\sum_{\mu,\nu=0}^{3} \eta_{\mu\nu} U^\mu U^\nu = c^2.
$$

Since $[P^\mu]$ is defined by $[P^\mu] = m[U^\mu]$, it follows that

$$
\sum_{\mu,\nu=0}^{3} \eta_{\mu\nu} P^\mu P^\nu = m^2 c^2. \tag{2.41}
$$

But using the Minkowski metric we also know that

$$
\sum_{\mu,\nu=0}^{3} \eta_{\mu\nu} P^\mu P^\nu = \frac{E^2}{c^2} - (p_x)^2 - (p_y)^2 - (p_z)^2 = \frac{E^2}{c^2} - p^2. \tag{2.42}
$$

Consequently, we can say that

$$
E^2 - c^2 p^2 = m^2 c^4.
$$

So, regardless of the frame of reference of an observer, the difference between the squared total energy and the squared momentum magnitude multiplied by the speed of light squared is proportional to the squared invariant mass. This extremely useful relationship is often called the **energy–momentum relation** and is usually written as follows.

Energy–momentum relation

$$E^2 = p^2c^2 + m^2c^4. \tag{2.43}$$

Taking the positive square root, we see that

$$E = \sqrt{m^2c^4 + c^2p^2}. \tag{2.44}$$

A plot of this relation can be seen in Figure 2.7. Apart from the presence of the distinctly non-Newtonian rest energy $E_0 = mc^2$, the behaviour at low speeds (when p is close to zero) is what would be expected in Newtonian mechanics, with (kinetic) energy increasing in proportion to p^2. However, as the momentum magnitude increases, the total energy becomes more and more nearly proportional to the momentum magnitude, as special relativity requires.

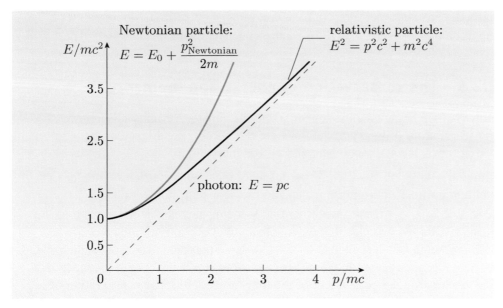

Figure 2.7 Plots of the energy–momentum relation for a Newtonian particle, a relativistic particle and a photon. Note that the energy is expressed in units of the massive particle's rest energy mc^2 and the momentum magnitude in units of mc.

● The electron has mass $m_e = 9.11 \times 10^{-31}$ kg. What is the energy of an electron that has a momentum of magnitude 1.00×10^{-22} kg m s^{-1}?

○ Making the substitutions $m^2c^4 = 6.72 \times 10^{-27}$ J^2 and $c^2p^2 = 9.00 \times 10^{-28}$ J^2, the energy–momentum relation shows that the energy is $E = \sqrt{6.72 \times 10^{-27} + 9.00 \times 10^{-28}}$ J $= 8.73 \times 10^{-14}$ J.

The energy–momentum relation has an important consequence with no analogue in Newtonian mechanics. A symmetry principle known as *gauge-invariance*, which is of great importance in particle physics, demands that the photon, which is often described as the 'particle of light', should be strictly massless with $m = 0$. It follows from the energy–momentum relation that for a photon, or any other massless particle,

$$p = E/c. \tag{2.45}$$

The photon carries energy, so even though it has no mass, it does have a momentum. This clearly shows the non-Newtonian nature of relativistic momentum.

It has been suggested that the momentum of the photon could be harnessed to make solar sails, a kind of propulsion system for spacecraft. A depiction of a solar-sail craft is shown in Figure 2.8.

Exercise 2.7   (a) The energy of a photon is hf, where $h = 6.63 \times 10^{-34}\,\mathrm{J\,s}$ is Planck's constant and f is the frequency of the photon. What is the magnitude of the momentum of a single photon belonging to a monochromatic beam of light with frequency $5.00 \times 10^{14}\,\mathrm{Hz}$?

(b) At what rate must such photons be absorbed by a solar sail if they are to cause a steady force of magnitude $10\,\mathrm{N}$ on the sail?

Exercise 2.8   You are told by a scientist of ill repute that a ficteron particle of mass m_f has been measured to have energy $E_\mathrm{f} = 3m_\mathrm{f}c^2$ and momentum of magnitude $p_\mathrm{f} = 7m_\mathrm{f}c$. Are those values consistent with special relativity?   ■

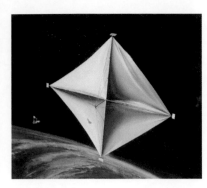

Figure 2.8   Solar sails have been proposed as a form of spacecraft propulsion. They are propelled by the momentum of photons.

2.2.6   The conservation of energy and momentum

Now that we know how the four-momentum transforms under a Lorentz transformation, it is easy to demonstrate the Lorentz covariance of the conservation laws of relativistic energy and momentum.

Imagine a collision in which N particles collide, and $\overline{N}$ particles emerge. Let incident particle i have mass m_i and an incident four-momentum $[P^\nu_{(i)}]$ $(i = 1, 2, 3, \ldots, N)$, and remember that some of the masses may be zero. Similarly, let the particles that emerge from the collision have masses $\overline{m}_j$ and four-momenta $[\overline{P}^\nu_{(j)}]$ $(j = 1, 2, 3, \ldots, \overline{N})$. Note that the index representing the particle has been placed in parentheses to avoid confusing it with the index that denotes a particular component of the four-momentum. The conservation of energy and momentum in an inertial frame S is represented by the relation

$$P^\nu_{(1)} + P^\nu_{(2)} + \cdots + P^\nu_{(N)} = \overline{P}^\nu_{(1)} + \overline{P}^\nu_{(2)} + \cdots + \overline{P}^\nu_{(\overline{N})}. \tag{2.46}$$

Note that ν is a free index in this expression, so this one line really represents four different equations, one for each possible value of ν.

What will be the energy and momentum involved in this collision as observed by some other inertial observer who uses frame S′? Performing the same Lorentz transformation on each side of the equation, we see that

$$\sum_{\nu=0}^{3} \Lambda^\mu{}_\nu (P^\nu_{(1)} + P^\nu_{(2)} + \cdots + P^\nu_{(N)}) = \sum_{\nu=0}^{3} \Lambda^\mu{}_\nu (\overline{P}^\nu_{(1)} + \overline{P}^\nu_{(2)} + \cdots + \overline{P}^\nu_{(\overline{N})}). \tag{2.47}$$

Since the transformation law of an individual four-momentum takes the form $[P'^{\mu}] = [\Lambda^{\mu}{}_{\nu}] [P^{\nu}]$, we know that each individual four-momentum in the sum will transform in the same way under the Lorentz transformation to frame S'. Consequently, transforming both sides of the conservation law, we get

$$P'^{\mu}_{(1)} + P'^{\mu}_{(2)} + \cdots + P'^{\mu}_{(N)} = \overline{P}'^{\mu}_{(1)} + \overline{P}'^{\mu}_{(2)} + \cdots + \overline{P}'^{\mu}_{(\overline{N})}. \qquad (2.48)$$

Apart from an irrelevant switch in the symbol used to represent the free index, from ν to μ, the only difference between the conservation law in frame S and that in frame S' is the addition of some primes.

The lesson is clear: by expressing the conservation laws of relativistic energy and relativistic momentum in terms of four-momenta that transform simply under Lorentz transformations, it has become obvious that the conservation laws can be written in the same form in all inertial frames without any need to carry out complicated transformations of E and $\boldsymbol{p}$. In such situations we say that the law is **manifestly covariant**. This is only a first glimpse of manifest covariance. We shall have much more to say on the subject later.

2.2.7 Four-force

The last major mechanics concept that we shall discuss is that of force. Recalling that in Newtonian particle mechanics, force may be defined by the rate of change of momentum, and taking the introduction of the four-momentum as a guide, a natural way to introduce a **four-force** in relativistic mechanics is via the manifestly covariant relation

$$[F^{\mu}] = \left[\frac{\mathrm{d}P^{\mu}}{\mathrm{d}\tau}\right] = \left(\frac{1}{c}\frac{\mathrm{d}E}{\mathrm{d}\tau}, \frac{\mathrm{d}p_x}{\mathrm{d}\tau}, \frac{\mathrm{d}p_y}{\mathrm{d}\tau}, \frac{\mathrm{d}p_z}{\mathrm{d}\tau}\right). \qquad (2.49)$$

Note that the differentiation is with respect to the invariant proper time τ. To make the link with Newtonian mechanics as close as possible, we identify the spatial components of the four-force with $\gamma(v)\boldsymbol{f}$, where $\boldsymbol{f}$ is a 'conventional' force vector: $\boldsymbol{f} = (f_x, f_y, f_z)$. (This is similar to our identification of the scaled velocity $\gamma(v)\boldsymbol{v}$ with the spatial components of $[U^{\mu}]$.) Making the usual identification $\Delta t = \gamma(v)\,\Delta\tau$, and taking the limit as $\Delta\tau$ tends to zero, gives

$$F^0 = \frac{1}{c}\frac{\mathrm{d}E}{\mathrm{d}\tau} = \frac{\gamma(v)}{c}\frac{\mathrm{d}E}{\mathrm{d}t}, \qquad (2.50)$$

and we can then identify $\mathrm{d}E/\mathrm{d}t$, the rate of change of total energy, with the rate at which the force $\boldsymbol{f}$ performs work, which is given by the scalar product $\boldsymbol{f} \cdot \boldsymbol{v}$. So we have

$$[F^{\mu}] = \left[\frac{\mathrm{d}P^{\mu}}{\mathrm{d}\tau}\right] = \left(\frac{\gamma}{c}\boldsymbol{f} \cdot \boldsymbol{v}, \gamma f_x, \gamma f_y, \gamma f_z\right) = \left(\frac{\gamma}{c}\boldsymbol{f} \cdot \boldsymbol{v}, \gamma \boldsymbol{f}\right). \qquad (2.51)$$

It's tempting to think that the 'conventional' force vector $\boldsymbol{f}$ must be the Newtonian force, but things are not quite so simple. Having defined the four-force as the derivative of the four-momentum with respect to the proper time, we can be sure that under a Lorentz transformation, the four-force will transform in a simple way, similar to that of the four-momentum and four-position. For the usual case of frames S and S′ in standard configuration, that means

$$F'^0 = \gamma(V)(F^0 - VF^1/c), \tag{2.52}$$

$$F'^1 = \gamma(V)(F^1 - VF^0/c), \tag{2.53}$$

$$F'^2 = F^2, \tag{2.54}$$

$$F'^3 = F^3. \tag{2.55}$$

This will automatically determine the way in which the force vector f must transform. It turns out that the electromagnetic Lorentz force that we consider in the next section does transform in just the required way, but the Newtonian gravitational force (which was shown to be form-invariant under the Galilean transformation in an earlier section) does *not* obey the required transformation. This means that it will be relatively simple to extend the ideas that we have been developing in this section to include electromagnetic forces, but we shall not be able to treat the Newtonian gravitational force as part of a four-force. In fact, we shall have to develop an entirely new theory of gravitation that will take us beyond special relativity and in which force will have almost no part to play at all. This is the role of general relativity.

Exercise 2.9   Given frames S and S′ in standard configuration with relative speed V, write down the expressions that relate the components of the three-force f' measured in frame S′ to the components of the same three-force f that would be measured in frame S.   ■

2.2.8   Four-vectors

You will have gathered by now that among the most important quantities in Lorentz-covariant mechanics are several four-component entities, including:

- the four-position $[x^\mu] = (ct, \boldsymbol{r})$
- the four-velocity $[U^\mu] = (\gamma c, \gamma \boldsymbol{v})$
- the four-momentum $[P^\mu] = (E/c, \boldsymbol{p})$
- the four-force $[F^\mu] = (\gamma \boldsymbol{f} \cdot \boldsymbol{v}/c, \gamma \boldsymbol{f})$.

To this list we may add the **four-displacement** $[\Delta x^\mu] = (c\,\Delta t, \Delta \boldsymbol{r})$. (The four-position is really a special case of the four-displacement in which the coordinate intervals are measured from the origin.) These quantities are all examples of a general class of four-component entities called **contravariant four-vectors**.

Four-vectors will play an important role in the next section, so we shall take this opportunity to introduce them properly and explain their mathematical properties. The defining characteristic that distinguishes the four-vectors introduced so far from other four-component objects is the way that they behave under a Lorentz transformation.

Given two inertial frames S and S' in standard configuration, the components A^μ of a contravariant four-vector $[A^\mu] \equiv (A^0, A^1, A^2, A^3)$ transform according to

$$A'^0 = \gamma(V)(A^0 - V A^1/c), \tag{2.56}$$
$$A'^1 = \gamma(V)(A^1 - V A^0/c), \tag{2.57}$$
$$A'^2 = A^2, \tag{2.58}$$
$$A'^3 = A^3, \tag{2.59}$$

which may be written more compactly in terms of matrices or as a summation:

$$[A'^\mu] = [\Lambda^\mu{}_\nu][A^\nu], \tag{2.60}$$

$$A'^\mu = \sum_{\nu=0}^{3} \Lambda^\mu{}_\nu A^\nu \quad (\mu = 0, 1, 2, 3). \tag{2.61}$$

To this extent, all contravariant four-vectors behave like four-displacements. However, not all four-component objects are four-vectors, nor, as you are about to see, are contravariant four-vectors the only kind of four-vectors.

Suppose that ϕ is some scalar function of x^0, x^1, x^2 and x^3 that is invariant under Lorentz transformations, so $\phi'(x'^0, x'^1, x'^2, x'^3) = \phi(x^0, x^1, x^2, x^3)$. Consider the behaviour of the derivative $\partial\phi/\partial x^0$, which we shall denote B_0. Under the usual Lorentz transformation from frame S to frame S', the function B_0 will become some new function B'_0, the form of which can be determined using the chain rule of partial differentiation:

$$B'_0 = \frac{\partial\phi'}{\partial x'^0} = \frac{\partial\phi}{\partial x^0}\frac{\partial x^0}{\partial x'^0} + \frac{\partial\phi}{\partial x^1}\frac{\partial x^1}{\partial x'^0} + \frac{\partial\phi}{\partial x^2}\frac{\partial x^2}{\partial x'^0} + \frac{\partial\phi}{\partial x^3}\frac{\partial x^3}{\partial x'^0}. \tag{2.62}$$

The partial derivatives $\partial x^0/\partial x'^0$, $\partial x^1/\partial x'^0$, $\partial x^2/\partial x'^0$ and $\partial x^3/\partial x'^0$ can each be easily determined from the inverse Lorentz transformations given in Chapter 1 as Equations 1.14–1.17, and turn out to be

$$\frac{\partial x^0}{\partial x'^0} = \gamma(V), \quad \frac{\partial x^1}{\partial x'^0} = \gamma(V)\frac{V}{c}, \quad \frac{\partial x^2}{\partial x'^0} = 0, \quad \frac{\partial x^3}{\partial x'^0} = 0.$$

Substituting these results into Equation 2.62, and representing $\partial\phi/\partial x^\mu$ by B_μ, you can see that under a Lorentz transformation,

$$B'_0 = \gamma(V)(B_0 + V B_1/c).$$

Performing similar calculations for all the other partial derivatives of ϕ leads to the following transformation rule for the four quantities B_μ:

$$B'_0 = \gamma(V)(B_0 + V B_1/c), \tag{2.63}$$
$$B'_1 = \gamma(V)(B_1 + V B_0/c), \tag{2.64}$$
$$B'_2 = B_2, \tag{2.65}$$
$$B'_3 = B_3. \tag{2.66}$$

Now, this is very similar to an inverse Lorentz transformation. In fact, if we use the symbol $[(\Lambda^{-1})_\mu{}^\nu]$ to represent the **inverse Lorentz transformation matrix**

$$[(\Lambda^{-1})_\mu{}^\nu] \equiv \begin{pmatrix} (\Lambda^{-1})_0{}^0 & (\Lambda^{-1})_0{}^1 & (\Lambda^{-1})_0{}^2 & (\Lambda^{-1})_0{}^3 \\ (\Lambda^{-1})_1{}^0 & (\Lambda^{-1})_1{}^1 & (\Lambda^{-1})_1{}^2 & (\Lambda^{-1})_1{}^3 \\ (\Lambda^{-1})_2{}^0 & (\Lambda^{-1})_2{}^1 & (\Lambda^{-1})_2{}^2 & (\Lambda^{-1})_2{}^3 \\ (\Lambda^{-1})_3{}^0 & (\Lambda^{-1})_3{}^1 & (\Lambda^{-1})_3{}^2 & (\Lambda^{-1})_3{}^3 \end{pmatrix}$$

$$= \begin{pmatrix} \gamma(V) & \gamma(V)V/c & 0 & 0 \\ \gamma(V)V/c & \gamma(V) & 0 & 0 \\ 0 & 0 & 1 & 0 \\ 0 & 0 & 0 & 1 \end{pmatrix}, \tag{2.67}$$

then we can write the transformation rule for the four-component entity $[B_\mu]$ in terms of matrices or components:

$$[B'_\mu] = [(\Lambda^{-1})_\mu{}^\nu][B_\nu], \tag{2.68}$$

$$B'_\mu = \sum_{\nu=0}^{3} (\Lambda^{-1})_\mu{}^\nu B_\nu \quad (\mu = 0, 1, 2, 3). \tag{2.69}$$

Any four-component entity that obeys this transformation law is said to be a **covariant four-vector**. Note that contravariant four-vectors transform like four-positions or four-displacements and are indicated by a raised index as in $[A^\mu]$, while covariant four-vectors transform like derivatives of scalar functions and are indicated by a lowered index as in $[B_\mu]$.

There are three important points to note concerning contravariant and covariant four-vectors.

I Raising and lowering four-vector indices

For every contravariant four-vector, a corresponding covariant four-vector can be formed, and vice versa. This is achieved by using the Minkowski metric introduced in Chapter 1:

$$[\eta_{\mu\nu}] \equiv \begin{pmatrix} \eta_{00} & \eta_{01} & \eta_{02} & \eta_{03} \\ \eta_{10} & \eta_{11} & \eta_{12} & \eta_{13} \\ \eta_{20} & \eta_{21} & \eta_{22} & \eta_{23} \\ \eta_{30} & \eta_{31} & \eta_{32} & \eta_{33} \end{pmatrix} = \begin{pmatrix} 1 & 0 & 0 & 0 \\ 0 & -1 & 0 & 0 \\ 0 & 0 & -1 & 0 \\ 0 & 0 & 0 & -1 \end{pmatrix}. \tag{Eqn 1.52}$$

If the four quantities A^0, A^1, A^2 and A^3 transform as a contravariant four-vector, then the four quantities defined by the sums

$$A_\mu = \sum_{\nu=0}^{3} \eta_{\mu\nu} A^\nu \quad (\mu = 0, 1, 2, 3) \tag{2.70}$$

will transform as a covariant four-vector. So the Minkowski metric can be used to lower the indices on four-vectors. Thanks to the very simple form of the Minkowski metric, it is easy to perform the necessary sums and to see that

if $\quad [A^\mu] = (a, b, c, d), \quad$ then $\quad [A_\mu] = (a, -b, -c, -d).$

This means that starting from the *contravariant* four-vectors that have already been introduced, we can now introduce a set of *covariant* counterparts simply by reversing the signs of the spatial components. This gives:

- the covariant four-displacement $[\Delta x_\mu] = (ct, -\Delta \boldsymbol{r})$
- the covariant four-velocity $[U_\mu] = (\gamma c, -\gamma \boldsymbol{v})$
- the covariant four-momentum $[P_\mu] = (E/c, -\boldsymbol{p})$
- the covariant four-force $[F_\mu] = (\gamma \boldsymbol{f} \cdot \boldsymbol{v}/c, -\gamma \boldsymbol{f})$.

Furthermore, if we introduce a new 16-component entity $[\eta^{\mu\nu}]$ with components $\eta^{\mu\nu}$ that can be identified from

$$[\eta^{\mu\nu}] \equiv \begin{pmatrix} \eta^{00} & \eta^{01} & \eta^{02} & \eta^{03} \\ \eta^{10} & \eta^{11} & \eta^{12} & \eta^{13} \\ \eta^{20} & \eta^{21} & \eta^{22} & \eta^{23} \\ \eta^{30} & \eta^{31} & \eta^{32} & \eta^{33} \end{pmatrix} = \begin{pmatrix} 1 & 0 & 0 & 0 \\ 0 & -1 & 0 & 0 \\ 0 & 0 & -1 & 0 \\ 0 & 0 & 0 & -1 \end{pmatrix}, \tag{2.71}$$

then we can use sums over those components to raise four-vector indices and convert covariant four-vectors into contravariant ones:

$$A^\mu = \sum_{\nu=0}^{3} \eta^{\mu\nu} A_\nu \quad (\mu = 0, 1, 2, 3). \tag{2.72}$$

Incidentally, it's worth noting for future reference that although $[\eta_{\mu\nu}]$ and $[\eta^{\mu\nu}]$ have identical components, the two quantities are actually inversely related, in the sense that

$$\sum_\nu \eta^{\alpha\nu} \eta_{\nu\beta} = \delta^\alpha{}_\beta, \tag{2.73}$$

where $[\delta^\alpha{}_\beta]$ is represented by the 4×4 matrix

$$[\delta^\alpha{}_\beta] = \begin{pmatrix} 1 & 0 & 0 & 0 \\ 0 & 1 & 0 & 0 \\ 0 & 0 & 1 & 0 \\ 0 & 0 & 0 & 1 \end{pmatrix}.$$

2 Forming invariants by contraction

The second point concerns invariants. We saw earlier that we could find invariants by considering sums of components such as

$$\sum_{\mu,\nu=0}^{3} \eta_{\mu\nu} U^\mu U^\nu = c^2. \tag{Eqn 2.31}$$

But it can now be seen that such a sum actually involves the corresponding components of a contravariant four-vector and its covariant counterpart:

$$\sum_{\nu=0}^{3} U_\nu U^\nu = U_0 U^0 + U_1 U^1 + U_2 U^2 + U_3 U^3. \tag{2.74}$$

Since the contravariant and covariant components transform in inversely related ways under a Lorentz transformation, it is really not surprising that this kind of sum is invariant. Other examples that you have already met include $\sum_{\nu=0}^{3} P_\nu P^\nu = m^2 c^2$ and even $\sum_{\nu=0}^{3} \Delta x_\nu \Delta x^\nu = (\Delta s)^2$.

It is very common to see expressions involving four-vectors in which a sum runs over one raised index and one lowered index. The process is often referred to as **contraction**, and is not limited to cases where the indices are on identical four-vectors. The contraction of $[A^\mu]$ with $[B_\nu]$, for example, would be the invariant quantity

$$\sum_{\nu=0}^{3} A^\nu B_\nu = A^0 B_0 + A^1 B_1 + A^2 B_2 + A^3 B_3. \tag{2.75}$$

The contraction of four-vectors is rather like the formation of a scalar product of ordinary (three-) vectors. Indeed, quantities that are invariant under Lorentz transformations are sometimes referred to as **Lorentz scalars**.

3 Transformation under arbitrary Lorentz transformation

The third point concerns the generality of the definition of four-vectors. So far, when considering Lorentz transformations, we have always considered the case where the frames S and S' are in standard configuration, though we have emphasized that there is no real loss of generality in doing this. Nonetheless, now that we are using behaviour under Lorentz transformation as the defining characteristic of four-vectors, we should make it clear that the definition applies to arbitrary Lorentz transformations and not just those that describe standard configuration. We shall have more to say about this later. The box below summarizes what has already been said.

Four-vectors and their transformation

The behaviour of momentum, energy and force under Lorentz transformation is most easily described in terms of four-vectors. Important contravariant four-vectors include the velocity four-vector $[U^\mu] = (\gamma c, \gamma \boldsymbol{v})$, the momentum four-vector $[P^\mu] = (E/c, \boldsymbol{p})$ and the force four-vector $[F^\mu] = (\gamma \boldsymbol{f} \cdot \boldsymbol{v}/c, \gamma \boldsymbol{f})$.

Under a Lorentz transformation in which $x'^\mu = \sum_{\nu=0}^{3} \Lambda^\mu{}_\nu x^\nu$, a contravariant four-vector $[A^\mu]$ transforms in the same way as a four-displacement:

$$A'^\mu = \sum_{\nu=0}^{3} \Lambda^\mu{}_\nu A^\nu. \tag{Eqn 2.61}$$

Under the same Lorentz transformation, a covariant four-vector $[B_\mu]$ transforms in the same way as a set of derivatives:

$$B'_\mu = \sum_{\nu=0}^{3} (\Lambda^{-1})_\mu{}^\nu B_\nu, \tag{Eqn 2.69}$$

where $[(\Lambda^{-1})_\mu{}^\nu]$ is the matrix inverse of $[\Lambda^\mu{}_\nu]$.

Indices on four-vectors may be lowered or raised using the Minkowski metric $\eta_{\mu\nu}$ or the related quantity $\eta^{\mu\nu}$ (defined by requiring that $\sum_\nu \eta^{\alpha\nu}\eta_{\nu\beta} = \delta^\alpha{}_\beta$). Thus

$$A_\mu = \sum_{\nu=0}^{3} \eta_{\mu\nu} A^\nu \tag{Eqn 2.70}$$

and

$$A^\mu = \sum_{\nu=0}^{3} \eta^{\mu\nu} A_\nu. \qquad \text{(Eqn 2.72)}$$

Lorentz invariants may be formed by the process of contraction (summing over one raised and one lowered index) as in $\sum_{\nu=0}^{3} U_\nu U^\nu = c^2$ and $\sum_{\nu=0}^{3} P_\nu P^\nu = m^2 c^2$ and, more generally,

$$\sum_{\nu=0}^{3} A^\nu B_\nu = A^0 B_0 + A^1 B_1 + A^2 B_2 + A^3 B_3. \qquad \text{(Eqn 2.75)}$$

Four-vectors may be used to formulate the laws of mechanics in a manifestly Lorentz covariant way, as in the relation $F^\mu = \mathrm{d}P^\mu/\mathrm{d}\tau$. However, the force described by Newton's inverse square law of gravitation fails to transform in the required way, so Newtonian gravitation is inconsistent with special relativity and must be replaced by a different theory of gravitation.

Exercise 2.10 $(c\rho, J_x, J_y, J_z)$ is a contravariant four-vector that you will meet in the next section. Even without knowing what the symbols represent, you should be able to write down the four equations that show how these quantities will transform under a Lorentz transformation. Do that for the case of frames S and S$'$ in standard configuration, then write down the four components of the counterpart covariant four-vector that will transform according to the corresponding inverse Lorentz transformation.

Exercise 2.11   If the four-vector given in the previous question is represented by $[J^\mu] = (c\rho, J_x, J_y, J_z)$, explain why you should expect the quantity $\sum_{\mu=0}^{3} J_\mu J^\mu$ to be invariant under a Lorentz transformation, but not the quantities $\sum_{\mu=0}^{3} J_\mu J_\mu$ or $\sum_{\mu=0}^{3} J^\mu J^\mu$.   ∎

2.3   The laws of electromagnetism

Turning to the laws of electromagnetism, the situation is rather different from that in mechanics. It turns out that the existing laws of electromagnetism are already consistent with special relativity. What is needed is a recasting of those laws so that the Lorentz covariance will be manifest. This involves identifying all the important electromagnetic quantities as components of four-vectors or other similar entities that behave simply under Lorentz transformations, and then expressing the laws of electromagnetism as relations between those entities. That is what we shall do in this section. To keep the discussion as simple as possible, we shall consider electromagnetism only in a vacuum.

2.3.1   The conservation of charge

One of the most fundamental laws of electromagnetism is the **conservation of electric charge**. Charge can be neither created nor destroyed. If particle physicists perform an experiment in which a positively-charged particle is produced, then an equal amount of negative charge must be produced at the same time. In less extreme circumstances, if the total amount of charge in some region changes, it must be because electric charge has been carried in or out of that region by electric currents. The law of electromagnetism that describes the conservation of electric charge is called the **equation of continuity** and is usually written as

$$\frac{\partial \rho}{\partial t} + \frac{\partial J_x}{\partial x} + \frac{\partial J_y}{\partial y} + \frac{\partial J_z}{\partial z} = 0, \tag{2.76}$$

where ρ represents the density of electric charge (measured in coulombs per cubic metre) and J_x, J_y and J_z are the three components of a vector that represents the electric current density (measured in amperes per square metre). When carefully examined, it turns out that under a Lorentz transformation the charge density and the current density transform as the components of a contravariant four-vector $[J^\mu] \equiv (J^0, J^1, J^2, J^3) = (c\rho, J_x, J_y, J_z)$, usually called the electric **four-current**, and the equation of continuity can be written as

$$\sum_{\nu=0}^{3} \frac{\partial J^\nu}{\partial x^\nu} = 0. \tag{2.77}$$

You will recall from our earlier discussion that derivatives transform like a covariant four-vector (the raised index in the denominator acts like a lowered index in the numerator). Consequently, the left-hand side of Equation 2.77 has the form of an invariant formed by contraction, and the right-hand side tells us that it is zero. The relationship is manifestly covariant — it is constructed from four-vectors, and there are no free indices on either side of the equation. So if experiment tells us — which it does — that the equation of continuity is true in some inertial frame S, then the theory of relativity tells us that it will also be true in any other inertial frame S′. We now have our first law of manifestly Lorentz-covariant electromagnetism.

The covariant equation of continuity

$$\sum_{\nu=0}^{3} \frac{\partial J^\nu}{\partial x^\nu} = 0. \tag{Eqn 2.77}$$

2.3.2   The Lorentz force law

The electrostatic force on a particle of charge q at position $\boldsymbol{r}$ due to another particle of charge Q at position $\boldsymbol{R}$ is given by Coulomb's law:

$$\boldsymbol{f} = \frac{Qq}{4\pi\varepsilon_0 d^2}\widehat{\boldsymbol{d}}, \tag{2.78}$$

where ε_0 is the permittivity of free space (an invariant constant with the value $8.854 \times 10^{-12} \, \mathrm{C^2 \, m^{-2} \, N^{-1}}$) and $\boldsymbol{d} = \boldsymbol{r} - \boldsymbol{R}$ is the displacement vector from the particle of charge Q to the particle of charge q (see Figure 2.9), so d is the distance between the two particles and $\widehat{\boldsymbol{d}}$ is a unit vector in the direction of $\boldsymbol{d}$.

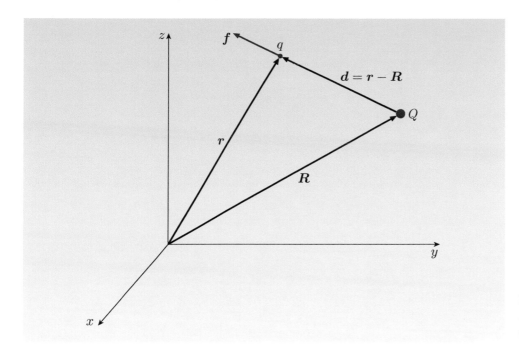

Figure 2.9 The electrostatic force $\boldsymbol{f}$ on a particle of charge q at position $\boldsymbol{r}$ due to a particle of charge Q at position $\boldsymbol{R}$.

Knowledge of this force is useful only in some highly specific cases. What is generally of much greater value is knowledge of the **electric field** $\boldsymbol{\mathcal{E}}(\boldsymbol{r})$. This is a *vector field*, meaning that it is a function of position that assigns a vector $\boldsymbol{\mathcal{E}}$ to each point $\boldsymbol{r}$ throughout some region. At any point $\boldsymbol{r}$, the assigned vector $\boldsymbol{\mathcal{E}}$ is the force per unit charge that would act on a test charge q placed at $\boldsymbol{r}$:

$$\boldsymbol{\mathcal{E}} - \boldsymbol{f}/q. \tag{2.79}$$

So, once the electric field throughout some region has been determined, the electrostatic force on any test charge q introduced at a point $\boldsymbol{r}$ can be predicted using

$$\boldsymbol{f} = q\boldsymbol{\mathcal{E}}(\boldsymbol{r}).$$

A similar approach may be taken to *magnetic forces*. This case is somewhat more complicated because the magnetic force on a charged particle generally depends on the particle's velocity as well as its position and charge. For example, the force on a charge q moving with velocity $\boldsymbol{v}$ through a point $\boldsymbol{r}$ that is at a perpendicular distance d from a long straight wire carrying a current I is given by

$$\boldsymbol{f} = q\boldsymbol{v} \times \frac{\mu_0 I}{2\pi d}\widehat{\boldsymbol{\theta}}, \tag{2.80}$$

where μ_0 is the permeability of free space (an invariant constant with the value $4\pi \times 10^{-7}\,\mathrm{T\,m\,A^{-1}}$) and $\widehat{\boldsymbol{\theta}}$ is a unit vector at right angles to the wire, as indicated in Figure 2.10 overleaf. Note that the symbol $\times$ in Equation 2.80 indicates a *vector product*, so directions are very important if it is to be correctly interpreted.

Once again, it is useful to have a more general prescription for the force, and this again involves the introduction of a vector field — in this case the **magnetic field** $\boldsymbol{B}(\boldsymbol{r})$, which is defined so that at the point $\boldsymbol{r}$,

$$\boldsymbol{f} = q\boldsymbol{v} \times \boldsymbol{B}(\boldsymbol{r}). \tag{2.81}$$

Once the magnetic field has been determined throughout some region, the force on any test charge moving through that region can be predicted.

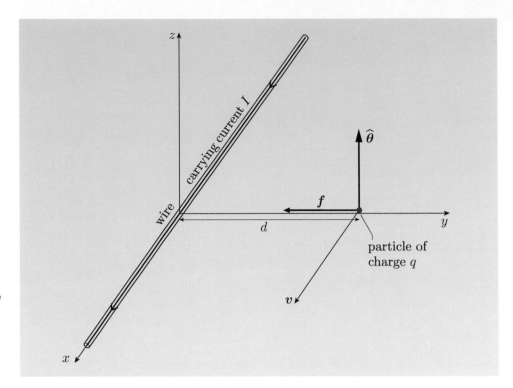

Figure 2.10   The magnetic force f on a particle of positive charge q moving with velocity v through a point at perpendicular distance d from a long straight wire carrying an electric current I.

Combining these descriptions of electric and magnetic forces, we see that in a region where there is both an electric field and a magnetic field, the electromagnetic force on a particle of charge q travelling with velocity v is given by the **Lorentz force law**

$$f = q(\mathcal{E} + v \times B). \tag{2.82}$$

The role of the vector product can be seen by writing out the individual components of the Lorentz force,

$$f_x = q(\mathcal{E}_x + v_y B_z - v_z B_y),$$
$$f_y = q(\mathcal{E}_y + v_z B_x - v_x B_z),$$
$$f_z = q(\mathcal{E}_z + v_x B_y - v_y B_x),$$

which can also be written in matrix form as

$$\begin{pmatrix} f_x \\ f_y \\ f_z \end{pmatrix} = q \begin{pmatrix} \mathcal{E}_x/c & 0 & B_z & -B_y \\ \mathcal{E}_y/c & -B_z & 0 & B_x \\ \mathcal{E}_z/c & B_y & -B_x & 0 \end{pmatrix} \begin{pmatrix} c \\ v_x \\ v_y \\ v_z \end{pmatrix}. \tag{2.83}$$

Our aim now is to find a way of rewriting the Lorentz force law in a manifestly covariant way. We should expect the final result to include four-vectors such as the four-force and the four-velocity, but the complexity of the above expressions suggests that something more will be required. The key extra ingredient is a new multi-component entity called the **electromagnetic four-tensor** or sometimes simply the **field tensor**. This can be denoted $[\mathsf{F}^{\mu\nu}]$ and will have 16 components $\mathsf{F}^{\mu\nu}$ that may be identified from the following:

$$[\mathsf{F}^{\mu\nu}] \equiv \begin{pmatrix} \mathsf{F}^{00} & \mathsf{F}^{01} & \mathsf{F}^{02} & \mathsf{F}^{03} \\ \mathsf{F}^{10} & \mathsf{F}^{11} & \mathsf{F}^{12} & \mathsf{F}^{13} \\ \mathsf{F}^{20} & \mathsf{F}^{21} & \mathsf{F}^{22} & \mathsf{F}^{23} \\ \mathsf{F}^{30} & \mathsf{F}^{31} & \mathsf{F}^{32} & \mathsf{F}^{33} \end{pmatrix}$$

$$= \begin{pmatrix} 0 & -\mathcal{E}_x/c & -\mathcal{E}_y/c & -\mathcal{E}_z/c \\ \mathcal{E}_x/c & 0 & -B_z & B_y \\ \mathcal{E}_y/c & B_z & 0 & -B_x \\ \mathcal{E}_z/c & -B_y & B_x & 0 \end{pmatrix}. \qquad (2.84)$$

It is unfortunate and potentially confusing that both the four-force and the field tensor are represented by an upper-case F, so we have used different typefaces for the two quantities. It may help that the field tensor will always have two indices while the four-force has only one. Nonetheless, you will need to take care not to confuse the two symbols.

Now, the truly remarkable thing about the electromagnetic four-tensor is that it behaves very simply under a Lorentz transformation. The positioning of the indices μ and ν in the raised contravariant location indicates the exact behaviour. If S and S' are two inertial frames in standard configuration, with coordinates related by $x'^{\mu} = \sum_{\nu=0}^{3} \Lambda^{\mu}{}_{\nu}\, x^{\nu}$, then the field tensor components $\mathsf{F}'^{\mu\nu}$ measured in frame S' will be related to those measured in frame S by

$$\mathsf{F}'^{\mu\nu} = \sum_{\alpha,\beta=0}^{3} \Lambda^{\mu}{}_{\alpha}\, \Lambda^{\nu}{}_{\beta}\, \mathsf{F}^{\alpha\beta}. \qquad (2.85)$$

Given that the fully contravariant field tensor $[\mathsf{F}^{\mu\nu}]$ does behave in this way, we can use the Minkowski metric to lower one of the indices, giving what is often referred to as the *mixed* version of the field tensor:

$$\mathsf{F}^{\mu}{}_{\beta} = \sum_{\nu=0}^{3} \eta_{\beta\nu}\, \mathsf{F}^{\mu\nu}. \qquad (2.86)$$

And then we can use the metric again to lower the remaining index, giving the fully covariant form:

$$\mathsf{F}_{\alpha\beta} = \sum_{\mu=0}^{3} \eta_{\alpha\mu}\, \mathsf{F}^{\mu}{}_{\beta}. \qquad (2.87)$$

Performing the sums is tedious and needs care, but the process is straightforward and leads to the result

$$[\mathsf{F}_{\mu\nu}] \equiv \begin{pmatrix} \mathsf{F}_{00} & \mathsf{F}_{01} & \mathsf{F}_{02} & \mathsf{F}_{03} \\ \mathsf{F}_{10} & \mathsf{F}_{11} & \mathsf{F}_{12} & \mathsf{F}_{13} \\ \mathsf{F}_{20} & \mathsf{F}_{21} & \mathsf{F}_{22} & \mathsf{F}_{23} \\ \mathsf{F}_{30} & \mathsf{F}_{31} & \mathsf{F}_{32} & \mathsf{F}_{33} \end{pmatrix}$$

$$= \begin{pmatrix} 0 & \mathcal{E}_x/c & \mathcal{E}_y/c & \mathcal{E}_z/c \\ -\mathcal{E}_x/c & 0 & -B_z & B_y \\ -\mathcal{E}_y/c & B_z & 0 & -B_x \\ -\mathcal{E}_z/c & -B_y & B_x & 0 \end{pmatrix}. \qquad (2.88)$$

Once again we see that, superficially at least, all that the index lowering has achieved is the reversal of some signs; but the real significance is that the fully covariant form of the field tensor transforms differently from the fully contravariant form. Under a Lorentz transformation implemented by the transformation matrix $[\Lambda^{\mu}{}_{\nu}]$, the fully covariant form transforms with the inverse Lorentz transformation matrix $[(\Lambda^{-1})_{\mu}{}^{\nu}]$ just as a derivative did. Specifically,

$$\mathsf{F}'_{\alpha\beta} = \sum_{\mu,\nu=0}^{3} (\Lambda^{-1})_{\alpha}{}^{\mu} (\Lambda^{-1})_{\beta}{}^{\nu} \, \mathsf{F}_{\mu\nu}. \tag{2.89}$$

These transformations are of great interest in their own right, and their implications for the electric and magnetic fields will be discussed in the next subsection. For the moment, however, we shall simply note that the transformations involve products of elements of the Lorentz transformation matrix or its inverse, and concentrate on the implications of this for the Lorentz force law.

Now consider the following equation:

$$F^{\mu} = q \sum_{\nu=0}^{3} \mathsf{F}^{\mu\nu} \, U_{\nu}. \tag{2.90}$$

On the left is a contravariant four-force; on the right is the product of an invariant (q), a four-tensor with two contravariant indices ($\mathsf{F}^{\mu\nu}$) and a covariant four-vector (U_{ν}) — there is a contraction over one raised index and the lowered one. So the right-hand side has only one free index, and that is raised, just like the one free index on the left-hand side. The upshot of all this is that the equation is expressed entirely in terms of entities that transform in simple ways under a Lorentz transformation, and those entities are combined in such a way that both sides of the equation will transform in the same manner. In other words, the given equation is manifestly covariant under Lorentz transformation. (Incidentally, note that in the last sentence we are using 'covariant' in the sense of 'form-invariant', not in the sense of 'transforming like a derivative'. It is unfortunate that the word is used in these two ways, but it is a customary practice.)

Of course, the real reason for our interest in Equation 2.90 is that it provides a covariant formulation of the Lorentz force law. You should convince yourself of this by actually performing the sum and checking the result, but the outcome can be more easily seen by interpreting the sum as the following matrix relationship (take the first index on any element to indicate the row):

$$\begin{pmatrix} (\gamma(v)/c)\boldsymbol{f} \cdot \boldsymbol{v} \\ \gamma(v)f_x \\ \gamma(v)f_y \\ \gamma(v)f_z \end{pmatrix} = q \begin{pmatrix} 0 & -\mathcal{E}_x/c & -\mathcal{E}_y/c & -\mathcal{E}_z/c \\ \mathcal{E}_x/c & 0 & -B_z & B_y \\ \mathcal{E}_y/c & B_z & 0 & -B_x \\ \mathcal{E}_z/c & -B_y & B_x & 0 \end{pmatrix} \begin{pmatrix} \gamma(v)c \\ -\gamma(v)v_x \\ -\gamma(v)v_y \\ -\gamma(v)v_z \end{pmatrix}. \tag{2.91}$$

Note that the v in this expression is the speed of the particle, the magnitude of the velocity $\boldsymbol{v} = (v_x, v_y, v_z)$. Also note that the negative signs in the right-hand column vector are there because it represents the *covariant* four-velocity. It is clear from the matrix expression that, after cancelling a $\gamma(v)$ on both sides, the last three rows reproduce the component expressions of the Lorentz force law that were given earlier.

What about the first row in the matrix equation? That is the equation

$$\boldsymbol{f} \cdot \boldsymbol{v} = q\boldsymbol{\mathcal{E}} \cdot \boldsymbol{v}. \tag{2.92}$$

It tells us the rate at which the Lorentz force does work and thereby increases the total energy. It does not contain any surprises, but it reflects the well-known fact that only the electric field is effective in doing work on the particle; this is because the magnetic part of the Lorentz force always acts at right angles to the particle's velocity.

So we now have a second law of Lorentz-covariant electromagnetism.

The covariant Lorentz force law

$$F^{\mu} = q \sum_{\nu=0}^{3} \mathsf{F}^{\mu\nu} U_{\nu}.$$

(Eqn 2.90)

Exercise 2.12   The Lorentz force law may also be expressed covariantly using the equation $F_{\mu} = q \sum_{\nu=0}^{3} \mathsf{F}_{\mu\nu} U^{\nu}$, but not $F_{\mu} = q \sum_{\nu=0}^{3} \mathsf{F}_{\nu\mu} U^{\nu}$. Why does the former work, but not the latter?   ∎

2.3.3   The transformation of electric and magnetic fields

The 'simple' transformation law of the electromagnetic four-tensor is vital for the successful formulation of the Lorentz-covariant force law, but it is also of great interest in itself. In particular, it shows that electric and magnetic fields become mixed together in relativity, in a way that is not unlike the mixing of energy and momentum seen earlier. What is an electric field to an observer in frame S will be observed as a combination of electric and magnetic fields by an observer in frame S′. In a relativistic universe, electric and magnetic phenomena are not completely separate. The existence of electric charge, combined with the requirements of special relativity, demands the existence of magnetism.

The transformation properties of electric and magnetic fields follow from the transformation properties of the field tensor. We already know that

$$[\mathsf{F}^{\mu\nu}] = \begin{pmatrix} 0 & -\mathcal{E}_x/c & -\mathcal{E}_y/c & -\mathcal{E}_z/c \\ \mathcal{E}_x/c & 0 & -B_z & B_y \\ \mathcal{E}_y/c & B_z & 0 & -B_x \\ \mathcal{E}_z/c & -B_y & B_x & 0 \end{pmatrix},$$

(Eqn 2.84)

and we know that in this fully contravariant case,

$$\mathsf{F}'^{\mu\nu} = \sum_{\alpha,\beta=0}^{3} \Lambda^{\mu}{}_{\alpha} \Lambda^{\nu}{}_{\beta} \mathsf{F}^{\alpha\beta}.$$

(Eqn 2.85)

In the case where the Lorentz transformation matrix is the usual one, relating frames S and S′ in standard configuration, the transformation is easier than it looks because many of the elements are zero. Even so, we shall not go through the details (you may do that if you wish), but we shall quote the result of a slightly more general Lorentz transformation in which frame S′ has an arbitrary velocity V (not necessarily in the x-direction) in frame S. In this case the transformation rules are usually expressed for field components that are parallel (indicated by $\parallel$) or perpendicular (indicated by $\perp$) to the direction of $\boldsymbol{V}$:

$$\mathcal{E}'_{\parallel} = \mathcal{E}_{\parallel}, \tag{2.93}$$

$$\boldsymbol{B}'_{\parallel} = \boldsymbol{B}_{\parallel}, \tag{2.94}$$

$$\mathcal{E}'_{\perp} = \gamma(V)\left[\mathcal{E}_{\perp} + \boldsymbol{V} \times \boldsymbol{B}_{\perp}\right], \tag{2.95}$$

$$\boldsymbol{B}'_{\perp} = \gamma(V)\left[\boldsymbol{B}_{\perp} - \boldsymbol{V} \times \mathcal{E}_{\perp}/c^2\right]. \tag{2.96}$$

These equations beautifully illustrate the blending of electricity and magnetism that relativity demands. Looking back at the covariant Lorentz force law, you can see the electromagnetic four-tensor as the mathematical entity required to allow a velocity-dependent force to be consistent with Lorentz covariance. From this point of view, electromagnetism is as simple as it could be.

Exercise 2.13   Using Equation 2.85 and taking $[\Lambda^{\mu}{}_{\nu}]$ to represent the usual Lorentz transformation between frames in standard configuration, show that $\mathcal{E}'_x = \mathcal{E}_x$.   ∎

2.3.4   The Maxwell equations

The remaining laws of vacuum electromagnetism are the **Maxwell equations**. These are the laws that determine the electric and magnetic fields in a given region. They relate the electric and magnetic fields to the charge and current densities that are their sources, and also to each other since a changing magnetic field can produce an electric field, and a changing electric field can produce a magnetic field.

In elementary treatments, the Maxwell equations are usually presented as a set of four differential equations written in the compact language of vector calculus, or sometimes as the equivalent set of eight component equations. This book does not assume any detailed familiarity with the Maxwell equations. The vector calculus versions are shown below, but all that matters mathematically is that the left-hand sides of the equations represent various combinations of partial derivatives of the electric and magnetic field components with respect to the spatial coordinates x, y and z:

$$\boldsymbol{\nabla} \cdot \mathcal{E} = \rho/\varepsilon_0, \qquad \text{(one-component equation)} \tag{2.97}$$

$$\boldsymbol{\nabla} \cdot \boldsymbol{B} = 0, \qquad \text{(one-component equation)} \tag{2.98}$$

$$\boldsymbol{\nabla} \times \mathcal{E} = -\frac{\partial \boldsymbol{B}}{\partial t}, \qquad \text{(three-component equation)} \tag{2.99}$$

$$\boldsymbol{\nabla} \times \boldsymbol{B} = \mu_0 \boldsymbol{J} + \frac{1}{c^2}\frac{\partial \mathcal{E}}{\partial t}, \qquad \text{(three-component equation)} \tag{2.100}$$

where $\boldsymbol{\nabla}$ represents the vector derivative

$$\boldsymbol{\nabla} = \left(\frac{\partial}{\partial x}, \frac{\partial}{\partial y}, \frac{\partial}{\partial z}\right). \tag{2.101}$$

The invariant constants that appear in these equations are not independent. They are linked by the equation $\mu_0\varepsilon_0 c^2 = 1$.

The charge and current densities were introduced earlier as components of the current four-vector. The fields, of course, are components of the electromagnetic four-tensor. The way to covariantly construct eight component equations from these ingredients is as follows.

The covariant Maxwell equations

$$\sum_{\mu=0}^{3} \frac{\partial F^{\mu\nu}}{\partial x^{\mu}} = \mu_0 J^{\nu}, \tag{2.102}$$

$$\frac{\partial F_{\lambda\mu}}{\partial x^{\nu}} + \frac{\partial F_{\nu\lambda}}{\partial x^{\mu}} + \frac{\partial F_{\mu\nu}}{\partial x^{\lambda}} = 0. \tag{2.103}$$

The first of these covariant equations has one free index and represents four component equations. These include references to the charge density and the current density, and reproduce Equations 2.97 and 2.100. The interpretation of the second covariant equation is less clear. It has three free indices, which indicates $64 \, (= 4 \times 4 \times 4)$ component equations. However, if any two of the indices are the same, the equation concerned is identically zero. Furthermore, in those cases where all the indices are different, permutations such as $\lambda = 1$, $\mu = 2$, $\nu = 3$ and $\lambda = 2$, $\mu = 3$, $\nu = 1$ lead to the same equation. Taking these symmetries into account, the original 64 component equations are reduced to just four independently meaningful equations. This second set of four component equations reproduces Equations 2.98 and 2.99. Thus, taken together, the two covariant equations reproduce the complete set of Maxwell equations and conclude our rewriting of the laws of vacuum electromagnetism in a manifestly covariant form. All that remains is to draw some lessons that will be of value in future chapters.

2.3.5 Four-tensors

Exposing the formal simplicity, almost the inevitability, of electromagnetism is one of the great triumphs of special relativity. However, from the point of view of relativity itself, the main development in this chapter has been the introduction of tensors. In this particular chapter the tensors have been called **four-tensors**. This indicates that they are specific to special relativity. You will meet a much more general class of tensors later, when we move on to general relativity, but a good understanding of four-tensors will be a valuable starting point for that more general experience.

The only four-tensor that we have formally introduced so far is the electromagnetic four-tensor $[F^{\mu\nu}]$ and its variants $[F^{\mu}{}_{\nu}]$ and $[F_{\mu\nu}]$, but you have already met some others. For instance, the vitally important Minkowski metric $[\eta_{\mu\nu}]$ is a fully covariant four-tensor, and the quantity $[\eta^{\mu\nu}]$ is a fully contravariant four-tensor. Moreover, the term four-tensor is used in such a general sense that these two-indexed examples represent only one particular class of four-tensors — technically referred to as four-tensors of **rank** 2. All four-vectors are also four-tensors, but they are of rank 1, and it is easy to define four-tensors of rank 3, rank 4, or any higher rank.

The defining characteristic of any four-tensor, whatever its rank, is its behaviour under Lorentz transformations. If S and S' are two inertial frames linked by a general Lorentz transformation (i.e. not necessarily in standard configuration), then we know that the coordinates in S will be related to those in S' by $[x'^\mu] = \sum_\nu [\Lambda^\mu{}_\nu][x^\nu]$. (Note that we are now using $\Lambda^\mu{}_\nu$ in a more general sense than before; we shall have to clarify this shortly.) Under such a general Lorentz transformation, a four-tensor $[T^{\mu_1,\mu_2,\ldots,\mu_m}]$ of contravariant rank m consists of 4^m components that transform according to

$$T'^{\mu_1,\mu_2,\ldots,\mu_m} = \sum_{\nu_1,\nu_2,\ldots,\nu_m} \Lambda^{\mu_1}{}_{\nu_1} \Lambda^{\mu_2}{}_{\nu_2} \cdots \Lambda^{\mu_m}{}_{\nu_m} T^{\nu_1,\nu_2,\ldots,\nu_m}. \quad (2.104)$$

Under the same Lorentz transformation, a covariant four-tensor of rank n is a collection of 4^n components that transform according to

$$T'_{\alpha_1,\alpha_2,\ldots,\alpha_n} = \sum_{\beta_1,\beta_2,\ldots,\beta_n} (\Lambda^{-1})_{\alpha_1}{}^{\beta_1} (\Lambda^{-1})_{\alpha_2}{}^{\beta_2} \cdots (\Lambda^{-1})_{\alpha_n}{}^{\beta_n} T_{\beta_1,\beta_2,\ldots,\beta_n},$$

$$(2.105)$$

where $[(\Lambda^{-1})_\mu{}^\nu]$ is the matrix inverse of $[\Lambda^\mu{}_\nu]$ in the usual sense that

$$\begin{pmatrix} \Lambda^0{}_0 & \Lambda^0{}_1 & \Lambda^0{}_2 & \Lambda^0{}_3 \\ \Lambda^1{}_0 & \Lambda^1{}_1 & \Lambda^1{}_2 & \Lambda^1{}_3 \\ \Lambda^2{}_0 & \Lambda^2{}_1 & \Lambda^2{}_2 & \Lambda^2{}_3 \\ \Lambda^3{}_0 & \Lambda^3{}_1 & \Lambda^3{}_2 & \Lambda^3{}_3 \end{pmatrix} \begin{pmatrix} (\Lambda^{-1})_0{}^0 & (\Lambda^{-1})_0{}^1 & (\Lambda^{-1})_0{}^2 & (\Lambda^{-1})_0{}^3 \\ (\Lambda^{-1})_1{}^0 & (\Lambda^{-1})_1{}^1 & (\Lambda^{-1})_1{}^2 & (\Lambda^{-1})_1{}^3 \\ (\Lambda^{-1})_2{}^0 & (\Lambda^{-1})_2{}^1 & (\Lambda^{-1})_2{}^2 & (\Lambda^{-1})_2{}^3 \\ (\Lambda^{-1})_3{}^0 & (\Lambda^{-1})_3{}^1 & (\Lambda^{-1})_3{}^2 & (\Lambda^{-1})_3{}^3 \end{pmatrix}$$

$$= \begin{pmatrix} 1 & 0 & 0 & 0 \\ 0 & 1 & 0 & 0 \\ 0 & 0 & 1 & 0 \\ 0 & 0 & 0 & 1 \end{pmatrix}. \quad (2.106)$$

A mixed four-tensor of contravariant rank m and covariant rank n consists of 4^{m+n} components that transform according to

$$T'^{\mu_1,\mu_2,\ldots,\mu_m}_{\alpha_1,\alpha_2,\ldots,\alpha_n} = \sum_{\nu_1,\nu_2,\ldots,\nu_m,\beta_1,\beta_2,\ldots,\beta_n} \Lambda^{\mu_1}{}_{\nu_1} \Lambda^{\mu_2}{}_{\nu_2} \cdots \Lambda^{\mu_m}{}_{\nu_m}$$

$$\times (\Lambda^{-1})_{\alpha_1}{}^{\beta_1} (\Lambda^{-1})_{\alpha_2}{}^{\beta_2} \cdots (\Lambda^{-1})_{\alpha_n}{}^{\beta_n}$$

$$\times T^{\nu_1,\nu_2,\ldots,\nu_m}_{\beta_1,\beta_2,\ldots,\beta_n}. \quad (2.107)$$

All that remains is to specify the elements of the general Lorentz transformation matrix that is the basis of this general definition of a four-tensor. We already know that if S and S' are in standard configuration, then $\Lambda^0{}_0 = \gamma(V)$, $\Lambda^0{}_1 = -\gamma(V)V/c$, $\Lambda^1{}_0 = -\gamma(V)V/c$ and $\Lambda^1{}_1 = \gamma(V)$, but what if the inertial frames S and S' are not in standard configuration? What if the axes are not aligned, for example, or the origin of S' never passes through the origin of S? What form do the matrix elements take under such general circumstances? We saw earlier, when deriving the Lorentz transformations in Chapter 1, that the primed coordinates have to be linear functions of the unprimed coordinates. In such circumstances, the constants that determine the transformation, the generalized analogues of $\gamma(V)$ and $\gamma(V)V/c$, can be represented by partial derivatives of the coordinates, so the elements of the general Lorentz transformation matrix can be written as

$$\Lambda^\mu{}_\nu = \frac{\partial x'^\mu}{\partial x^\nu}, \quad (2.108)$$

and the elements of the corresponding inverse transformation will be

$$(\Lambda^{-1})_\mu{}^\nu = \frac{\partial x^\nu}{\partial x'^\mu}. \tag{2.109}$$

Substituting these expressions into Equation 2.107 gives

$$
\begin{aligned}
T'^{\mu_1,\mu_2,\ldots,\mu_m}_{\quad\alpha_1,\alpha_2,\ldots,\alpha_n} = \sum_{\nu_1,\nu_2,\ldots,\nu_m,\beta_1,\beta_2,\ldots,\beta_n} & \frac{\partial x'^{\mu_1}}{\partial x^{\nu_1}} \frac{\partial x'^{\mu_2}}{\partial x^{\nu_2}} \cdots \frac{\partial x'^{\mu_m}}{\partial x^{\nu_m}} \\
\times\; & \frac{\partial x^{\beta_1}}{\partial x'^{\alpha_1}} \frac{\partial x^{\beta_2}}{\partial x'^{\alpha_2}} \cdots \frac{\partial x^{\beta_n}}{\partial x'^{\alpha_n}} \\
\times\; & T^{\nu_1,\nu_2,\ldots,\nu_m}_{\quad\beta_1,\beta_2,\ldots,\beta_n}.
\end{aligned}
\tag{2.110}
$$

This is the form of the general tensor transformation rule that you will meet later. The main difference is that in the case of four-tensors and special relativity, the partial derivatives are all constants that are independent of spacetime position. This will not always be the case in general relativity, as will soon become clear.

Exercise 2.14   You are told that the 256-component object $[H_{\mu\nu\rho\eta}]$ with elements $H_{\mu\nu\rho\eta}$ is a fully covariant four-tensor of rank 4. Write down the general rule for transforming its components from frame S to frame S'. ∎

Summary of Chapter 2

1. Invariants that take the same value in all inertial frames include the speed of light in a vacuum, the spacetime separation between events, the proper time between time-like separated events, the charge of a particle and the mass of a particle.

2. The principle of relativity demands that the laws of physics should be form-invariant under Lorentz transformations. Such laws are said to be Lorentz-covariant.

3. The relativistic momentum of a particle of mass m and velocity $\boldsymbol{v}$ is
$$\boldsymbol{p} = \gamma(v)m\boldsymbol{v}. \tag{Eqn 2.16}$$

4. The relativistic kinetic energy of a particle of mass m and speed v is
$$E_\mathrm{K} = (\gamma(v) - 1)mc^2. \tag{Eqn 2.22}$$

5. The total relativistic energy of a particle of mass m and speed v is
$$E = \gamma(v)mc^2 = E_\mathrm{K} + E_0, \tag{Eqn 2.24}$$
where $E_0 = mc^2$ is the mass energy of the particle.

6. In the absence of external forces, relativistic total energy is conserved, but neither kinetic energy nor mass energy is necessarily conserved. This establishes an 'equivalence' of mass and energy, with many important consequences.

7. The four-momentum $[P^\mu] = (E/c, p_x, p_y, p_z)$ brings together momentum and energy. It transforms in the same way as a four-displacement:
$$E' = \gamma(V)(E - Vp_x), \tag{Eqn 2.34}$$
$$p'_x = \gamma(V)(p_x - VE/c^2), \tag{Eqn 2.35}$$
$$p'_y = p_y, \tag{Eqn 2.36}$$
$$p'_z = p_z. \tag{Eqn 2.37}$$

8. The energy–momentum relation for a particle of mass m is
$$E^2 = p^2 c^2 + m^2 c^4, \qquad \text{(Eqn 2.43)}$$
showing that for a massless particle $p = E/c$.

9. Laws of conservation of total energy and momentum are combined in a manifestly covariant law of four-momentum conservation.

10. The four-force $[F^\mu] = ((\gamma/c)\boldsymbol{f} \cdot \boldsymbol{v}, \gamma \boldsymbol{f})$ determines the rate of change of a particle's four-momentum with respect to proper time. It transforms like the four-momentum, placing restrictions on the acceptable expressions for the three-force $\boldsymbol{f}$. The electromagnetic Lorentz force meets these requirements; Newton's gravitational force does not.

11. Under a Lorentz transformation in which $x'^\mu = \sum_{\nu=0}^{3} \Lambda^\mu{}_\nu \, x^\nu$, a contravariant four-vector $[A^\mu]$ transforms in the same way as a four-displacement:
$$A'^\mu = \sum_{\nu=0}^{3} \Lambda^\mu{}_\nu \, A^\nu. \qquad \text{(Eqn 2.61)}$$

Under the same Lorentz transformation, a covariant four-vector $[B_\mu]$ transforms in the same way as a set of derivatives:
$$B'_\mu = \sum_{\nu=0}^{3} (\Lambda^{-1})_\mu{}^\nu \, B_\nu, \qquad \text{(Eqn 2.69)}$$

where $[(\Lambda^{-1})_\mu{}^\nu]$ is the matrix inverse of $[\Lambda^\mu{}_\nu]$. In the case of two frames in standard configuration,
$$[\Lambda^\mu{}_\nu] = \begin{pmatrix} \gamma(V) & -\gamma(V)V/c & 0 & 0 \\ -\gamma(V)V/c & \gamma(V) & 0 & 0 \\ 0 & 0 & 1 & 0 \\ 0 & 0 & 0 & 1 \end{pmatrix}, \qquad \text{(Eqn 1.12)}$$
$$[(\Lambda^{-1})_\mu{}^\nu] = \begin{pmatrix} \gamma(V) & \gamma(V)V/c & 0 & 0 \\ \gamma(V)V/c & \gamma(V) & 0 & 0 \\ 0 & 0 & 1 & 0 \\ 0 & 0 & 0 & 1 \end{pmatrix}. \qquad \text{(Eqn 2.67)}$$

12. Indices on four-vectors may be lowered or raised using the Minkowski metric $\eta_{\mu\nu}$ or the related inverse quantity $\eta^{\mu\nu}$ defined by $\sum_\nu \eta^{\alpha\nu} \eta_{\nu\beta} = \delta^\alpha{}_\beta$:
$$A_\mu = \sum_{\nu=0}^{3} \eta_{\mu\nu} A^\nu \qquad \text{(Eqn 2.70)}$$
and
$$A^\mu = \sum_{\nu=0}^{3} \eta^{\mu\nu} A_\nu. \qquad \text{(Eqn 2.72)}$$

13. Contraction involves summing over one raised and one lowered index, and may be used to form invariants as in
$$\sum_{\nu=0}^{3} A^\nu B_\nu = A^0 B_0 + A^1 B_1 + A^2 B_2 + A^3 B_3. \qquad \text{(Eqn 2.75)}$$

14. The Lorentz-covariant laws of electromagnetism are:

the covariant equation of continuity

$$\sum_{\nu=0}^{3} \frac{\partial J^{\nu}}{\partial x^{\nu}} = 0;$$ (Eqn 2.77)

the covariant Lorentz force law

$$F^{\mu} = q \sum_{\nu=0}^{3} \mathsf{F}^{\mu\nu} U_{\nu};$$ (Eqn 2.90)

the covariant Maxwell equations

$$\sum_{\mu=0}^{3} \frac{\partial \mathsf{F}^{\mu\nu}}{\partial x^{\mu}} = \mu_0 J^{\nu},$$ (Eqn 2.102)

$$\frac{\partial \mathsf{F}_{\lambda\mu}}{\partial x^{\nu}} + \frac{\partial \mathsf{F}_{\nu\lambda}}{\partial x^{\mu}} + \frac{\partial \mathsf{F}_{\mu\nu}}{\partial x^{\lambda}} = 0,$$ (Eqn 2.103)

where $[J^{\mu}] = (c\rho, J_x, J_y, J_z)$ is the contravariant current four-vector, and $[\mathsf{F}^{\mu\nu}]$ is the fully contravariant electromagnetic four-tensor given by

$$[\mathsf{F}^{\mu\nu}] = \begin{pmatrix} 0 & -\mathcal{E}_x/c & -\mathcal{E}_y/c & -\mathcal{E}_z/c \\ \mathcal{E}_x/c & 0 & -B_z & B_y \\ \mathcal{E}_y/c & B_z & 0 & -B_x \\ \mathcal{E}_z/c & -B_y & B_x & 0 \end{pmatrix}.$$ (Eqn 2.84)

15. Under a Lorentz transformation, the electromagnetic four-tensor transforms according to

$$\mathsf{F}'^{\mu\nu} = \sum_{\alpha,\beta=0}^{3} \Lambda^{\mu}{}_{\alpha} \Lambda^{\nu}{}_{\beta} \mathsf{F}^{\alpha\beta}.$$ (Eqn 2.85)

This leads to the following transformation laws for the electric and magnetic fields:

$$\mathcal{E}'_{\parallel} = \mathcal{E}_{\parallel},$$ (Eqn 2.93)
$$B'_{\parallel} = B_{\parallel},$$ (Eqn 2.94)
$$\mathcal{E}'_{\perp} = \gamma(V)\left[\mathcal{E}_{\perp} + V \times B_{\perp}\right],$$ (Eqn 2.95)
$$B'_{\perp} = \gamma(V)\left[B_{\perp} - V \times \mathcal{E}_{\perp}/c^2\right].$$ (Eqn 2.96)

16. Under a general Lorentz transformation, the components of a four-tensor transform according to

$$T'^{\mu_1,\mu_2,...,\mu_m}_{\alpha_1,\alpha_2,...,\alpha_n} = \sum_{\nu_1,\nu_2,...,\nu_m,\beta_1,\beta_2,...,\beta_n} \frac{\partial x'^{\mu_1}}{\partial x^{\nu_1}} \frac{\partial x'^{\mu_2}}{\partial x^{\nu_2}} \cdots \frac{\partial x'^{\mu_m}}{\partial x^{\nu_m}}$$
$$\times \frac{\partial x^{\beta_1}}{\partial x'^{\alpha_1}} \frac{\partial x^{\beta_2}}{\partial x'^{\alpha_2}} \cdots \frac{\partial x^{\beta_n}}{\partial x'^{\alpha_n}}$$
$$\times T^{\nu_1,\nu_2,...,\nu_m}_{\beta_1,\beta_2,...,\beta_n}.$$ (Eqn 2.110)

Chapter 3 Geometry and curved spacetime

Introduction

Einstein's 1905 theory of special relativity concerns relationships between observations made by inertial observers in uniform relative motion. As you learned in the previous chapter, the theory is inconsistent with Newtonian gravitation. In 1907, in what he later described as 'the happiest thought of my life', Einstein realized that a theory of general relative motion — one that included relationships between observations made by accelerated observers — would also shed light on the problem of gravitation. It was not long after this that Minkowski introduced his four-dimensional spacetime approach to special relativity, which revealed the geometric basis of the theory. Under these influences, Einstein's own thinking took on an increasingly geometric flavour, and by the middle of 1912 he realized that to make further progress in relativity and gravitation, he needed to find out what mathematicians knew about certain problems concerning invariants in geometry. At that point he asked his friend, the mathematician Marcel Grossman (1878–1936), to help him to find the required information. Grossman was soon able to tell Einstein that what he was looking for was contained in the subject known as *Riemannian geometry* — a branch of mathematics particularly concerned with the study of *curved spaces*.

Geometry is the study of shape and spatial relationships. The kind of geometry taught in high schools is known as **Euclidean geometry**, after Euclid of Alexandria who collected together the main results of the field in around 300 BC. Among the best known of those results (see Figure 3.1) are:

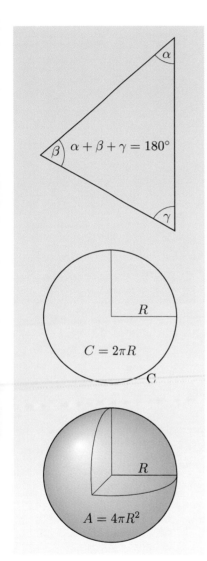

Figure 3.1 Some well-known results of Euclidean geometry.

- the internal angles of a triangle add up to $180°$
- a circle of radius R has a circumference of length $C = 2\pi R$
- a sphere of radius R has a surface area $A = 4\pi R^2$.

It was long thought that Euclidean geometry was the only kind of geometry, and that these results would therefore apply to all triangles, circles and spheres. However, in the first half of the nineteenth century, three mathematicians, János Bolyai (1802–1860), Nikolai Lobachevsky (1792–1856), and Carl Friedrich Gauss (1777–1855; Figure 3.2), independently established that it was possible to formulate a kind of geometry that made mathematical sense but was quite different from traditional Euclidean geometry. In **non-Euclidean geometry**, none of the Euclidean results quoted above is necessarily true.

The realization that there was more than one kind of geometry meant that determining the geometric properties of the space around us was an experimental question, not just a mathematical one. Lobachevsky considered the possibility of using astronomical measurements to determine the true geometry of space, but concluded that they would not be sufficiently accurate. Gauss became involved in a land survey and examined the angles of the large triangle between three mountain tops. He failed to find any sign of non-Euclidean geometry, but he too realized that this might simply reflect the limited sensitivity of the technique that he was using.

Gauss was one of the greatest of all mathematicians. His many discoveries included several important contributions to the development of geometry. Not least was his part in helping to found **differential geometry**, the branch of mathematics that applies the techniques of calculus to the analysis of geometric problems. It was in furthering this subject that Gauss's assistant Bernhard Riemann (1826–1866; Figure 3.3) introduced the geometry that now bears his name.

Figure 3.2 Carl Friedrich Gauss (1777–1855) was one of the founders of non-Euclidean geometry, sometimes described as the 'prince of geometers'.

Figure 3.3 Bernhard Riemann (1826–1866), a protegé of Gauss, was a great mathematician in his own right and the founder of Riemannian geometry.

The purpose of this chapter is to introduce you to some of the tools and techniques of Riemannian geometry. We shall not attempt a complete or rigorous development of the subject; rather, our aim is to motivate and introduce those concepts that will be needed when general relativity is discussed in the next chapter. What will become apparent as you work through this chapter is the immense importance of a quantity known as the *metric*, the components of which are usually represented by the symbol $g_{\mu\nu}$. This is a generalization of the Minkowski metric $\eta_{\mu\nu}$ that you have already met. Using the metric, initially in spaces of only two or three dimensions and then later in four-dimensional spacetime, we shall successively introduce methods of measuring the *length of a curve*, defining the *parallel transport* of a vector, finding *geodesics* (the curved space analogues of straight lines) and quantifying the *curvature* of a space or spacetime.

It is not until the last of these steps — the quantification of spacetime curvature — has been completed that we can formally define a *curved spacetime*. At that stage you will see that the four-dimensional Minkowski spacetime of special relativity has zero curvature and is therefore described as a 'flat' spacetime. Until curvature

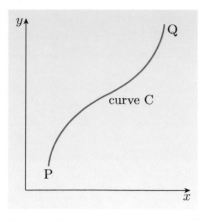

Figure 3.4   A smooth curve C in a Euclidean plane.

has been properly explained, it will be sufficient to think of a 'flat' space as one in which the conventional Euclidean geometrical results hold true, and a 'curved' space as one in which they fail. Note that the terms 'flat' and 'curved' are used to describe geometric properties and may be applied to spaces with any number of dimensions. They do not simply mean 'curved like a bow' or 'flat like a pancake'.

3.1   Line elements and differential geometry

3.1.1   Line elements in a plane

In order to analyze the geometry of curved space, we need to clarify what we mean by the length of a curve. Figure 3.4 shows a smooth curve C linking two points P and Q in an ordinary (Euclidean) plane. The plane is equipped with Cartesian coordinates so that each point on the curve can be assigned coordinates (x, y). The length of the curve can be approximately determined by dividing it into n short segments, each of which can be regarded as a straight line of length Δl_i ($i = 1, 2, \ldots, n$), and then adding together the lengths of all those short straight lines. The approximate length of the curve C from P to Q will then be given by

$$L_{\mathrm{C}}(\mathrm{P}, \mathrm{Q}) \approx \sum_{i=1}^{n} \Delta l_i. \tag{3.1}$$

According to Pythagoras's theorem, which is one of the fundamental results of Euclidean geometry, the length Δl of the straight line linking two points separated by the coordinate intervals Δx and Δy (see Figure 3.5) is given by

$$(\Delta l)^2 = (\Delta x)^2 + (\Delta y)^2. \tag{3.2}$$

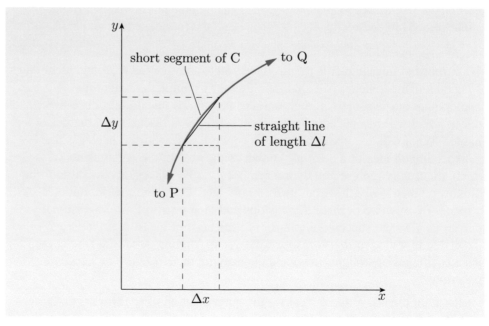

Figure 3.5   Each short segment of a curve C can be approximated by a straight line of length Δl.

Decreasing the length of those short segments will increase their number and improve the accuracy of the approximation to the total length of C. Taking the limit as $\Delta x \to 0$ and $\Delta y \to 0$, the sum will become an integral, and we can write the length of curve C from P to Q as

$$L_{\mathrm{C}}(\mathrm{P},\mathrm{Q}) = \int_{\mathrm{P}}^{\mathrm{Q}} \mathrm{d}l, \qquad (3.3)$$

where the **line element**, $\mathrm{d}l$, is defined by

$$\mathrm{d}l^2 = \mathrm{d}x^2 + \mathrm{d}y^2, \qquad (3.4)$$

or

$$\mathrm{d}l = (\mathrm{d}x^2 + \mathrm{d}y^2)^{1/2}. \qquad (3.5)$$

Unfortunately, this is not enough to let us actually work out the length of C; we need to know how to perform such an integral. In particular, in order to add up all the line elements along the curve, we need to take account of their differing directions, which will cause each element $\mathrm{d}l$ to correspond to differently-sized increments in the x- and y-directions.

One powerful way of taking the shape of C into account involves representing it as a **parameterized curve**. This requires that every point on the curve should be identified with a unique value of some continuously varying parameter, u say, so that the x- and y-coordinates of any particular point on the curve represent specific values of two **coordinate functions** $x(u)$ and $y(u)$ that effectively define the curve. So, for example:

- the parabola $y = x^2$ can be described in terms of a parameter u by the functions $x(u) = u$, $y(u) = u^2$
- the circle $x^2 + y^2 = 1$ can be described in terms of a parameter u by the functions $x(u) = \cos(u)$, $y(u) = \sin(u)$.

(Notice how in the first example, it is easy to parameterize a single-valued function $y = f(x)$: we just write $x(u) = u$ and $y(u) = f(u)$.)

Adopting this parametric approach, it's clear that any two points on the curve C that are separated by coordinate intervals Δx and Δy, will also be separated by some corresponding parameter interval Δu, and we can say that

$$\Delta x = \frac{\Delta x}{\Delta u} \Delta u$$

and

$$\Delta y = \frac{\Delta y}{\Delta u} \Delta u.$$

As $\Delta u \to 0$ (so that $\Delta x \to 0$ and $\Delta y \to 0$), the fractions $\Delta x/\Delta u$ and $\Delta y/\Delta u$ become the derivatives $\mathrm{d}x/\mathrm{d}u$ and $\mathrm{d}y/\mathrm{d}u$ of $x(u)$ and $y(u)$ with respect to u, and it follows that

$$\mathrm{d}x = \frac{\mathrm{d}x}{\mathrm{d}u}\,\mathrm{d}u, \quad \mathrm{d}y = \frac{\mathrm{d}y}{\mathrm{d}u}\,\mathrm{d}u,$$

and hence, from Equation 3.5,

$$\mathrm{d}l = \left(\left(\frac{\mathrm{d}x}{\mathrm{d}u}\right)^2 \mathrm{d}u^2 + \left(\frac{\mathrm{d}y}{\mathrm{d}u}\right)^2 \mathrm{d}u^2 \right)^{1/2} = \left(\left(\frac{\mathrm{d}x}{\mathrm{d}u}\right)^2 + \left(\frac{\mathrm{d}y}{\mathrm{d}u}\right)^2 \right)^{1/2} \mathrm{d}u.$$

So, finally, the length of the curve C from $\mathrm{P} = (x(u_\mathrm{P}), y(u_\mathrm{P}))$ to $\mathrm{Q} = (x(u_\mathrm{Q}), y(u_\mathrm{Q}))$ is given by the following.

Length of a curve in a Euclidean plane

$$L_\mathrm{C}(\mathrm{P}, \mathrm{Q}) = \int_\mathrm{P}^\mathrm{Q} \mathrm{d}l = \int_{u_\mathrm{P}}^{u_\mathrm{Q}} \left(\left(\frac{\mathrm{d}x}{\mathrm{d}u} \right)^2 + \left(\frac{\mathrm{d}y}{\mathrm{d}u} \right)^2 \right)^{1/2} \mathrm{d}u. \qquad (3.6)$$

Once we know the functions $x(u)$ and $y(u)$ that parameterize the curve C, and the values of u that correspond to the points P and Q, this expression for the length of a curve between two points in a Euclidean plane really can be evaluated. It is our first major result in this chapter.

Worked Example 3.1

(a) Parameterize the straight line $y = 2(\frac{6}{5}x + 1)$.

(b) Using the line element method described above, calculate the length of the line from $(0, 2)$ to $(5, 14)$. Check your result using Pythagoras's theorem.

Solution

(a) This is a single-valued function, so a suitable parameterization is $x = u$, $y = 2(\frac{6}{5}u + 1)$.

(b) Differentiating with respect to u, we obtain

$$\frac{\mathrm{d}x}{\mathrm{d}u} = 1 \quad \text{and} \quad \frac{\mathrm{d}y}{\mathrm{d}u} = \frac{12}{5}.$$

Since $x = u$, we have $u(0, 2) = 0$ and $u(5, 14) = 5$, so Equation 3.6 gives

$$L_\mathrm{C}((0, 2), (5, 14)) = \int_0^5 \left((1)^2 + \left(\frac{12}{5} \right)^2 \right)^{1/2} \mathrm{d}u = \left[\frac{13}{5} u \right]_0^5 = 13.$$

Pythagoras's theorem gives the same answer:

$$L_\mathrm{C}((0, 2), (5, 14)) = ((5 - 0)^2 + (14 - 2)^2)^{1/2} = 13.$$

Worked Example 3.2

Parameterize the circle $x^2 + y^2 = R^2$, and find the length of the circumference in terms of the (constant) radius R.

Solution

The simplest way to parameterize the circle is to set $x(u) = R\cos(u)$ and $y(u) = R\sin(u)$, as given earlier. Differentiating with respect to u, we obtain

$$\frac{\mathrm{d}x}{\mathrm{d}u} = -R\sin(u) \quad \text{and} \quad \frac{\mathrm{d}y}{\mathrm{d}u} = R\cos(u).$$

To get the circumference C, we need to let u vary from 0 to 2π, so using $\sin^2(u) + \cos^2(u) = 1$, we have

$$C = \int_0^{2\pi} ((R^2 \sin^2(u) + R^2 \cos^2(u))^{1/2}\,\mathrm{d}u = R\,[u]_0^{2\pi} = 2\pi R.$$

Exercise 3.1   (a) Sketch the curve parameterized by $x = 3u^2$, $y = 4u^2$. (b) Calculate the length L of the curve from $u = 0$ to $u = 3$. ∎

There are always many ways to parameterize a curve, but it is usually best to choose the simplest. For example, in Exercise 3.1 we used the parameterization $x = 3u^2$, $y = 4u^2$, but this gives us no particular benefit and it would be simpler to use $x = 3u$, $y = 4u$. For the circle in Worked Example 3.2, another possibility is $x = u$, $y = \pm(R^2 - u^2)^{1/2}$, but this would make the calculations much more difficult.

When dealing with a general curve in the plane, instead of Cartesian coordinates, it is often more convenient to use **plane polar coordinates** (r, ϕ), which can be defined in terms of (x, y) by

$$x = r \cos \phi,$$
$$y = r \sin \phi,$$

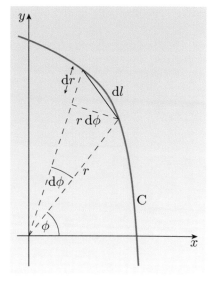

Figure 3.6   A line segment in plane polar coordinates.

as shown in Figure 3.6. Note that r is now a variable (not the constant radius R of Worked Example 3.2), so we can define any point in the plane by the coordinates (r, ϕ), where r is the distance from the origin measured along a line that makes an angle ϕ with the x-axis.

Using the rule for differentiating a product, it follows from the above definitions that

$$\mathrm{d}x = \cos \phi\,\mathrm{d}r - r \sin \phi\,\mathrm{d}\phi,$$
$$\mathrm{d}y = \sin \phi\,\mathrm{d}r + r \cos \phi\,\mathrm{d}\phi,$$

and so, from Equation 3.4, the line element in a Euclidean plane is also given by

$$\mathrm{d}l^2 = \mathrm{d}r^2 + r^2\,\mathrm{d}\phi^2. \tag{3.7}$$

This too is indicated in Figure 3.6, in which $\mathrm{d}r$ and $\mathrm{d}\phi$ are infinitesimal.

Exercise 3.2   Use the parameterization $r = R$ (a constant) and $\phi = u$ (a variable parameter) together with Equation 3.7 to again find the circumference C of a circle of radius R. ∎

3.1.2 Curved surfaces

The differential approach to curved surfaces geometry that we have just been using can be generalized to higher dimensions. In three-dimensional Euclidean space with Cartesian coordinates, the definition of the line element in Equation 3.4 generalizes to

$$\mathrm{d}l^2 = \mathrm{d}x^2 + \mathrm{d}y^2 + \mathrm{d}z^2. \tag{3.8}$$

In **spherical coordinates**, as illustrated in Figure 3.7, x, y, z can be written as

$$x = r\sin\theta\cos\phi,$$
$$y = r\sin\theta\sin\phi,$$
$$z = r\cos\theta.$$

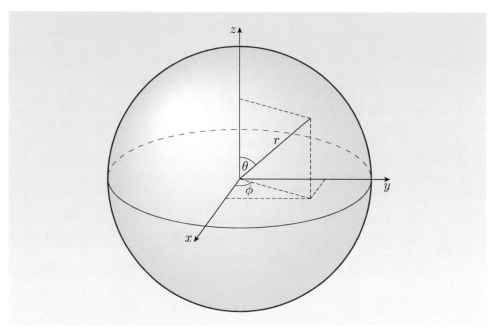

Figure 3.7   Spherical coordinates.

Applying the rule for differentiating a product, we see that

$$\mathrm{d}x = \sin\theta\cos\phi\,\mathrm{d}r + r\cos\theta\cos\phi\,\mathrm{d}\theta - r\sin\theta\sin\phi\,\mathrm{d}\phi,$$
$$\mathrm{d}y = \sin\theta\sin\phi\,\mathrm{d}r + r\cos\theta\sin\phi\,\mathrm{d}\theta + r\sin\theta\cos\phi\,\mathrm{d}\phi,$$
$$\mathrm{d}z = \cos\theta\,\mathrm{d}r - r\sin\theta\,\mathrm{d}\theta,$$

which leads, after some algebra, to

$$\mathrm{d}l^2 = \mathrm{d}r^2 + r^2\,\mathrm{d}\theta^2 + r^2\sin^2\theta\,\mathrm{d}\phi^2. \tag{3.9}$$

Using these alternative expressions for the line element, we can give meaning to the length of a curve in three-dimensional Euclidean space, and from there we could start to build up the whole of three-dimensional Euclidean geometry, just as we started to do in the two-dimensional case. As Gauss realized, these line elements are really the key to unlocking an entire geometry.

One topic that we can investigate is the geometry of two-dimensional surfaces in three-dimensional space. If, in Equation 3.9, we set r equal to a constant, R, then we are restricting ourselves to the surface of a sphere of radius R, and the equation for the line element reduces to

$$\mathrm{d}l^2 = R^2\,\mathrm{d}\theta^2 + R^2\sin^2\theta\,\mathrm{d}\phi^2. \tag{3.10}$$

There are just two variables in Equation 3.10, θ and ϕ, so it really does describe the geometry of a two-dimensional space. But the geometry of this two-dimensional space — the surface of the sphere — differs significantly from that of the plane, as the following example shows.

Worked Example 3.3

Figure 3.8 shows a sphere of radius R and a spherical coordinate system. Suppose that we draw a circle on the sphere by sweeping round the 'north pole' at a fixed angle θ. Starting from Equation 3.10, find the length of the circumference C of the circle.

Solution

Since θ is constant, Equation 3.10 tells us that a line element along the circle's circumference is given by $dl^2 = R^2 \sin^2\theta \, d\phi^2$. Adding together (i.e. integrating) all the line elements around the circle is easy in this case, since each one points in the direction of increasing ϕ, so the circumference is

$$C = \int_0^{2\pi} R \sin\theta \, d\phi = R \sin\theta \, [\phi]_0^{2\pi} = 2\pi R \sin\theta.$$

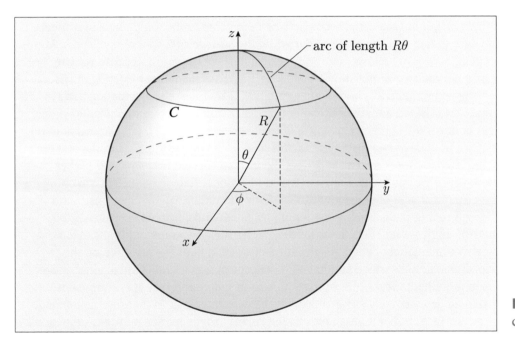

Figure 3.8 The geometry of a circle on the sphere.

If the geometry of a spherical surface were the same as that of a plane, we would expect the circumference C to be 2π times the radius of the circle, with both the circumference and the radius *measured in the spherical surface*. The radius measured in the spherical surface is $R\theta$, so the geometry of a plane would lead us to expect $C = 2\pi R\theta$. However, as the worked example showed, the circumference of the circle on the sphere is actually $C = 2\pi R \sin\theta$, which is less than plane geometry implies. So the geometry of a spherical surface is different from that of a plane. This has been well known to mathematicians and navigators for a long time. (Euclid used spherical geometry in his writings on astronomy.) But its real significance was not properly appreciated until the discovery of non-Euclidean geometry (now sometimes called *hyperbolic geometry*) caused mathematicians to reconsider the nature of geometry in general.

We shall not try to formulate spherical geometry here, but it is worth noting some

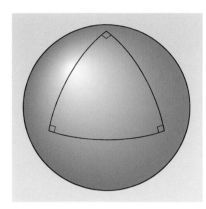

Figure 3.9 The angles of a triangle on a sphere can all be right angles.

key points that will be of significance later. A topic of great interest in spherical geometry is the behaviour of triangles. Obviously, there are no straight lines on a spherical surface, so before we can discuss spherical triangles, we need to know what are the spherical analogues of straight lines from which such triangles can be constructed. On a spherical surface this special role is played by the arcs of *great circles*. A **great circle** is a curve on the surface of a sphere created by the intersection of the sphere and a plane that passes through its centre. (On the Earth, the equator is an example of a great circle, and so are the meridian circles that pass through the North and South Poles.) In a Euclidean plane, the shortest path between any two points is the straight line that joins them. Similarly, on the surface of a sphere, the shortest path between any two points is the minor (i.e. shorter) arc of the great circle that passes through those points.

Figure 3.9 shows a spherical triangle constructed from the minor arcs of three great circles. In this case the spherical triangle is a rather special one since each of the interior angles is a right angle, but this illustrates another important difference between spherical geometry and plane geometry: the sum of the interior angles of a spherical triangle is greater than $180°$.

What lies behind the differences between the geometries of a plane and a sphere is the simple fact that the plane is *flat* while the surface of a sphere is *curved*. At this stage it is easy to believe that the spherical surface is curved because we can 'see' it as a curved two-dimensional surface in a three-dimensional Euclidean space, but this is not generally a reliable guide nor is such visual information always obtainable. Later, a mathematical definition of curvature will be introduced that will confirm the curvature of the spherical surface. However, it's important to note that we now have two tests for the presence of curvature that do not depend on being able to 'see', or even imagine, the curved surface in a space of higher dimension. Using the appropriate two-dimensional line element, we can compare the circumference of a circle with 2π times the radius, or we can construct a triangle (using paths of shortest length as sides) and compare the sum of the interior angles with $180°$. Each of these tests for curvature could be carried out by two-dimensional beings — traditionally called bugs — who live on the two-dimensional surface and have no concept of any higher-dimensional space. From a mathematical point of view this is an indication that curvature is an **intrinsic** property of a surface that can be determined from measurements made in the surface, rather than an **extrinsic** property that depends on measurements made in some higher dimension.

It is important to be aware of the intrinsic nature of curvature and our ability to detect it for at least three reasons. First, unlike spherical surfaces, not all surfaces that are of mathematical interest can be reproduced (the proper term is *embedded*) in three-dimensional Euclidean space. The 'hyperbolic' surface of the original non-Euclidean geometry is of this kind. The geometry exists, but the two-dimensional surface to which it applies cannot be embedded in three-dimensional Euclidean space. Second, when we come to deal with the curvature of the physical four-dimensional spacetime in which we live, it's very hard to imagine that we might successfully visualize it as existing within some other space or spacetime of even higher dimension. Third, not everything that appears curved in three dimensions really is curved in the mathematical sense. This last point is illustrated by the example of the cylinder given below.

A cylinder is formed by taking a strip of a plane, say the xy-plane from $x = a$ to $x = b$, and rolling it up so that the line $x = a$ becomes identified with the line $x = b$, as shown in Figure 3.10. Before rolling up the strip, we can draw on it a circle with radius r and circumference $2\pi r$. We can also draw a triangle whose interior angles add up to $180°$. These two features don't change when we roll up the strip of the plane, so our two-dimensional bugs carrying out local measurements of distances and angles would not be able to detect what we see as extrinsic curvature due to the rolling up in a third dimension. The process of 'rolling up' is what enables us to embed the cylindrical surface in three-dimensional space, but it does not produce any intrinsic curvature at all. In fact, the geometry of the cylinder is intrinsically flat.

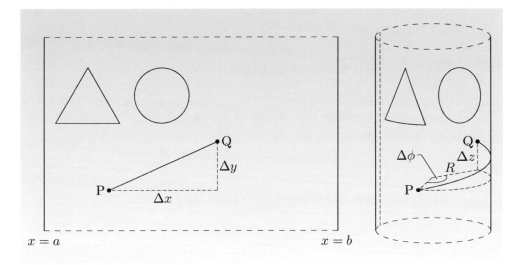

Figure 3.10 Geometry on a cylinder.

We can approach this idea more mathematically by using the appropriate two-dimensional line elements. The length L of the straight line from P to Q in the plane is given by $L^2 = (\Delta x)^2 + (\Delta y)^2$, reminding us that the line element in a plane, expressed in Cartesian coordinates, is $\mathrm{d}l^2 = \mathrm{d}x^2 + \mathrm{d}y^2$. Using the **cylindrical coordinates** (z, ϕ) shown in Figure 3.10, where z is measured parallel to the axis of the cylinder and ϕ is an angle measured in the plane perpendicular to the axis, we see that the distance from P to Q in the cylindrical surface is given by $L^2 = (\Delta z)^2 + R^2(\Delta \phi)^2$, where $R = (a - b)/2\pi$ is the radius of the cylinder. This shows that the line element in the cylindrical surface will be $\mathrm{d}l^2 = \mathrm{d}z^2 + R^2\,\mathrm{d}\phi^2$. However, if we make the change of variables $x = R\phi$, $y = z$, we see that these two line elements are actually the same.

As a final example of the importance of intrinsic curvature, consider a hotplate consisting of a circular region of the plane with a heat source at the centre point. The heat diffuses through the disc so that it gets cooler as the distance from the heat source increases. The two-dimensional bugs and their measuring sticks expand with the heat, so from our point of view they are bigger towards the centre of the disc (see Figure 3.11), although this is not noticeable to the bugs themselves. As a result of the temperature distribution, the shortest distance from P to Q as measured by the bugs will appear to us to curve in towards the centre, where fewer measuring sticks are needed to cover the distance (this too is shown in Figure 3.11). Hence the angles of the triangle PQR in Figure 3.11 add up to less than $180°$, and so, despite looking like a part of a flat plane to us, the hotplate

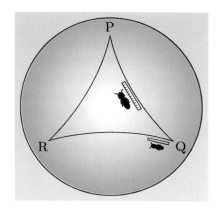

Figure 3.11 A circular hotplate with a source of heat at the centre.

89

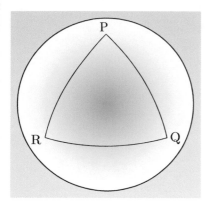

Figure 3.12   A circular hotplate heated uniformly around the edge.

has an intrinsically curved geometry according to the bugs that inhabit it.

It was Gauss who first recognized the intrinsic nature of the curvature of surfaces, but, as you will see in the next section, it was Riemann who enthusiastically embraced the idea and extended it to spaces of higher dimension.

Exercise 3.3   Using the same sort of informal arguments as in the above examples, investigate the curvature of the following spaces.

(a) A cone, excluding the point at its apex. Note that this means that you shouldn't consider circles and triangles drawn *around* the apex, as they are not completely contained in the space.

(b) A circular 'hotplate' where the heat source is around the edge of the disc, so that it cools towards the centre (Figure 3.12). ∎

3.2   Metrics and connections

Having informally introduced the idea of a curved space, we now focus on the branch of differential geometry known as *Riemannian geometry* that is mainly used to analyze such spaces. As we shall see in Chapter 4, it is Riemannian geometry that is particularly relevant to Einstein's theory of general relativity.

3.2.1   Metrics and Riemannian geometry

In the previous section we saw that in the differential approach to geometry, line elements hold the key to determining lengths of curves and paths of shortest distance, and through them to the properties of circles and triangles, and hence to the whole geometry of Euclidean space or the surface of a sphere. Several line elements were written down for two- and three-dimensional spaces, flat and curved, using a variety of coordinate systems (Equations 3.4, 3.7, 3.8, 3.9, 3.10). In each case, by analogy with Pythagoras's theorem, the line element was expressed as a sum of squares of **coordinate differentials**, such as dx, dy, dr and $d\theta$. In all those cases the line element was deduced from the known geometrical properties of the space concerned. Riemann's great insight was to recognize that line elements could be used not merely to summarize a geometry but rather as the starting point for the consideration of a geometry. He realized that by constructing line elements in accordance with certain simple general principles, it would be possible to develop a whole family of geometries that could describe flat and curved spaces with any desired number of dimensions. This is the basis of Riemannian geometry.

An n-dimensional **Riemann space** is a space in which the line element takes the general form

$$dl^2 = \sum_{i,j=1}^{n} g_{ij}\, dx^i\, dx^j, \tag{3.11}$$

where $dx^1, dx^2, \ldots, dx^n$ are the differentials of the n coordinates that describe the space, and the various g_{ij} are functions of the coordinates known as **metric coefficients** that are required to be symmetric in the sense that $g_{ij} = g_{ji}$.

Each of the line elements that we examined in the previous section was a special case of this general Riemannian line element. In the case of the Euclidean plane described by plane polar coordinates, for example, we saw in Equation 3.7 that

$$\mathrm{d}l^2 = \mathrm{d}r^2 + r^2 \,\mathrm{d}\phi^2,$$

which corresponds to the choices $n = 2$, $x^1 = r$, $x^2 = \phi$ and the metric coefficients $g_{11} = 1$, $g_{22} = r^2$ and $g_{12} = g_{21} = 0$.

In an n-dimensional Euclidean space described by n Cartesian coordinates $(x^1, x^2, x^3, \ldots, x^n)$, the line element is

$$\mathrm{d}l^2 = (\mathrm{d}x^1)^2 + (\mathrm{d}x^2)^2 + \cdots + (\mathrm{d}x^n)^2,$$

and the metric coefficients can be written as $g_{ij} = \delta_{ij}$, where δ_{ij} is the **Kronecker delta** defined by

$$\delta_{ij} = \begin{cases} 1 & \text{if } i = j, \\ 0 & \text{if } i \neq j. \end{cases}$$

In general, the metric coefficients can be regarded as forming an $n \times n$ array with n^2 elements, though due to the symmetry requirement $g_{ij} = g_{ji}$, the number of *independent* elements is only $n(n + 1)/2$, i.e. half the number of off-diagonal elements, plus the n diagonal ones. The complete set of metric coefficients $[g_{\mu\nu}]$ is called the **metric** or sometimes the **metric tensor**. (We shall not be much concerned with coordinate transformations in this chapter, but you will see later that the metric does transform in the way required of a rank 2 covariant tensor.) Consequently, the metric tensor for the three-dimensional Euclidean space defined by the line element of Equation 3.8 can be written as

$$[g_{ij}] = \begin{pmatrix} 1 & 0 & 0 \\ 0 & 1 & 0 \\ 0 & 0 & 1 \end{pmatrix},$$

where the i, j simply indicate the positions of the indices and have no other significance. (In much of the literature on general relativity, *no* explicit distinction is made between a tensor and its components. Rather than follow this potentially confusing practice, we use brackets [ ] to indicate the full tensor.)

Note that, in general, the metric coefficients are not constants, but are functions of the coordinates x^i. Once the coordinates being used to describe a space have been specified, it is the metric coefficients that perform the crucially important task of relating the coordinate differentials to lengths and thereby determine the geometry of the space. This point is so important to all that follows that it deserves special emphasis. *Once you know the metric, the geometry of the space is entirely determined.* However, the converse is not true. The geometry does not uniquely determine the metric; this is simply because there are many possible coordinate systems and hence many different ways of writing the metric.

Exercise 3.4 Writing $x^1 = \theta$, $x^2 = \phi$, find the metric that defines the curved geometry of the surface of a sphere of radius R with the line element given by Equation 3.10. ∎

We have now seen that both flat and curved spaces can be represented by metrics that are diagonal arrays. In fact, diagonal metrics occur whenever we have **orthogonal** coordinate systems, in which the different sets of grid lines corresponding to the directions of the x^i are at right angles to each other. All the coordinate systems that we have used so far, Cartesian, plane polar and spherical, have been of this kind. It turns out that the metrics of interest in general relativity and cosmology are usually orthogonal, so most of the examples of metrics that we use in this book will be diagonal, but non-diagonal arrays are possible.

Exercise 3.5   Here we consider the metric of three-dimensional Euclidean space in spherical coordinates. With $x^1 = r$, $x^2 = \theta$, $x^3 = \phi$, write down the metric coefficients g_{ij} that correspond to Equation 3.9, i.e.

$$dl^2 = dr^2 + r^2\,d\theta^2 + r^2\sin^2\theta\,d\phi^2.$$

Exercise 3.6   The metric coefficients for a plane in polar coordinates have already been given. Rewrite them as an array using appropriate notation.   ∎

Notice that in both of these exercises, the metric is a function of one or more of the coordinates, even though the spaces are certainly flat. This demonstrates that simply observing that the metric is a function of the coordinates is *not* sufficient to conclude that the space is curved; we may merely have a flat space in a non-Cartesian coordinate system.

We can summarize the main results of this subsection as follows.

Metrics

In an n-dimensional Riemann space, the line element is given by

$$dl^2 = \sum_{i,j=1}^{n} g_{ij}\,dx^i\,dx^j, \qquad\qquad \text{(Eqn 3.11)}$$

where the n^2 metric coefficients g_{ij} that define the geometry of the space are symmetric in the sense that $g_{ij} = g_{ji}$, and transform as the components of a rank 2 covariant tensor $[g_{ij}]$ called the metric tensor.

3.2.2   Connections and parallel transport

The main purpose of this subsection is to introduce an important set of quantities known as *connection coefficients*. In an n-dimensional Riemannian space there are n^3 such coefficients, usually denoted $\Gamma^i{}_{jk}$ ($i, j, k = 1, 2, \ldots, n$), though due to symmetry they are not all independent. Despite the indices, the connection coefficients are *not* the components of a tensor; under a coordinate transformation they do not transform in the way that tensor components must. The connection coefficients are directly related to the metric coefficients and are important in several contexts, including differentiation in curved space and a related process known as *parallel transport*. We shall start with a physical discussion of parallel transport and then go on to a more mathematical discussion that includes the connection coefficients.

Imagine a scientist studying the distribution of wind velocity in the Earth's atmosphere. The scientist might well want to compare the wind velocity $\boldsymbol{v}_P$ at some point P with the wind velocity $\boldsymbol{v}_Q$ at some other point Q. To do this, the scientist really needs to convey a copy of $\boldsymbol{v}_P$ along some chosen path C to the point Q, preserving the direction of the original $\boldsymbol{v}_P$ throughout each infinitesimal step. This is the process of **parallel transport**. It is illustrated in Figure 3.13, where the copy of $\boldsymbol{v}_P$ that has been parallel transported to Q is denoted $\boldsymbol{v}_{\parallel Q}$.

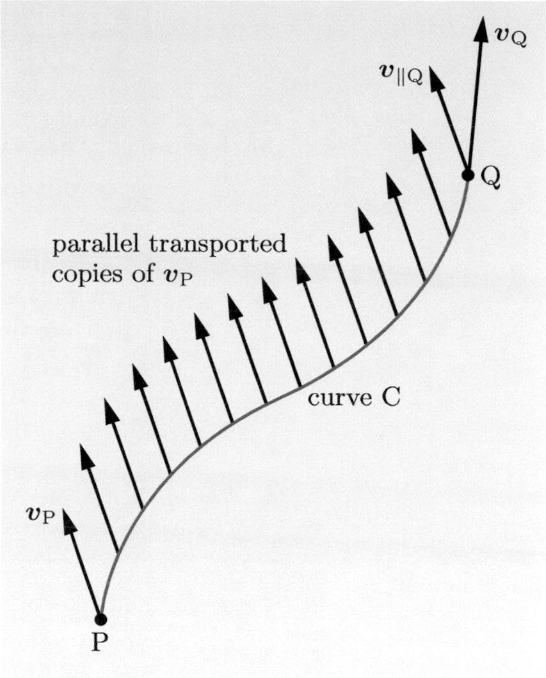

The mathematical difficulty of performing such a parallel transport of a vector along a curve depends very much on the nature of the space and coordinates involved. If the space is Euclidean and the coordinates Cartesian, the process is very simple. The wind velocity at any point can be written as $\boldsymbol{v} = v^1\boldsymbol{i} + v^2\boldsymbol{j} + v^3\boldsymbol{k}$, where the unit vectors $\boldsymbol{i}$, $\boldsymbol{j}$ and $\boldsymbol{k}$ in the $x^1 = x$, $x^2 = y$ and $x^3 = z$ directions are said to be **coordinate basis vectors**, since they point in the direction of increasing coordinate values, and v^1, v^2 and v^3 are the components of $\boldsymbol{v}$ in the coordinate basis. Since we are using Cartesian coordinates in Euclidean space, a vector may be parallel transported by simply keeping its components constant, so the components of $\boldsymbol{v}_{\parallel Q}$ will be $v^1_{\parallel Q} = v^1_P$, $v^2_{\parallel Q} = v^2_P$ and $v^3_{\parallel Q} = v^3_P$.

The situation is not so simple if the Cartesian coordinates are replaced by spherical coordinates with $x^1 = r$, $x^2 = \theta$ and $x^3 = \phi$. The reason for the extra complexity is easy to see and is illustrated in Figure 3.14. Spherical coordinates belong to the family of **curvilinear coordinates**. That means that the coordinate basis vectors $\widehat{r}$, $\widehat{\theta}$ and $\widehat{\phi}$ change their direction from place to place. As a

Figure 3.13 The parallel transport of a vector along a curve from P to Q, so that it can be compared with a vector already at Q.

consequence, in these coordinates, the components of the parallel transported vector at Q, $\boldsymbol{v}_{\parallel Q}$, will be different from those of the original vector $\boldsymbol{v}_P$ at P. So, in order to parallel transport a vector in this case, we need to know exactly how the components must change during each infinitesimal displacement along the curve.

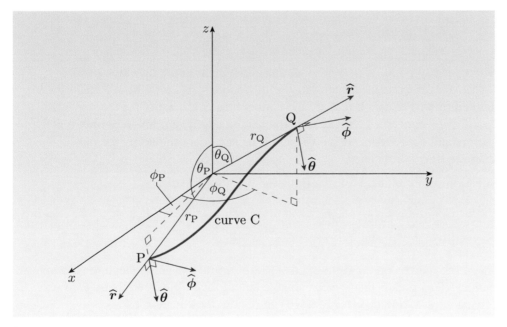

Figure 3.14 When spherical coordinates are used, the coordinate basis vectors point in different directions at different points.

Now let's generalize this problem to a three-dimensional Riemann space (which may be intrinsically curved) with coordinates x^1, x^2, x^3 and metric $[g_{ij}]$ $(i, j = 1, 2, 3)$ in which we want to parallel transport a vector specified at point P along a curve C to point Q. We shall suppose that positions along the curve are described by a parameter u and that the curve is therefore described by three coordinate functions $x^1(u)$, $x^2(u)$ and $x^3(u)$. If we denote the coordinate basis vectors (the analogues of $\boldsymbol{i}, \boldsymbol{j}, \boldsymbol{k}$ or $\widehat{\boldsymbol{r}}, \widehat{\boldsymbol{\theta}}, \widehat{\boldsymbol{\phi}}$) by $\boldsymbol{e}_1, \boldsymbol{e}_2, \boldsymbol{e}_3$, then at any point on C that corresponds to the parameter value u, we can write the local value of an arbitrary vector field $\boldsymbol{v}(u)$ in terms of its components in the coordinate basis and the coordinate basis vectors at that point. Thus

$$\boldsymbol{v}(u) = \sum_j v^j(u)\, \boldsymbol{e}_j(u). \tag{3.12}$$

Applying the rule for differentiating a product, we see that the rate of change of the vector field with respect to u as we move along the curve is given by

$$\frac{\mathrm{d}\boldsymbol{v}}{\mathrm{d}u} = \sum_j \left(\frac{\mathrm{d}v^j}{\mathrm{d}u} \boldsymbol{e}_j + v^j \frac{\mathrm{d}\boldsymbol{e}_j}{\mathrm{d}u} \right),$$

where the first term on the right represents the effect of changing the components, while the second term represents the effect of the changing basis vectors. Using the chain rule we can express the last term as a sum of terms, giving

$$\frac{\mathrm{d}\boldsymbol{v}}{\mathrm{d}u} = \sum_j \left(\frac{\mathrm{d}v^j}{\mathrm{d}u} \boldsymbol{e}_j + \sum_k v^j \frac{\partial \boldsymbol{e}_j}{\partial x^k} \frac{\mathrm{d}x^k}{\mathrm{d}u} \right). \tag{3.13}$$

Consider the term $\partial \boldsymbol{e}_j / \partial x^k$ — note that this is a vector quantity. It represents the rate of change of $\boldsymbol{e}_j$ with respect to x^k and will have components in the direction of each of the basis vectors. This means that we can write it as a sum:

$$\frac{\partial \boldsymbol{e}_j}{\partial x^k} = \sum_i \Gamma^i{}_{jk}\, \boldsymbol{e}_i, \tag{3.14}$$

where, at any point, $\Gamma^i{}_{jk}$ represents the component in the direction of basis vector $\boldsymbol{e}_i$ of the rate of change of $\boldsymbol{e}_j$ with respect to x^k. It is the n^3 quantities $\Gamma^i{}_{jk}$ defined by this equation that are the **connection coefficients** for the space and coordinates concerned. Since each connection coefficient involves only unit vectors and coordinates, it is clear that it must be determined by the metric; we shall see how a little later.

Substituting Equation 3.14 into Equation 3.13, we see that

$$\frac{\mathrm{d}\boldsymbol{v}}{\mathrm{d}u} = \sum_j \left(\frac{\mathrm{d}v^j}{\mathrm{d}u} \boldsymbol{e}_j + \sum_{i,k} \Gamma^i{}_{jk}\, \boldsymbol{e}_i v^j \frac{\mathrm{d}x^k}{\mathrm{d}u} \right). \tag{3.15}$$

All of the indices on the right-hand side are summed over, so they are all dummy indices. This means that we can change any of them, provided that we do so consistently. Using this freedom we can rewrite the equation as

$$\frac{\mathrm{d}\boldsymbol{v}}{\mathrm{d}u} = \sum_i \left(\frac{\mathrm{d}v^i}{\mathrm{d}u} + \sum_{j,k} \Gamma^i{}_{jk}\, v^j \frac{\mathrm{d}x^k}{\mathrm{d}u} \right) \boldsymbol{e}_i. \tag{3.16}$$

If we now require that the vector field that we have been discussing represents the same parallel transported vector at every point, then we can say that its rate of change must be zero. So the condition that must be satisfied if the vector v is actually being parallel transported along the curve is that

$$\sum_i \left(\frac{\mathrm{d}v^i}{\mathrm{d}u} + \sum_{j,k} \Gamma^i{}_{jk}\, v^j \frac{\mathrm{d}x^k}{\mathrm{d}u} \right) e_i = 0. \tag{3.17}$$

Thus, even in the case of a curved space, where the geometric interpretation is not simple, we can ensure the parallel transport of a vector by requiring that for each component,

$$\frac{\mathrm{d}v^i}{\mathrm{d}u} = -\sum_{j,k} \Gamma^i{}_{jk}\, v^j \frac{\mathrm{d}x^k}{\mathrm{d}u}. \tag{3.18}$$

So, given the components $v^i(u)$ of a vector at some point on the curve, the components of the parallel transported vector at a neighbouring point are

$$v^i(u + \mathrm{d}u) = v^i(u) + \frac{\mathrm{d}v^i}{\mathrm{d}u}\,\mathrm{d}u = v^i(u) - \sum_{j,k} \Gamma^i{}_{jk}\, v^j \frac{\mathrm{d}x^k}{\mathrm{d}u}\,\mathrm{d}u. \tag{3.19}$$

All that remains is to determine the expression for the connection coefficient $\Gamma^i{}_{jk}$ in terms of the metric.

If we consider two nearby points, we can write their infinitesimal vector separation as

$$\mathrm{d}l = \sum_i e_i\,\mathrm{d}x^i,$$

and consequently

$$\mathrm{d}l^2 = \mathrm{d}l \cdot \mathrm{d}l = \sum_i e_i\,\mathrm{d}x^i \cdot \sum_j e_j\,\mathrm{d}x^j = \sum_{i,j} (e_i \cdot e_j)\,\mathrm{d}x^i\,\mathrm{d}x^j.$$

Comparing this with the original line element (Equation 3.11)

$$\mathrm{d}l^2 = \sum_{i,j} g_{ij}\,\mathrm{d}x^i\,\mathrm{d}x^j,$$

we see that

$$e_i \cdot e_j = g_{ij}. \tag{3.20}$$

So the basis vectors are directly related to the metric coefficients.

Now, if we partially differentiate Equation 3.20 with respect to x^k, we see that

$$\frac{\partial e_i}{\partial x^k} \cdot e_j + e_i \cdot \frac{\partial e_j}{\partial x^k} = \frac{\partial g_{ij}}{\partial x^k}. \tag{3.21}$$

Using Equation 3.14 again, this can be rewritten as

$$\sum_l \Gamma^l{}_{ik}\, e_l \cdot e_j + e_i \cdot \sum_l \Gamma^l{}_{jk}\, e_l = \frac{\partial g_{ij}}{\partial x^k}. \tag{3.22}$$

After several lines of additional algebra, this leads to the final result

$$\Gamma^i{}_{jk} = \frac{1}{2} \sum_l g^{il} \left(\frac{\partial g_{lk}}{\partial x^j} + \frac{\partial g_{jl}}{\partial x^k} - \frac{\partial g_{jk}}{\partial x^l} \right), \tag{3.23}$$

where g^{il} is a component of the contravariant form of the metric tensor $[g^{ij}]$. This latter quantity is the inverse of $[g_{ij}]$ regarded as a matrix; that is, $[g^{ij}][g_{ij}]$ is equal to the identity matrix or, more explicitly,

$$\sum_k g^{ik} g_{kj} = \delta^i{}_j. \tag{3.24}$$

Since $[g^{ij}]$ is the inverse of the metric $[g_{ij}]$, it too must contain all the information about the geometry of the space. It is sometimes referred to as the **dual metric**.

Our findings regarding parallel transport can now be summarized as follows.

Parallel transport and connection coefficients

Given the components v^i of a vector at some point on a curve specified by $x^i(u)$ in a Riemann space with coordinates $x^1, \ldots, x^n$ and metric $[g_{ij}]$, the components of the parallel transported vector at some neighbouring point on the curve are given by

$$v^i(u + du) = v^i(u) - \sum_{j,k} \Gamma^i{}_{jk}\, v^j \frac{dx^k}{du}\, du, \tag{Eqn 3.19}$$

where the connection coefficient $\Gamma^i{}_{jk}$ is given by

$$\Gamma^i{}_{jk} = \frac{1}{2} \sum_l g^{il} \left(\frac{\partial g_{lk}}{\partial x^j} + \frac{\partial g_{jl}}{\partial x^k} - \frac{\partial g_{jk}}{\partial x^l} \right), \tag{Eqn 3.23}$$

and the dual metric $[g^{ij}]$, the matrix inverse of $[g_{ij}]$, is defined by the requirement that

$$\sum_k g^{ik} g_{kj} = \delta^i{}_j. \tag{Eqn 3.24}$$

Exercise 3.7 Calculate the connection coefficients $\Gamma^i{}_{jk}$ for:
(a) a two-dimensional Euclidean space using Cartesian coordinates;
(b) the surface of a sphere of radius $R = 1$, using polar coordinates. ■

As mentioned earlier, connection coefficients and parallel transport are important in several contexts, particularly in connection with differentiation in curved spaces. However, as Exercise 3.7 shows, they also provide an important indicator of the curvature of a space. Two-dimensional surfaces provide some easily visualized examples of this. In the case of the cylinder shown in Figure 3.15, parallel transport does exactly what it says: if we transport a vector v around a closed curve, it stays parallel to itself all the way around and gets back to the initial point exactly as it started out. That's because the surface of a cylinder is actually a flat space in terms of its intrinsic geometry.

However, as shown in Figure 3.16, there are no parallel lines in the curved geometry of a spherical surface, so we can't really expect that even a vector that is parallel transported over infinitesimal steps will manage to stay 'parallel' to itself when transported around a loop of finite size. And indeed, after being moved around a closed spherical triangle by 'parallel' transport, the vector in Figure 3.16 arrives back at its starting position pointing in a different direction.

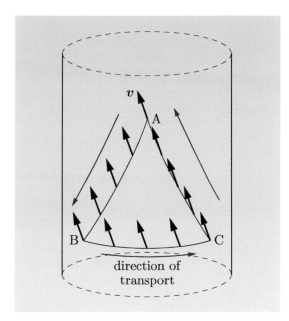

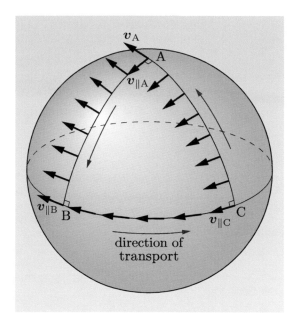

Figure 3.15 Parallel transport of the vector v around the triangle ABC drawn on the surface of a cylinder. Because the cylinder is intrinsically flat, the direction of a vector is preserved under parallel transport.

Figure 3.16 Parallel transport of the vector v around the triangle ABC drawn on the surface of a sphere. Under parallel transport, the original vector v_A becomes $v_{\|B}$, then $v_{\|C}$, then $v_{\|A}$, which points in a different direction from v_A.

Thus parallel transport of a vector around a closed curve or path gives us another test for whether a particular geometry is intrinsically flat or curved. Indeed, as you will see later, the difference between the initial and final directions of the vector gives us a measure of just how curved the geometry is in the vicinity of the closed path.

3.3 Geodesics

In a flat space, straight lines are of particular importance. A straight line represents the most direct route between two points and also the path of shortest distance between those points. Great circles play a similar role in the curved surface of a sphere. The analogues of straight lines and great circles in a general Riemannian space are referred to as *geodesics*. In this section we generalize the notions of 'most direct path' and 'shortest distance' in order to present two different derivations of the equations that are used to determine geodesics.

3.3.1 Most direct route between two points

One way of defining a straight line in Euclidean space is as a curve that always goes in the same direction. In order to extend this definition to the more general spaces of Riemannian geometry, we need to analyze the concept of 'direction

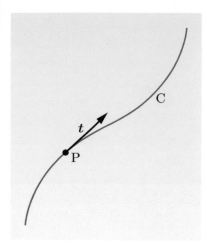

Figure 3.17   The tangent vector t to the curve C at the point P.

along a curve' and what it means to 'always go in the same direction'. At any point on a curve parameterized by u and defined by the coordinate functions $x^i(u)$ ($i = 1, \ldots, n$), we can define the **tangent vector t** to be the vector that points along the curve, as shown in Figure 3.17. The components of such a vector are $t^i = \mathrm{d}x^i/\mathrm{d}u$. If the curve is always going to go in the same direction, then the tangent vector should not change its direction as the parameter u varies and the tangent vector travels along the curve. In other words, if we parallel transport the tangent vector at u along the curve C to the point specified by $u + \mathrm{d}u$, the resulting vector should be proportional to the tangent vector at $u + \mathrm{d}u$. This means that

$$\frac{\mathrm{d}\boldsymbol{t}}{\mathrm{d}u} = f(u)\,\boldsymbol{t}, \tag{3.25}$$

where $f(u)$ is some function of u. It then follows from the condition for parallel transport that for each component of $\boldsymbol{t}$,

$$\frac{\mathrm{d}t^i}{\mathrm{d}u} + \sum_{j,k} \Gamma^i{}_{jk}\, t^j\, \frac{\mathrm{d}x^k}{\mathrm{d}u} = f(u)\, t^i. \tag{3.26}$$

Recalling that $t^i = \mathrm{d}x^i/\mathrm{d}u$, this gives

$$\frac{\mathrm{d}^2 x^i}{\mathrm{d}u^2} + \sum_{j,k} \Gamma^i{}_{jk}\, \frac{\mathrm{d}x^j}{\mathrm{d}u}\, \frac{\mathrm{d}x^k}{\mathrm{d}u} = f(u)\, \frac{\mathrm{d}x^i}{\mathrm{d}u}.$$

Now this can be simplified by choosing the parameter u in such a way that the function $f(u)$ is equal to zero. When the parameter is chosen in this particular way, it is said to be an **affine parameter** and will be denoted by the symbol λ. (This choice ensures that the tangent vector will preserve its magnitude as well as its direction as we move along the curve.) So, provided that we choose to use an affine parameter λ, the condition for a parameterized curve defined by a set of coordinate functions $x^i(\lambda)$ to always 'go in the same direction', is that

$$\frac{\mathrm{d}^2 x^i}{\mathrm{d}\lambda^2} + \sum_{j,k} \Gamma^i{}_{jk}\, \frac{\mathrm{d}x^j}{\mathrm{d}\lambda}\, \frac{\mathrm{d}x^k}{\mathrm{d}\lambda} = 0. \tag{3.27}$$

These are called the **geodesic equations**. Any parameterized pathway defined by a set of n functions $x^i(\lambda)$, $i = 1, \ldots, n$, that satisfies these differential equations is said to be a **geodesic** in the n-dimensional Riemannian space with metric $[g_{ij}]$ and connection coefficients $\Gamma^i{}_{jk}$. This is the analogue of a straight line in the curved space.

3.3.2   Shortest distance between two points

We saw earlier that in a two-dimensional Euclidean space, the length of a curve, parameterized by u and defined by the functions $x(u)$ and $y(u)$, between the points P and Q is given by integrating the line element $\mathrm{d}l$ from P to Q (Equation 3.6) so that

$$L(\mathrm{P}, \mathrm{Q}) = \int_{u=u_\mathrm{P}}^{u=u_\mathrm{Q}} \mathrm{d}l = \int_{u_\mathrm{P}}^{u_\mathrm{Q}} \left(\left(\frac{\mathrm{d}x}{\mathrm{d}u}\right)^2 + \left(\frac{\mathrm{d}y}{\mathrm{d}u}\right)^2 \right)^{1/2} \mathrm{d}u.$$

In an n-dimensional Riemannian space, Equation 3.11 extends the definition of the line element to include the metric via

$$dl^2 = \sum_{i,j}^{n} g_{ij}\, dx^i\, dx^j,$$

so we must correspondingly extend the formula for the length of the curve to

$$L(\mathrm{P},\mathrm{Q}) = \int_{u_\mathrm{P}}^{u_\mathrm{Q}} \left(\sum_{i,j} g_{ij} \frac{dx^i}{du} \frac{dx^j}{du} \right)^{1/2} du. \tag{3.28}$$

What we want to find is the parameterized curve $(x^1(u), x^2(u), \ldots, x^n(u))$ between P and Q that gives the smallest value for $L(\mathrm{P},\mathrm{Q})$, i.e. the shortest distance between the two points. Such a curve would be the analogue of a straight line, and therefore a geodesic. We use a method that is analogous to finding the minimum of a function $f(x)$ by differentiating it and looking for points at which $df/dx = 0$. The full mathematical treatment uses the *calculus of variations*, which is beyond the scope of this book, although a flavour of it is sketched below. You are not expected to follow the details, unless you have prior knowledge of the calculus of variations.

We can see from Figure 3.18 that since the geodesic between P and Q is the path of shortest length L, the curves that are close to it are of almost the same length. Now, if we consider all possible curves linking P and Q, and in each case we imagine changing the curve very slightly by making an infinitesimal variation of the form $x^i(u) \to x^i(u) + \delta x^i(u)$, then in each case the length of the curve will change by a small amount δL. However, in the case of the true geodesic, where the length is a minimum, we will find that δL is zero. So, writing

$$F = \left(\sum_{i,j} g_{ij} \frac{dx^i}{du} \frac{dx^j}{du} \right)^{1/2}, \tag{3.29}$$

it can be shown that

$$\delta L = \delta \int_{u_\mathrm{P}}^{u_\mathrm{Q}} F\, du$$

$$= \int_{u_\mathrm{P}}^{u_\mathrm{Q}} \sum_{m} \left(\frac{\partial F}{\partial x^m} \delta x^m + \frac{\partial F}{\partial \left(\frac{dx^m}{du} \right)} \delta \left(\frac{dx^m}{du} \right) \right) du.$$

Integrating the second part of the sum by parts, and noting that

$$\frac{d}{du}(\delta x^m) = \delta \left(\frac{dx^m}{du} \right),$$

leads to

$$\delta L = \left[\sum_{m} \frac{\partial F}{\partial \left(\frac{dx^m}{du} \right)} \delta x^m \right]_{u_\mathrm{P}}^{u_\mathrm{Q}} + \int_{u_\mathrm{P}}^{u_\mathrm{Q}} \sum_{m} \left(\frac{\partial F}{\partial x^m} - \frac{d}{du} \left(\frac{\partial F}{\partial \left(\frac{dx^m}{du} \right)} \right) \right) \delta x^m\, du.$$

However, $\delta x^m = 0$ at P and Q for all m, so the first bracket is zero. Consequently, for $\delta L = 0$, we have

$$\delta \int_{u_\mathrm{P}}^{u_\mathrm{Q}} F\, du = \int_{u_\mathrm{P}}^{u_\mathrm{Q}} \sum_{m} \left(\frac{\partial F}{\partial x^m} - \frac{d}{du} \left(\frac{\partial F}{\partial \left(\frac{dx^m}{du} \right)} \right) \right) \delta x^m\, du = 0.$$

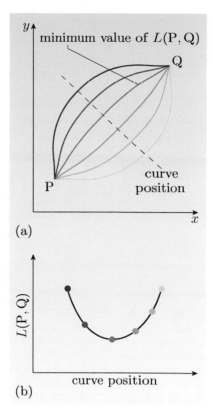

Figure 3.18 (a) In general there are many curves between P and Q; the shortest is the geodesic. (b) Distances along the curves shown in (a).

Since this is true for arbitrary variation δx^m, we obtain

$$\frac{\mathrm{d}}{\mathrm{d}u}\left(\frac{\partial F}{\partial\left(\frac{\mathrm{d}x^m}{\mathrm{d}u}\right)}\right) - \frac{\partial F}{\partial x^m} = 0, \qquad (m = 0, 1, 2, 3). \tag{3.30}$$

These are known as the **Euler–Lagrange equations** and are of fundamental importance to the study of the calculus of variations. If we substitute the expression for F (Equation 3.29) into the Euler–Lagrange equations, and choose u so that it is an affine parameter λ, it can be shown that

$$\frac{\mathrm{d}^2 x^i}{\mathrm{d}\lambda^2} + \sum_{j,k}\Gamma^i{}_{jk}\frac{\mathrm{d}x^j}{\mathrm{d}\lambda}\frac{\mathrm{d}x^k}{\mathrm{d}\lambda} = 0.$$

These are just the geodesic equations again (Equation 3.27), which shows that both methods of generalizing the definition of a straight line lead to the same concept of the geodesic.

So, to summarize, we have the following.

Geodesics and the geodesic equations

In an n-dimensional Riemannian space, the analogues of straight lines are known as geodesics. A geodesic can be represented by a curve parameterized by an affine parameter λ and defined by a set of n coordinate functions $x^i(\lambda)$ that satisfy the geodesic equations

$$\frac{\mathrm{d}^2 x^i}{\mathrm{d}\lambda^2} + \sum_{j,k}\Gamma^i{}_{jk}\frac{\mathrm{d}x^j}{\mathrm{d}\lambda}\frac{\mathrm{d}x^k}{\mathrm{d}\lambda} = 0. \tag{Eqn 3.27}$$

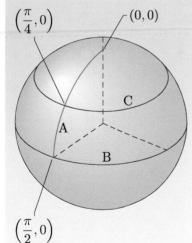

Exercise 3.8   Solve the geodesic equations for two-dimensional Euclidean space and verify that the geodesics are indeed straight lines.

Exercise 3.9   Figure 3.19 shows three curves on the surface of a sphere:

- a portion of a meridian A, with end-points $\left(\frac{\pi}{2}, 0\right)$ and $(0, 0)$
- the equator B, defined by $\theta = \frac{\pi}{2}$ and $0 \le \phi < 2\pi$
- a line of latitude C, defined by $\theta = \frac{\pi}{4}$ and $0 \le \phi < 2\pi$.

Starting from the geodesic equations (Equation 3.27), show that for the curves A, B and C in Figure 3.19:

(a) curve A is a geodesic;

(b) curve B is also a geodesic;

(c) the line of latitude C is *not* a geodesic.   ∎

Figure 3.19   Three curves, A, B and C, on the surface of a sphere, with coordinates of certain points.

3.4   Curvature

In this section we formalize and quantify the notion of the curvature of space. In particular we learn how to measure the curvature in an intrinsic way that does not depend on being able to embed the space being studied in some other space of higher dimension.

3.4.1 Curvature of a curve in a plane

We start with the comparatively simple idea of a curved line in a plane. Looking at the curve AH in Figure 3.20, it makes sense to say that the section around BCD is 'more curved' than the section around EFG. Our first objective is to associate a quantity k with this curvature at each point, such that $k_C > k_F$.

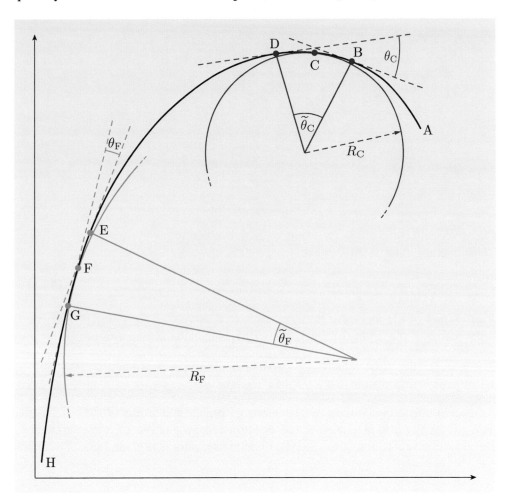

Figure 3.20   A curve ABCDEFGH in the plane and the approximating circles for the sections BCD and EFG.

First consider the section BCD of the curve, with mid-point C and of length l. This short section can be approximated by an arc of a circle of radius R_C, as shown in Figure 3.20. The tangent swings through an angle θ_C as it moves from point B to point D. It is this change in the direction of the tangent that gives our measure of curvature at C. Because the arc of the circle approximates the curve between these points, the angle $\widetilde{\theta}_C$ between the radii at points B and D is approximately equal to θ_C. This means that we have

$$l \approx R_C \widetilde{\theta}_C \approx R_C \theta_C,$$

and as l gets smaller, the approximations get better. We can do the same thing with the section EFG of the curve, also of length l, although this time the angle θ_F between the tangents is smaller than θ_C, and the radius R_F of the approximating circle is larger than R_C. This gives the relation

$$l \approx R_F \theta_F.$$

Using the angles between the tangents as the measure of curvature, we can say that

$$\text{curvature at } C > \text{curvature at } F$$

because

$$\theta_C > \theta_F.$$

But

$$\theta_C \approx \frac{l}{R_C} \quad \text{and} \quad \theta_F \approx \frac{l}{R_F},$$

so

$$\frac{l}{R_C} > \frac{l}{R_F}$$

and therefore

$$\frac{1}{R_C} > \frac{1}{R_F}.$$

Consequently, the quantity

$$k_X = \frac{1}{R_X} \tag{3.31}$$

is a measure of the curvature at any point X of a curve C in the plane, where R_X is the radius of the circle that best approximates C in the region close to X.

Exercise 3.10   (a) What is the curve of constant curvature $k = 0.2\,\mathrm{cm}^{-1}$?
(b) What is the curvature $k = 1/R$ of a straight line?   ■

For more complicated curves, a better way of finding the radius of the approximating circle is needed. It can be shown that if a curve is parameterized by the coordinate functions $(x(\lambda), y(\lambda))$, then its curvature k is given by

$$k = \frac{|\dot{x}\ddot{y} - \dot{y}\ddot{x}|}{(\dot{x}^2 + \dot{y}^2)^{3/2}}, \tag{3.32}$$

where a single dot over a function indicates the first derivative of that function with respect to λ, and a double dot indicates the second derivative.

Exercise 3.11   Find the curvature of the parabola $y = x^2$ at $x = 0$. Where is the centre of the circle that best approximates the parabola in the region close to $x = 0$?

Exercise 3.12   Find the curvature at any point on an ellipse parameterized by $x = a\cos\lambda$, $y = b\sin\lambda$. Use your answer to show that it leads to the expected result for the curvature of a circle of radius R.   ■

3.4.2   Gaussian curvature of a two-dimensional surface

We now consider the curvature of a two-dimensional surface embedded in three-dimensional Euclidean space. Suppose that we want to measure the curvature at a point A on the two-dimensional surface. From the point of view of the three-dimensional space, we can construct a vector N, normal to the surface at A, and this partly defines a plane PL containing N. As shown in Figure 3.21, the plane intersects the two-dimensional surface to give a curve C. (In the neighbourhood of A, this curve C will be a geodesic of the surface.) The curvature of C can be measured, as in the previous subsection, by finding the circle that best approximates the curve at A and then taking the reciprocal of the radius of that circle to obtain the curvature k. However, the plane PL is not completely defined since it can have any orientation with respect to N: different orientations will give different curves C and hence different curvatures k. We can get a measure of the curvature of the two-dimensional surface (rather than just a single curve C) at A by letting the plane PL rotate about N and picking the largest and smallest values of k, which we can denote by $k_{\max}$ and $k_{\min}$. The curvature of the two-dimensional surface at A is then characterized by what is known as the **Gaussian curvature**, which is defined by

$$K = k_{\max}k_{\min}. \tag{3.33}$$

One important subtlety is that for different curves at the same point A on a surface, the approximating circles may lie on opposite sides of the surface: for instance, this occurs in Figure 3.22(a) but not in Figure 3.22(b). In order to distinguish between these situations, we define the curvature k to be positive if the centre of the approximating circle is on the opposite side to the arrowhead of the normal vector N, and negative if it is on the same side. To ensure a unique result, negative curvatures are always taken as being smaller than positive ones in the search for $k_{\max}$ and $k_{\min}$. (Of course, the orientation of N is arbitrary, but this doesn't matter.)

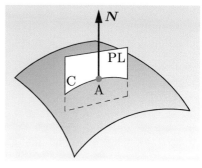

Figure 3.21 The curve C is the intersection between the surface and the plane PL through N.

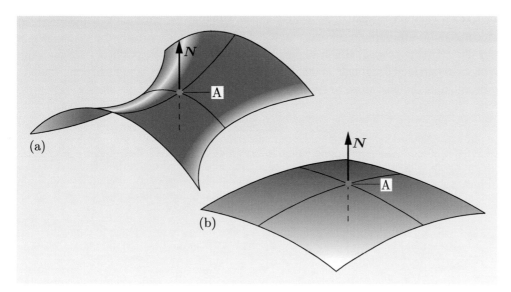

Figure 3.22 (a) A surface containing curves with curvature of opposite signs. (b) A surface only containing curves with curvature of the same sign.

● What is the Gaussian curvature for the surface of a two-dimensional sphere of radius R?

○ For the sphere,

$$k_{\mathrm{max}} = k_{\mathrm{min}} = \frac{1}{R},$$

so the Gaussian curvature is

$$K = k_{\mathrm{max}}k_{\mathrm{min}} = \frac{1}{R^2}.$$

So far, our arguments have depended on being able to embed the surface being studied in a three-dimensional space so that it has an obvious extrinsic curvature. In 1828 Gauss discovered a result regarding the curvature of surfaces that surprised him so much that he called it the 'remarkable theorem' (*theorema egregium*). The theorem provided a formula for working out the Gaussian curvature K of a two-dimensional surface, but the remarkable aspect of the result was that it showed K to be an invariant, independent of the coordinate system used. This was one of the inspirations for Riemann's work, and is now seen as an indication that curvature is an *intrinsic* property; it can be calculated directly from the metric $[g_{ij}]$ and does not require any embedding in a space of higher dimension. We shall not prove the *theorema egregium* here — even an outline proof requires four pages of dense mathematics — but we shall return to its main outcome once we have considered the curvature of spaces with three or more dimensions, in the next subsection.

3.4.3   Curvature in spaces of higher dimensions

Now consider an n-dimensional Riemann space with metric $[g_{ij}]$ that can be used to determine the space's connection coefficients $\Gamma^i{}_{jk}$ ($i = 1, \ldots, n$). Suppose that we have a vector $\boldsymbol{v}$ specified at some point P and that we parallel transport that vector around an infinitesimal rectangle PQRS with sides specified by $\mathrm{d}x^j$ and $\mathrm{d}x^k$. This process is illustrated in Figure 3.23, where the parallel transported vector that arrives back at P is denoted $\boldsymbol{v}_{\|\mathrm{P}}$ and is shown as being different from $\boldsymbol{v}$ because of the curvature of the space. We should expect the difference

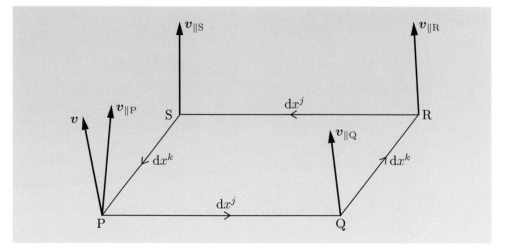

Figure 3.23   A vector $\boldsymbol{v}$ at point P is parallel transported around an infinitesimal rectangle PQRS to produce another vector $\boldsymbol{v}_{\|\mathrm{P}}$ at point P.

between $v_{\|\mathrm{P}}$ and v to have components that are proportional to the infinitesimal displacements and to the components of the original vector, so we can write any given component of the diffcrence as

$$v^l_{\|\mathrm{P}} - v^l = \sum_{i,j,k} R^l_{\ ijk}\, v^i\, \mathrm{d}x^j\, \mathrm{d}x^k, \tag{3.34}$$

where $R^l_{\ ijk}$ will be some measure of the curvature. (In a flat space we know that $v_{\|\mathrm{P}} = v$, so in that case we know that $R^l_{\ ijk} = 0$ for all choices of i, j, k and l.) When the parallel transport is actually carried out, it can be shown that

$$R^l_{\ ijk} \equiv \frac{\partial \Gamma^l_{\ ik}}{\partial x^j} - \frac{\partial \Gamma^l_{\ ij}}{\partial x^k} + \sum_m \Gamma^m_{\ ik}\, \Gamma^l_{\ mj} - \sum_m \Gamma^m_{\ ij}\, \Gamma^l_{\ mk}. \tag{3.35}$$

It turns out that under a general coordinate transformation, the quantity $R^l_{\ ijk}$ transforms in the manner required of a rank 4 tensor with one contravariant index and three covariant indices. Consequently, $R^l_{\ ijk}$ is known as the **Riemann curvature tensor** or the **Riemann tensor**. It is possible to show that the vanishing of the Riemann tensor $[R^l_{\ ijk}]$ at all points in a space is a necessary and sufficient condition for a space to be **flat**, i.e. not **curved**. So we finally have a quantitative measure of curvature, and — since it is related directly to the metric, albeit in a complicated way — it is clearly an intrinsic quantity that does not require any embedding in higher dimensions.

In n dimensions the Riemann tensor has n^4 components, giving $2^4 = 16$ components in two dimensions and $3^4 = 81$ in three dimensions. However, because of the definition of the connection (Equation 3.23) and Equation 3.35 itself, the Riemann tensor has many symmetries involving its indices. For example,

$$R^l_{\ ijk} = -R^l_{\ ikj}. \tag{3.36}$$

These symmetries reduce the number of *independent* components to 6 in three-dimensional spaces and only one in two-dimensional spaces. In two dimensions, the single independent component can be related to the Gaussian curvature K. From the point of view of Riemannian geometry, this is the explanation of Gauss's *theorema egregium*, with its implication that Gaussian curvature is intrinsic.

Exercise 3.13   Use Equation 3.35 to show that $R^l_{\ ijk} = -R^l_{\ ikj}$.

Exercise 3.14   Find the Riemann tensor for two-dimensional Euclidean space with the line element given by Equation 3.4. Extend your result to an n-dimensional Euclidean space. (*Hint*: Use Equation 3.35 and the results of Exercise 3.7(a).)

Exercise 3.15   Find the component $R^1_{\ 212}$ of the Riemann tensor for a two-dimensional sphere of radius R with the line element given in Equation 3.10. (*Hint*: Use Equation 3.35 and the results of Exercise 3.7(b).)

Exercise 3.16   The Gaussian curvature K for a two-dimensional surface is related to the Riemann tensor by

$$K = \frac{R_{1212}}{g},$$

where

$$g = \det[g_{ij}]$$

and the first index of R_{1212} is 'lowered' by means of the metric tensor (see Chapter 2). Use the result of the earlier in-text question concerning the Gaussian curvature for the surface of a two-dimensional sphere and the results of Exercises 3.7 and 3.15 to verify this relationship for a two-dimensional sphere of radius a. (We use a for the radius of the sphere in order to avoid possible confusion with the Riemann tensor.) ∎

So, to summarize, we have the following.

The Riemann tensor

In an n-dimensional Riemannian space, the curvature is described by the rank 4 Riemann tensor

$$R^l_{\ ijk} \equiv \frac{\partial \Gamma^l_{\ ik}}{\partial x^j} - \frac{\partial \Gamma^l_{\ ij}}{\partial x^k} + \sum_m \Gamma^m_{\ ik}\, \Gamma^l_{\ mj} - \sum_m \Gamma^m_{\ ij}\, \Gamma^l_{\ mk}. \qquad \text{(Eqn 3.35)}$$

The necessary and sufficient condition for a space to be flat (i.e. not curved) is that all the components of this tensor should vanish at every point.

3.4.4   Curvature of spacetime

So far, we have considered curved spaces that are Riemannian. In a strict mathematical sense, such spaces are defined by a line element taking the form

$$\mathrm{d}l^2 = \sum_{i,j} g_{ij}\, \mathrm{d}x^i\, \mathrm{d}x^j, \qquad \text{(Eqn 3.11)}$$

where $\mathrm{d}l^2 > 0$. As you will see in the next chapter, Einstein's general theory of relativity is a geometric theory of gravity that makes essential use of the Riemann tensor. However, in searching for a geometric theory of gravity, Einstein needed to generalize the Minkowski spacetime of special relativity, which is defined by a line element of the form

$$\mathrm{d}s^2 = \sum_{\mu,\nu=0}^{3} \eta_{\mu\nu}\, \mathrm{d}x^\mu\, \mathrm{d}x^\nu, \qquad (3.37)$$

where

$$\eta_{\mu\nu} = \begin{cases} 1 & \text{if } \mu = \nu = 0, \\ -1 & \text{if } \mu = \nu = 1,2,3, \\ 0 & \text{otherwise.} \end{cases} \qquad (3.38)$$

More explicitly, the line element in Minkowski spacetime is

$$\mathrm{d}s^2 = c^2\, \mathrm{d}t^2 - \mathrm{d}\boldsymbol{x} \cdot \mathrm{d}\boldsymbol{x} = c^2\, \mathrm{d}t^2 - \mathrm{d}x^2 - \mathrm{d}y^2 - \mathrm{d}z^2. \qquad (3.39)$$

This is the infinitesimal generalization of the *spacetime separation* $(\Delta s)^2$ that was introduced in Chapter 1. It is clearly possible to choose the differentials so that $\mathrm{d}s^2$ is negative, breaking the $\mathrm{d}l^2 > 0$ requirement of a Riemannian geometry.

Spaces for which the squared line element can be positive, negative or zero (null) are called **pseudo-Riemannian spaces** by mathematicians. However, physicists often don't bother to make this distinction and use the term *Riemannian space* to cover any space (or spacetime) defined by a metric as in Equation 3.11.

The generalization from the flat Minkowski spacetime of special relativity to the curved spacetime of general relativity is made by replacing the Minkowski spacetime metric coefficients $\eta_{\mu\nu}$, which are constants, with metric coefficients $g_{\mu\nu}$ that are function of the coordinates, so that

$$ds^2 = \sum_{\mu,\nu=0}^{3} g_{\mu\nu}\,dx^\mu\,dx^\nu. \qquad (3.40)$$

Notice that it is traditional to use Greek letters for the indices of four-dimensional Minkowski spacetime and for its extension to the curved spacetime of general relativity, with 0 representing the time coordinate. Latin letters are reserved for indices relating to space coordinates, usually taking the values $1, 2, 3$.

Many of the properties of Riemannian spaces carry over to pseudo-Riemannian ones. Most importantly, the vanishing of the Riemann tensor $R^\delta{}_{\alpha\beta\gamma}$ is a necessary and sufficient condition for a spacetime to be flat. For flat spacetimes, it is possible to choose a coordinate system so that the metric reduces to that of Minkowski spacetime at every point. For a curved spacetime, it is possible to choose a coordinate system so that the metric reduces to the Minkowski metric in the vicinity of any specific point P, but it is not generally possible to find a coordinate system in which this happens everywhere. Thus in general relativity we shall find that the results of special relativity will continue to hold true in the neighbourhood of any point but cannot be relied on generally. Special relativity will apply locally but not globally. This is similar to the finding that any small part of the Earth's surface can be treated as flat, but any extensive investigation will soon show that the Earth is actually curved.

One important property of a pseudo-Riemannian space is that it is possible to have curves for which ds^2 is zero at all points along the curve. Such curves are known as **null curves** since they have zero 'length' in the generalized sense of length residing in Equation 3.40. An important example of a null curve is a **null geodesic**. A null geodesic cannot therefore be defined as the shortest distance between the end-points of the curve (as in Subsection 3.3.2), but the definition as a curve along which the tangent always points in the same direction (as in Subsection 3.3.1) *is* still valid. Null geodesics are important in general relativity since, as you will see in the next chapter, they represent the possible paths of light rays in curved spacetime.

Exercise 3.17   (a) Find the connection coefficients for the Minkowski metric of Equation 3.37.

(b) Find the component $R^1{}_{212}$ of the Riemann tensor for the Minkowski metric of Equation 3.37.

Exercise 3.18   A two-dimensional Minkowski spacetime has the metric
$$ds^2 = c^2\,dt^2 - f^2(t)\,dx^2.$$

(a) Setting $x^0 = t$ and $x^1 = x$, find the connection coefficients.

(b) Hence find the component $R^0{}_{101}$ of the Riemann tensor.   ∎

Summary of Chapter 3

1. The line element for a Riemannian space is given by

    $$dl^2 = \sum_{i,j} g_{ij}\, dx^i\, dx^j,\qquad\text{(Eqn 3.11)}$$

    where the g_{ij} are the metric coefficients and the array $[g_{ij}]$ represents the metric tensor.

2. The metric tensor completely defines the geometry of the space. The converse is not true due to the freedom to choose different coordinates.

3. The line element in Cartesian coordinates for a plane is given by

    $$dl^2 = dx^2 + dy^2.\qquad\text{(Eqn 3.4)}$$

4. The line element in spherical coordinates for the surface of a sphere with radius R is given by

    $$dl^2 = R^2\, d\theta^2 + R^2 \sin^2\theta\, d\phi^2.\qquad\text{(Eqn 3.10)}$$

5. On a parameterized curve, each point corresponds to a unique value of a single parameter u. The curve can be described in an n-dimensional space by specifying a set of n coordinate functions $x^i(u)$ that assign to each point coordinates $x^1, x^2, \ldots, x^n$ that depend on the value of u.

6. A vector $\boldsymbol{v}$ that is moved along a curve while remaining parallel to its original direction is said to undergo parallel transport.

7. When a vector $\boldsymbol{v}$ is parallel transported along a curve specified by the coordinate functions $x^i(u)$, its components in the coordinate basis must change (to compensate for any changes in the coordinate basis vectors) at the rate

    $$\frac{dv^i}{du} = -\sum_{j,k} \Gamma^i{}_{jk}\, v^j\, \frac{dx^k}{du}.\qquad\text{(Eqn 3.18)}$$

8. The connection coefficient $\Gamma^i{}_{jk}$ describes the component in the direction of basis vector $\boldsymbol{e}_i$ of the rate of change of the basis vector $\boldsymbol{e}_j$ with respect to changes in the coordinate x^k. It is directly related to the metric by the expression

    $$\Gamma^i{}_{jk} = \frac{1}{2}\sum_l g^{il}\left(\frac{\partial g_{lk}}{\partial x^j} + \frac{\partial g_{jl}}{\partial x^k} - \frac{\partial g_{jk}}{\partial x^l}\right).\qquad\text{(Eqn 3.23)}$$

9. $[g^{ij}]$ is called the dual metric, and is the inverse of $[g_{ij}]$ regarded as a matrix, i.e. $\sum_{i,j}[g^{ij}][g_{ij}]$ is equal to the identity matrix. Or, more explicitly,

    $$\sum_k g^{ik} g_{kj} = \delta^i{}_j,\qquad\text{(Eqn 3.24)}$$

    where $\delta^i{}_j$ is known as the Kronecker delta, which is defined by

    $$\delta^i{}_j = \begin{cases} 1 & \text{if } i = j, \\ 0 & \text{if } i \neq j. \end{cases}$$

10. In a curved space, the geodesic between two points is the most direct path between those points (its tangent vector always points in the same direction) and also the path of shortest distance between them. Geodesics are analogous to straight lines in Euclidean space and minor arcs of great circles on the surface of a sphere. Affinely parameterized geodesics are described by coordinate functions $x^i(\lambda)$ that satisfy the geodesic equations

$$\frac{\mathrm{d}^2 x^i}{\mathrm{d}\lambda^2} + \sum_{j,k} \Gamma^i{}_{jk} \frac{\mathrm{d}x^j}{\mathrm{d}\lambda} \frac{\mathrm{d}x^k}{\mathrm{d}\lambda} = 0. \qquad \text{(Eqn 3.27)}$$

11. The curvature k at a point P of a curve in the plane is defined by

$$k = \frac{1}{R}, \qquad \text{(Eqn 3.31)}$$

where R is the radius of the circle that best approximates the curve in the region of P.

12. The Gaussian curvature K of a two-dimensional surface at a point P is defined by

$$K = k_{\mathrm{max}} k_{\mathrm{min}}, \qquad \text{(Eqn 3.33)}$$

where k_{max} and k_{min} are the maximum and minimum curvatures obtained by considering all possible geodesics through P.

13. The (intrinsic) curvature of an n-dimensional Riemannian space is characterized by the n^4 components of the Riemann tensor, which are directly related to the metric by the expression

$$R^l{}_{ijk} \equiv \frac{\partial \Gamma^l{}_{ik}}{\partial x^j} - \frac{\partial \Gamma^l{}_{ij}}{\partial x^k} + \sum_m \Gamma^m{}_{ik}\, \Gamma^l{}_{mj} - \sum_m \Gamma^m{}_{ij}\, \Gamma^l{}_{mk}. \quad \text{(Eqn 3.35)}$$

14. The Riemann tensor has many symmetries with respect to interchanging its indices, and this considerably restricts the number of independent components. In four dimensions there are 20 independent components, in three dimensions 6, and in two dimensions only 1.

15. The vanishing of the Riemann tensor is a necessary and sufficient condition for a space to be flat.

16. Strictly speaking, one requirement for a Riemannian space is that the line element satisfies $\mathrm{d}l^2 > 0$. Spaces for which the line element can be positive, negative or zero (null) are technically known as pseudo-Riemannian. The four-dimensional Minkowski spacetime of special relativity in which $\mathrm{d}s^2 = c^2\,\mathrm{d}t^2 - \mathrm{d}x^2 - \mathrm{d}y^2 - \mathrm{d}z^2$ is a pseudo-Riemannian space, as is its generalization to the curved spacetime of general relativity.

17. In pseudo-Riemannian spaces, a geodesic for which $\mathrm{d}s^2$ vanishes at all points along the curve is known as a null geodesic. It remains true that the tangent vector at any point along a null geodesic always points in the same direction.

Chapter 4   General relativity and gravitation

Introduction

Figure 4.1 Pierre-Simon Laplace (1749–1827), was born in Turin, but is regarded as one of the greatest of French mathematical physicists.

Figure 4.2 Siméon-Denis Poisson (1781–1840), a protegé of Laplace, made a number of significant contributions to mathematics, including the theory of probability.

Gravitation is an observable phenomenon; unsupported objects have a general tendency to fall downwards. In the Aristotelian physics of ancient Greece this was explained in terms of the composition of a body and the idea that objects had a 'natural place' in an Earth-centred universe. An apple released from a tree would fall downwards because its earthy composition gave it a natural place below the ground and its 'gravity' was the result of its tendency to move towards that place when free to do so. Likewise, smoke from a fire rose upwards because its airy composition gave it a natural place above the Earth that its innate 'levity' (the opposite of gravity) caused it to seek. Newton wrote scathingly of these ancient ideas. He offered a more mechanistic explanation of the phenomenon. Gravitation, according to Newton, was the result of a *gravitational force* that acted between massive bodies. In the case of two massive particles separated by a distance r, the gravitational force acting on each particle varied in proportion to $1/r^2$, so the Newtonian law that described this force became known as the inverse square law.

Neither Newton nor any of his followers was ever able to give a convincing explanation of the origin of this force. Newton tried to do so using ideas that were in vogue at the time, but he found that they did not work, so he said instead that he would 'feign no hypothesis' as to the origin of the gravitational force. The inverse square law of Newtonian gravitation simply described the way things were — it was a **phenomenological law**, based on experience, with no deeper justification than the fact that it worked. But it worked supremely well.

Over the generations that followed, innumerable scientists and engineers used the Newtonian concept of a gravitational force to explain a vast array of phenomena. Nowhere was this more true than in the field of celestial mechanics — the application of mechanical principles to the study of the motion of celestial bodies. Newton himself had shown that his notion of a gravitational force could explain the gross features of the Moon's motion but it fell to others, particularly French investigators such as Pierre-Simon Laplace (Figure 4.1), his pupil Siméon-Denis Poisson (Figure 4.2), and later still Charles Delaunay (1816–1872) to develop powerful ways of exploiting Newton's insights and working out their detailed consequences. That line of work continues to this day, particularly among the astrodynamicists who devise the trajectories of interplanetary spacecraft. These often include several 'gravity assist' manoeuvres in which a probe is helped on its way to a distant target by energy that it gathers from the planets that it encounters en route (Figure 4.3).

The Newtonian approach to gravitation has been so successful that many confuse Newton's proposed explanation of gravitation with the phenomenon itself. The question 'What is gravitation?' deserves an answer that speaks of the general tendency of massive bodies to draw together, yet even today a common answer is that it is an attractive force described by an inverse square law. Newton's brilliant and highly successful concept of a gravitational force has taken over gravitation in much the same way that the term 'Hoover' has replaced 'vacuum cleaner'.

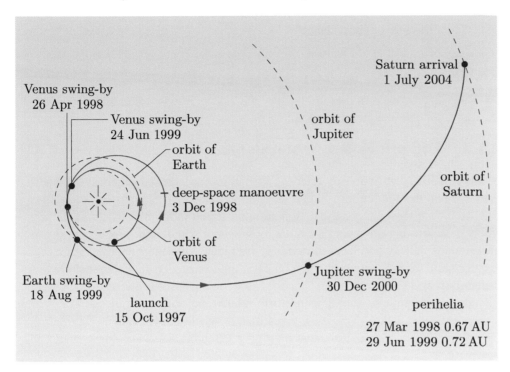

Figure 4.3 The trajectory that took the Cassini spacecraft to Saturn using a VVEJ manoeuvre that involved gravity assists from Venus, Venus again, Earth and Jupiter.

However, as Chapter 2 started to show, the development of Einsteinian relativity exposed problems deep in the heart of the Newtonian approach to gravitation. Under a change of inertial reference frame, a force described by an inverse square law does not transform in the way that a (three-) force should according to special relativity. Perhaps even more seriously, the Newtonian requirement that for every action there is an equal and opposite reaction implies that the gravitational forces linking two widely separated bodies should act instantly, irrespective of the distance between the two bodies. This is clearly at odds with the special relativistic requirement that such effects should not travel faster than the speed of light. Such arguments showed that Newtonian gravitation was not consistent with special relativity, and it soon became clear that no minor modification would make the two consistent.

The aim of this chapter is to introduce the core ideas of **general relativity** — Einstein's relativistic theory of gravity. We start with the principles that guided Einstein in his search for the theory, then go on to examine the basic mathematical ingredients of the theory, and finally present the Einstein field equations that relate those ingredients and use them to provide a new explanation of gravitation that does not require the existence of any gravitational force.

4.1   The founding principles of general relativity

Formulating a new theory in fundamental physics is not an entirely logical process. The search usually involves some general fundamental principles, consistency with known experimental facts, elegance and economy of ideas, and, inevitably, some guesswork. Of course, the ultimate test of any theory is provided by confronting its predictions with new experiments, and we shall come to this in Chapter 7; first we have to formulate the theory. Einstein was motivated in his search by three basic principles:

1. The principle of equivalence
2. The principle of covariance
3. The principle of consistency.

We shall discuss each of these in turn.

4.1.1 The principle of equivalence

It was in 1907, just two years after the formulation of special relativity, that Einstein had the sudden insight that he later described as 'the happiest thought of my life'. That thought was the realization that for an individual who was falling freely, accelerating downwards from a roof, say, or some other high place, it was almost as if gravity had been turned off. This idea, linking gravitation and acceleration, gave Einstein his start on extending relativity theory to include gravitation and showed him that a theory of general relative motion — one that included accelerations as well as uniform relative motions — could also be a theory of gravitation. This idea, that Einstein would later formalize as the *principle of equivalence*, also shed light on a troubling aspect of Newtonian mechanics.

The equality of gravitational and inertial mass

Newtonian mechanics involves two different concepts of mass:

1. **Inertial mass**, m, which describes a particle's resistance to being accelerated by a force. The inertial mass of a particle is defined, according to Newton's second law, by the ratio of the magnitude of the force on the particle to the magnitude of the acceleration it produces, $m = |\boldsymbol{F}|/|\boldsymbol{a}|$.

2. **Gravitational mass**, μ, which determines the force that a given particle experiences due to, or exerts on, another particle as a result of gravity. The gravitational mass is defined through Newton's law of gravitation for the force $\boldsymbol{F}_{12}$ on particle 1 of gravitational mass μ_1 due to particle 2 of gravitational mass μ_2. The magnitude of this force can be written as

$$|\boldsymbol{F}_{12}| = G\frac{\mu_1\mu_2}{|\boldsymbol{x}_1 - \boldsymbol{x}_2|^2}. \tag{4.1}$$

Now, as will be discussed later, it is a well-established experimental fact that the ratio μ/m is the same for all bodies, to an accuracy of at least one part in 10^{11}. In Newtonian mechanics, this is simply an extraordinary coincidence with no explanation. However, for Einstein it was something that cried out for a fundamental explanation. Of course, once we accept that the ratio of gravitational to inertial mass is a constant, then we can (and do) choose to use units of measurement that make the two masses for any body equal, so that $\mu/m = 1$. This is why we can ignore the distinction between gravitational and inertial masses for almost all practical purposes.

Freely falling frames are locally inertial frames

In Newtonian physics, the equality of inertial and gravitational mass implies that the acceleration of any body due to a gravitational force is independent of the mass of the body.

● Prove the above statement.

○ The equality of inertial and gravitational mass implies that μ_i in Equation 4.1 may be replaced by m_i, so

$$|\boldsymbol{F}_{12}| = G\frac{m_1 m_2}{|\boldsymbol{x}_1 - \boldsymbol{x}_2|^2}.$$

The acceleration $\boldsymbol{a}_1$ of particle 1 due to this force is given by

$$\boldsymbol{F}_{12} = m_1\boldsymbol{a}_1,$$

and hence

$$m_1|\boldsymbol{a}_1| = G\frac{m_1 m_2}{|\boldsymbol{x}_1 - \boldsymbol{x}_2|^2}.$$

Clearly, the mass m_1 cancels, and consequently the acceleration of any body subject only to gravitational forces will be independent of the mass of the body.

This result leads us to consider a famous 'thought experiment' in which it is supposed that a frictionless (non-rotating) lift is falling freely down an airless lift shaft (see Figure 4.4). The acceleration of the lift or any object in the vicinity of the lift is independent of its mass. Consequently, for an observer inside the lift, an object released from rest (relative to the observer) would remain stationary; that is, according to the freely falling observer, the object would be free of any force and would continue in its state of rest. Moreover, if the observer were to exert a force on the object, it would move according to Newton's laws of motion. In other words, from the point of view of the observer in the freely falling lift, a frame of reference fixed in the lift is an inertial frame of reference.

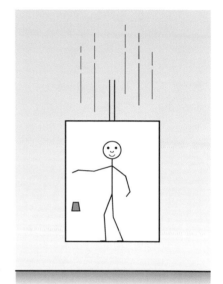

Figure 4.4 A freely-falling lift.

Such a frame is properly described as a *locally* inertial frame (as opposed to a *globally* inertial frame) because we need to restrict our measurements to sufficiently small regions of space and sufficiently small intervals of time if we are not to observe departures from inertial behaviour. This is because the gravitational field in which the lift and its contents are located is not uniform. Two objects released simultaneously from the same height on opposite sides of the lift will each fall towards the centre of the Earth, so instead of falling along parallel paths and maintaining a constant separation, as they would in a uniform gravitational field, they will in fact fall along converging paths and gradually approach each other. The horizontal forces responsible for this non-inertial behaviour are examples of the **tidal forces** that cause neighbouring particles in any non-uniform gravitational field to have different accelerations. Such effects are usually small but they can have observable consequences (such as the tides in the Earth's oceans!), and even within a freely falling lift they would be observable if experiments were performed with sufficient precision or over a sufficiently long period of time. Nonetheless, the point remains that a freely falling frame in a gravitational field is a locally inertial frame where the laws of special relativity will hold true.

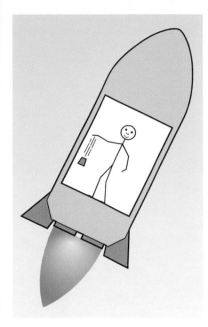

Figure 4.5 A uniformly accelerating rocket.

Exercise 4.1 Two objects are 2.00 m apart in a freely falling lift near to the surface of the Earth (which has a radius of 6.38×10^6 m).

(a) Calculate the magnitude of their acceleration towards each other when their separation is horizontal.

(b) Calculate the magnitude of their acceleration towards each other when their separation is vertical. ■

Of course, you might well ask what is meant by 'sufficiently small' for a frame to be locally inertial. The answer is that we assume that having decided on limits to the accuracy of a particular experiment, we can always choose a small enough region and a short enough time interval so that a freely falling frame will appear to be inertial to within this accuracy.

Another thought experiment involves a rocket in a region in which there is no gravitational field. If the rocket is accelerated with a uniform acceleration of magnitude g, no sufficiently localized experiment within the rocket can distinguish between the consequences of the acceleration and the gravitational field on the surface of the Earth. An object released from rest within the rocket would accelerate downwards, just as an object on Earth would do (see Figure 4.5).

Principle of equivalence

In 1907, Einstein elevated to a formal principle the idea that locally one cannot distinguish between gravity and acceleration. What is now known as the **weak equivalence principle** can be stated as follows.

> **Weak equivalence principle**
>
> Within a sufficiently localized region of spacetime adjacent to a concentration of mass, the motion of bodies subject to gravitational effects alone cannot be distinguished by any experiment from the motion of bodies within a region of appropriate uniform acceleration.

The weak equivalence principle is a direct consequence of the fact that the acceleration of freely falling objects does not depend on their composition, and it is therefore sometimes referred to as the **principle of universality of free fall**. Note that this does not apply to very massive objects that would substantially change the gravitational field in their vicinity. Moreover, it only relates to gravitational effects, so experiments involving electromagnetic forces or nuclear interactions are specifically excluded.

The restriction to gravitational effects does not apply to the **strong equivalence principle**.

> **Strong equivalence principle**
>
> Within a sufficiently localized region of spacetime adjacent to a concentration of mass, the physical behaviour of bodies cannot be distinguished by *any* experiment from the physical behaviour of bodies within a region of appropriate uniform acceleration.

This statement (which is often simply referred to as the **equivalence principle**) clearly goes beyond the universality of free fall, although that is included as a special case.

Both versions of the equivalence principle have been subject to many direct experimental tests. Galileo is often said to have demonstrated the universality of free fall by dropping different objects from the leaning tower of Pisa. It is unlikely that he actually performed such an experiment, but the experiments that he did perform — rolling bodies down inclined planes — should have made him aware of the outcome to expect. The first high-precision tests were carried out over many years with steadily improving sensitivity, eventually reaching better than one part in 10^8, by the Hungarian scientist Loränd Eötvös (pronounced 'ert-vos') in the late nineteenth and early twentieth centuries. These results were quoted by Einstein in his first complete formulation of general relativity. Currently, the most rigorous test of the weak equivalence principle is provided by the Eöt-Wash experiments, which provide agreement to better than one part in 10^{12} (see Figure 4.6). Projected satellite experiments could provide even more stringent tests. For instance, the proposed Satellite Test of the Equivalence Principle (STEP), a space mission that is still in the design stage, could provide an accuracy of one part in 10^{18}.

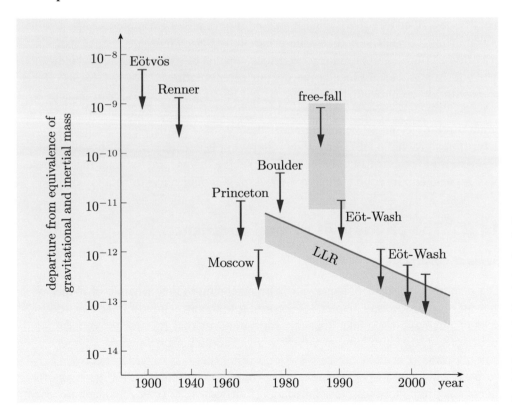

Figure 4.6 Tests of the weak equivalence principle. Most use torsion balances to seek tiny differences in the gravitational and inertial mass of a body, but the green region represents the results of experiments in drop towers, and LLR indicates lunar ranging experiments that compare the acceleration of the Earth and the Moon in the gravitational field of the Sun.

Experimental tests of the strong equivalence principle are much less clear-cut, but most theories that violate it predict that the locally measured value of the gravitational constant, G, may vary with time. Current constraints on the rate of change of G are approaching one part in 10^{13} year^{-1}. Einstein's theory of general relativity is thought to be the only theory of gravity that is consistent with the strong equivalence principle.

Although the strong equivalence principle is certainly in need of additional tests, the weak equivalence principle alone was sufficient to lead Einstein to predict two new effects that eventually became part of general relativity. First, consider a horizontally travelling beam of light that enters and crosses the interior of a rocket that is accelerating vertically upwards — at right angles to the beam of light. From the point of view of the accelerated observer travelling with the rocket, the light ray follows a downward-curving path. The local equivalence of gravitation and acceleration therefore led Einstein to predict that one effect of gravitation would be the deflection of light rays towards concentrations of mass. The second effect was based on the fact that an observer in an upward-accelerating rocket would find that the frequency of light waves emitted from the floor of the rocket would be redshifted (i.e. their frequency would be decreased) as successive wave peaks took longer and longer to reach the ceiling. (These effects are illustrated in Figure 4.7.) This led Einstein to predict that light escaping from a concentration of mass should exhibit a redshift due to gravity. As you will see later, these two predicted effects, the **gravitational deflection of light** and the **gravitational redshift of light**, both became the subject of refined calculations in the full theory of general relativity and both eventually became important tests of the theory.

The equivalence principle calls into question the reality of gravitational forces. The centrifugal force experienced by an observer in a bus turning a corner is usually described as a *fictitious force* since it only arises from the use of a non-inertial frame. The equivalence principle indicates that, locally at least, gravitational forces are similarly 'fictitious' results of using a frame that is not freely falling and so not locally inertial. In a freely falling frame the (local) effects of gravitation would vanish and there would be no gravitational force. This does not mean that gravitation is not real but suggests that in its final formulation general relativity might be a 'geometric theory' in which spacetime curvature determines gravitational effects (including what is a locally inertial frame), and gravitational forces will not be needed. In such a theory gravitational mass would play no role and its equivalence to inertial mass would cease to be a mystery.

4.1.2   The principle of general covariance

General covariance

The *principle of general covariance* is an extension of the principle of relativity that was introduced in Chapter 1. According to the principle of relativity, the laws of physics should take the same form in all inertial frames. As you saw in Chapter 2, that implied that physical laws should be form-invariant under Lorentz transformations, and a way of ensuring that was to write the laws as properly balanced four-tensor relations. We saw how to do that for the laws of electromagnetism using scalar invariants (four-tensors of rank zero), contravariant and covariant four-vectors (four-tensors of rank 1), and some four-tensors of rank 2 — specifically, the contravariant field four-tensor $[\mathsf{F}^{\mu\nu}]$, the mixed field four-tensor $[\mathsf{F}^{\mu}{}_{\nu}]$, and the covariant field four-tensor $[\mathsf{F}_{\mu\nu}]$. You will also recall that it was the principle of relativity that excluded Newtonian gravitation from being a viable relativistic theory of gravity; the Newtonian gravitational force cannot be described as part of a four-vector, because it does not transform in the right way.

Remember that when we enclose a tensor component in square brackets, it indicates that we are discussing the entire tensor, not just the individual component.

The **principle of general covariance** extends the principle of relativity by requiring the physical equivalence of all frames, including non-inertial ones.

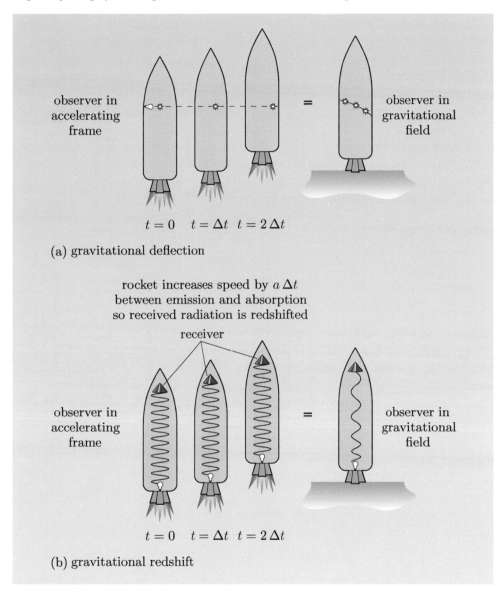

(a) gravitational deflection

(b) gravitational redshift

Figure 4.7 The effect of observer acceleration on the behaviour of light, and the equivalent gravitational deflection and gravitational redshift of light.

There is still debate about the significance of this principle and the extent to which Einstein was successful in implementing it in general relativity. However, what he did in practice was to require that physical laws should retain their form under a broad class of coordinate transformations, and he did this by requiring that the laws should be expressed in terms of mathematical objects called **general tensors**, or more often just **tensors**. Most of this section will be devoted to making clear what tensors are, how they differ from the more restricted four-tensors that you met in Chapter 2, and how they may be combined to form tensor equations that might describe generally covariant laws of physics, including gravitation.

Defining general tensors

The study of tensors can be approached in several ways, but for our purposes tensors are multi-component mathematical objects that can be recognized and classified according to the way their components behave under **general coordinate transformations** — that is, under coordinate transformations in which the new coordinates x'^μ are functions of the old coordinates x^ν, as in $x'^\mu = x'^\mu(x^\nu)$ for $\mu, \nu = 0, 1, 2, 3$. These functions are required to be sufficiently well-behaved that they can be differentiated, but they are still more general than the Lorentz transformations of special relativity, which were restricted to linear functions. In the case of the Lorentz transformations, the linearity ensured that derivatives such as $\partial x'^\mu/\partial x^\nu$ would be constants (such as c, V, γ or combinations of those parameters). In the case of a general coordinate transformation $x'^\mu = x'^\mu(x^\nu)$, the sixteen functions $\partial x'^\mu/\partial x^\nu$ ($\mu, \nu = 0, 1, 2, 3$) and the sixteen functions $\partial x^\beta/\partial x'^\alpha (\alpha, \beta = 0, 1, 2, 3)$ are free of such restrictions. Having explained what is meant by a general coordinate transformation, we can say that a tensor of contravariant rank m and covariant rank n has components $T'^{\mu_1 \mu_2 \dots \mu_m}_{\alpha_1 \alpha_2 \dots \alpha_n}$ that transform according to

$$T'^{\mu_1 \mu_2 \dots \mu_m}_{\alpha_1 \alpha_2 \dots \alpha_n} = \sum_{\nu_1, \nu_2, \dots, \nu_m, \beta_1, \beta_2, \dots, \beta_n} \frac{\partial x'^{\mu_1}}{\partial x^{\nu_1}} \frac{\partial x'^{\mu_2}}{\partial x^{\nu_2}} \cdots \frac{\partial x'^{\mu_m}}{\partial x^{\nu_m}}$$

$$\times \frac{\partial x^{\beta_1}}{\partial x'^{\alpha_1}} \frac{\partial x^{\beta_2}}{\partial x'^{\alpha_2}} \cdots \frac{\partial x^{\beta_n}}{\partial x'^{\alpha_n}}$$

$$\times T^{\nu_1 \nu_2 \dots \nu_m}_{\beta_1 \beta_2 \dots \beta_n}. \tag{Eqn 2.110}$$

Expressed in such general terms this looks very complicated, but the simple fact is that you have already met many of the most important tensor quantities that will be needed in this book. In particular, you are already familiar with the notion of a scalar invariant, S say, that remains unchanged under a general coordinate transformation. And you are also familiar with the infinitesimal displacement $[\mathrm{d}x^\mu] = (\mathrm{d}x^0, \mathrm{d}x^1, \mathrm{d}x^2, \mathrm{d}x^3)$. This is actually a contravariant tensor of rank 1 with components that transform according to

$$\mathrm{d}x'^\mu = \sum_{\alpha=0}^{3} \frac{\partial x'^\mu}{\partial x^\alpha} \, \mathrm{d}x^\alpha. \tag{4.2}$$

You have also met the vastly important rank 2 metric tensor $[g_{\mu\nu}]$. In its contravariant (dual) form its components transform according to

$$g'^{\mu\nu} = \sum_{\alpha=0}^{3} \frac{\partial x'^\mu}{\partial x^\alpha} \sum_{\beta=0}^{3} \frac{\partial x'^\nu}{\partial x^\beta} \, g^{\alpha\beta}, \tag{4.3}$$

and in the covariant form they transform according to

$$g'_{\mu\nu} = \sum_{\alpha=0}^{3} \frac{\partial x^\alpha}{\partial x'^\mu} \sum_{\beta=0}^{3} \frac{\partial x^\beta}{\partial x'^\nu} \, g_{\alpha\beta}. \tag{4.4}$$

The metric tensor components satisfy the useful relationship

$$\sum_{\gamma=0}^{3} g^{\alpha\gamma} \, g_{\gamma\beta} = \delta^\alpha{}_\beta, \tag{Eqn 3.24}$$

where $\delta^\alpha{}_\beta$ is a four-dimensional version of the Kronecker delta and is itself defined by

$$\delta^\alpha{}_\beta = \begin{cases} 1 & \text{if } \alpha = \beta, \\ 0 & \text{if } \alpha \neq \beta. \end{cases}$$

You have even met the Riemann curvature tensor $[R^{\alpha}{}_{\beta\gamma\delta}]$, a mixed tensor of contravariant rank 1 and covariant rank 3. In four-dimensional spacetime this tensor has 256 components, though due to symmetries, only 20 are independent. Each component transforms according to

$$R'^{\alpha}{}_{\beta\gamma\delta} = \sum_{\mu=0}^{3} \frac{\partial x'^{\alpha}}{\partial x^{\mu}} \sum_{\nu=0}^{3} \frac{\partial x^{\nu}}{\partial x'^{\beta}} \sum_{\rho=0}^{3} \frac{\partial x^{\rho}}{\partial x'^{\gamma}} \sum_{\sigma=0}^{3} \frac{\partial x^{\sigma}}{\partial x'^{\delta}} R^{\mu}{}_{\nu\rho\sigma}. \tag{4.5}$$

A final point to note — or rather to recall, since it was mentioned in Chapter 3 — is that not all multi-component objects are tensors. It was pointed out earlier that the 64 connection coefficients $\Gamma^{\alpha}{}_{\beta\gamma}$ of a four-dimensional spacetime do not satisfy the appropriate transformation law for a mixed rank 3 tensor, so they simply do not form a tensor.

Exercise 4.2   Suppose that in a two-dimensional Euclidean space with coordinates x^{μ} ($\mu = 1, 2$) the coordinates x^1 and x^2 correspond to the polar coordinates r and θ. Also suppose that the coordinates x'^{μ} correspond to the usual Cartesian coordinates x, y.

(a) If A^{μ} is a general tensor component in r, θ coordinates, and A'^{μ} is the corresponding tensor component in x, y coordinates, find the transformation that expresses A'^{μ} in terms of A^{μ} for each value of μ.

(b) Confirm that this transformation law is satisfied by the two-dimensional infinitesimal displacement vector that has components $(\mathrm{d}x^1, \mathrm{d}x^2) = (\mathrm{d}r, \mathrm{d}\theta)$ and $(\mathrm{d}x'^1, \mathrm{d}x'^2) = (\mathrm{d}x, \mathrm{d}y)$.   ∎

Raising and lowering general tensor indices

It is the metric tensor that relates contravariant and covariant tensor components via

$$A_{\mu} = \sum_{\alpha=0}^{3} g_{\mu\alpha} A^{\alpha} \tag{4.6}$$

and

$$A^{\mu} = \sum_{\alpha=0}^{3} g^{\mu\alpha} A_{\alpha}. \tag{4.7}$$

In other words, the contravariant metric tensor 'raises' indices and the covariant metric tensor 'lowers' them.

Exercise 4.3   Show that if we use the covariant metric tensor to 'lower' the index on A^{μ} and then we use the contravariant metric tensor to 'raise' the index again, we get back to A^{μ}.   ∎

● If we have a mixed tensor with some indices up and some down, it is usually important to leave spaces so that, for example, we write $R^{\alpha}{}_{\beta\gamma\delta}$ rather than $R^{\alpha}_{\beta\gamma\delta}$. Explain why.

○ Suppose that we start with $R_{\alpha\beta\gamma\delta}$, then use the contravariant metric tensor to raise the α index without paying attention to the order of the indices. We will obtain the result $R^{\alpha}_{\beta\gamma\delta}$. The problem is that the individual indices are just placeholders and have no special significance. This means that if we subsequently use the covariant metric tensor to lower the α index, it is impossible to tell if the lowered index should be put in the first or second 'slot', i.e. whether the result should be $R_{\alpha\beta\gamma\delta}$ or $R_{\beta\alpha\gamma\delta}$. Unless the tensor happens to be symmetric with respect to interchange of the first two indices, the two possible results will be different. It is therefore usually important to preserve the order of the indices despite any raising or lowering that may be performed. That's why we should generally write $R^{\alpha}{}_{\beta\gamma\delta}$ rather than $R^{\alpha}_{\beta\gamma\delta}$.

The rules of tensor algebra

Einstein's aim was to use tensors to write down a theory of gravity in a generally covariant form — in other words, following the rules of general tensor algebra for multiplying tensors by scalars, adding and subtracting tensors, multiplying tensors together and reducing the rank of a tensor through contraction. These rules are similar to those that we have already used to manipulate four-tensors in special relativity, but to make them completely clear, we now list them in their general forms.

1. **Scaling**   A tensor $[T^{\mu_1\mu_2\ldots\mu_m}_{\alpha_1\alpha_2\ldots\alpha_n}]$ of contravariant rank m and covariant rank n may be multiplied by a scalar S to produce a new tensor $[U^{\mu_1\mu_2\ldots\mu_m}_{\alpha_1\alpha_2\ldots\alpha_n}]$ of the same rank. Each component of the new tensor is obtained by multiplying the corresponding component of the original tensor by the same scalar S. So, for example, for all values of μ and α,
$$S\,T^{\mu}{}_{\alpha} \equiv U^{\mu}{}_{\alpha}.$$

2. **Addition and subtraction**   Tensors may be added or subtracted to form new tensors, but those being added or subtracted must be of the same type, i.e. with the same contravariant rank and the same covariant rank. Again the addition or subtraction is carried out component by component. So, for example, for all values of μ and α,
$$S^{\mu}{}_{\alpha} + T^{\mu}{}_{\alpha} \equiv U^{\mu}{}_{\alpha}.$$

3. **Multiplication**   Tensors may be multiplied together by forming products of their components. So, for example, given three tensors $[X^{\mu}]$, $[Y_{\alpha}]$ and $[Z_{\beta}]$, we can form a new tensor $[A^{\mu}{}_{\alpha\beta}]$ with components
$$A^{\mu}{}_{\alpha\beta} \equiv X^{\mu}\,Y_{\alpha}\,Z_{\beta}.$$
The rank of the new tensor is then the sum of the ranks of the tensors being multiplied together (e.g. $A^{\mu}{}_{\alpha\beta}$ has rank 3). The tensors being multiplied together may even be the same, as in
$$A^{\mu\nu} \equiv U^{\mu}\,U^{\nu}.$$

4. **Contraction**   In the case of a single tensor with contravariant rank m and covariant rank n, or in the case of a product of tensors with combined contravariant rank m and covariant rank n, it is possible to form another tensor, of contravariant rank $m-1$ and covariant rank $n-1$, by summing over one raised index and one lowered index. So, for example,
$$B_{\gamma} \equiv \sum_{\sigma=0}^{3} A^{\sigma}{}_{\sigma\gamma} \equiv \sum_{\sigma=0}^{3} X^{\sigma}\,Y_{\sigma}\,Z_{\gamma}.$$

These rules imply that tensors can appear in expressions only in certain well-defined ways. In order to illustrate this, consider the following (fairly arbitrary) equation involving tensors:

$$A^\mu{}_\nu = S\,B^\mu{}_\nu + \sum_{\alpha=0}^{3} C^\mu{}_\alpha\,E^\alpha{}_\nu + \sum_{\alpha=0}^{3}\sum_{\beta=0}^{3} X^{\mu\alpha}{}_\beta\,Y_{\nu\alpha}{}^\beta. \tag{4.8}$$

The right-hand side of Equation 4.8 consists of the sum of three 'terms', which we can use to emphasize some important general properties of tensor equations.

- The only indices that are not 'summed over' are μ and ν. These are the *free indices*. They exhibit the following properties:

  (a) The μ and ν indices are consistently 'up' (contravariant) or 'down' (covariant).

  (b) The μ and ν indices appear once and only once in every term on each side of the equation.

  (c) The letters μ and ν have no special significance. We can replace either (or both) of them with a different (Greek) letter provided that we carry out the replacement in every term (on both sides of the equation) and the new letter does not clash with one that is already in use. For example, we could replace μ with λ, but replacing μ with α would cause confusion.

- Some indices (α and β in this example) appear precisely twice in a term. These are the *dummy indices*.

  (a) Such indices are always summed over.

  (b) One appearance is always 'up' and the other is 'down'.

  (c) The letter used has no special significance and can always be replaced with another (Greek) letter provided that we replace *both* occurrences within any one term and the new letter doesn't clash with one that is already in use. For example, α in the third term on the right-hand side could be replaced with γ, but not with β.

As you can see, the indices within a covariant equation form very distinct patterns that you will soon become adept at spotting. Expressions such as Equation 4.8 are said to be **generally covariant** or, more simply, in **covariant** form. This means that the equation will take the same form in any coordinate system; it does *not*, of course, mean that the numerical values of the components are necessarily the same. It is worth noticing how the word 'covariant' is a bit over-used. A rank 1 'covariant tensor' is one with components that transform according to

$$A'_\alpha = \sum_{\beta=0}^{3} \frac{\partial x^\beta}{\partial x'^\alpha} A_\beta, \tag{4.9}$$

and is denoted by having the indices 'down'. A 'covariant equation' is an equation that takes the same form in different coordinate systems, and may or may not involve covariant tensors. Indeed, a covariant equation may involve contravariant tensors.

- What is the analogous equation to Equation 4.9 that describes how the components of a rank 1 contravariant tensor transform?

○ From Equation 2.110 or from the rank 1 example that follows it in Equation 4.2, the required transformation rule is

$$A'^{\alpha} = \sum_{\beta=0}^{3} \frac{\partial x'^{\alpha}}{\partial x^{\beta}} A^{\beta}. \tag{4.10}$$

Exercise 4.4 Explain why each of the following is not a generally covariant tensor equation.

(a) $A^{\mu} = B_{\mu} + K$ (b) $X^{\mu} = \sum_{\nu} Y^{\mu\nu} Z^{\nu}$ (c) $A_{\mu} = \sum_{\nu} W_{\mu}{}^{\nu} X_{\nu} Y^{\nu}$ ∎

The rules of covariant differentiation

When we wrote down the laws of Lorentz-covariant electromagnetism in Chapter 2, in addition to scaling, adding, multiplying and contracting four-tensors, we also formed four-tensors by taking partial derivatives of existing tensors. Being able to represent derivatives of four-tensors was important because the basic laws of electromagnetism (the Maxwell equations and the equation of continuity) were differential equations. We should expect the generally covariant theory of gravitation to involve differential equations, so we need to know how to differentiate a general tensor in a covariant way. This turns out to be more complicated in general relativity than it was in special relativity because simple partial derivatives of tensors are *not* generally covariant.

Defining the derivative of a function involves evaluating the function at some point, x say, and at a nearby point, $x + \delta x$ say, and then taking the difference. In a flat space this does not present any particular problem. Nor is it particularly complicated in a curved space as long as we are only considering functions. However, we know from Chapter 3 that transporting a vector $[v^{\alpha}]$ (i.e. a rank 1 tensor) requires some care since the parallel transport of a vector generally involves the connection coefficients

$$\Gamma^{\alpha}{}_{\beta\gamma} = \frac{1}{2} \sum_{\delta} g^{\alpha\delta} \left(\frac{\partial g_{\delta\gamma}}{\partial x^{\beta}} + \frac{\partial g_{\beta\delta}}{\partial x^{\gamma}} - \frac{\partial g_{\beta\gamma}}{\partial x^{\delta}} \right). \tag{Eqn 3.23}$$

For a vector with components v^{α}, the expression

$$\frac{\partial v^{\alpha}}{\partial x^{\beta}}$$

simply does not transform in the right way under general coordinate transformations to be a component of a rank 2 tensor. Nor does the expression

$$\sum_{\lambda} \Gamma^{\alpha}{}_{\lambda\beta} v^{\lambda}.$$

However, sums of the form

$$\frac{\partial v^{\alpha}}{\partial x^{\beta}} + \sum_{\lambda} \Gamma^{\alpha}{}_{\lambda\beta} v^{\lambda}$$

arise when considering the limit of a difference in a vector and its parallel transported version, and this quantity does transform as a component of a rank 2 tensor. Expressions of this kind occur so frequently in general relativity that it is useful to give them a name and a symbol. Consequently, we write

$$\nabla_\beta v^\alpha \equiv \frac{\partial v^\alpha}{\partial x^\beta} + \sum_\lambda \Gamma^\alpha{}_{\lambda\beta}\, v^\lambda \qquad (4.11)$$

and say that $\nabla_\beta v^\alpha$ represents the **covariant derivative** of v^α. In effect, the non-tensorial behaviour of $\partial v^\alpha / \partial x^\beta$ is cancelled by the non-tensorial behaviour of $\sum_\lambda \Gamma^\alpha{}_{\lambda\beta}\, v^\lambda$. At this stage, you should regard $\nabla_\beta v^\alpha$ as no more than a shorthand for the right-hand side of Equation 4.11. Of course, we don't just want to differentiate rank 1 contravariant tensors. We also need to know how to covariantly differentiate rank 1 covariant tensors and tensors of higher rank, so that the result is a tensor in each case. It can be shown that Equation 4.11 implies that the covariant derivative of a covariant tensor v_α can be expressed as

$$\nabla_\beta v_\alpha = \frac{\partial v_\alpha}{\partial x^\beta} - \sum_\lambda \Gamma^\lambda{}_{\alpha\beta}\, v_\lambda. \qquad (4.12)$$

Note that in this case the final term is subtracted from the partial derivative, whereas in the case of a contravariant vector it was added. The covariant derivatives of higher-rank tensors are direct generalizations of Equations 4.11 and 4.12, as appropriate. For instance,

$$\nabla_\lambda T^{\mu\nu} = \frac{\partial T^{\mu\nu}}{\partial x^\lambda} + \sum_\rho \Gamma^\mu{}_{\rho\lambda}\, T^{\rho\nu} + \sum_\rho \Gamma^\nu{}_{\rho\lambda}\, T^{\mu\rho}.$$

● Write down the expression for $\nabla_\lambda T^\mu{}_\nu$ in terms of the connection coefficients.

○ From Equations 4.11 and 4.12, we have

$$\nabla_\lambda T^\mu{}_\nu = \frac{\partial T^\mu{}_\nu}{\partial x^\lambda} + \sum_\rho \Gamma^\mu{}_{\rho\lambda}\, T^\rho{}_\nu - \sum_\rho \Gamma^\rho{}_{\nu\lambda}\, T^\mu{}_\rho. \qquad (4.13)$$

This is a good point at which to restate the principle of general covariance and summarize its significance in the formulation of general relativity.

General covariance, tensors and covariant differentiation

According to the principle of general covariance, the laws of physics should take the same form in all frames of reference. In practice this means that they should be expressed as balanced tensor relationships that are covariant under general coordinate transformations.

Legitimate algebraic operations involving tensors include scaling, addition and subtraction (provided that the types are identical), multiplication and contraction. The partial differentiation of a tensor does not generally produce another tensor, but the process of covariant differentiation does. This may be applied to a tensor of any rank and is exemplified by

$$\nabla_\lambda T^\mu{}_\nu = \frac{\partial T^\mu{}_\nu}{\partial x^\lambda} + \sum_\rho \Gamma^\mu{}_{\rho\lambda}\, T^\rho{}_\nu - \sum_\rho \Gamma^\rho{}_{\nu\lambda}\, T^\mu{}_\rho. \qquad (\text{Eqn } 4.13)$$

Exercise 4.5   What is the covariant derivative of the invariant scalar function $S(ct, x, y, z)$? (*Hint*: This is a tensor of rank 0.) ■

Figure 4.8   Isaac Newton (1642–1727) was the founding genius of natural philosophy as we know it today.

4.1.3   The principle of consistency

The **principle of consistency** asserts that a new theory that aims to replace or supersede earlier theories should account for the successful predictions of those earlier theories. In the particular case of general relativity, we should expect consistency with the successes of Einstein's own special relativity and Newtonian gravitation. The former requirement is guaranteed by using a spacetime that is locally equivalent to Minkowski spacetime; the latter provides a useful constraint on the kinds of tensor equations that can be used in the formulation of general relativity.

For the purposes of establishing consistency with Newtonian predictions, it is helpful to first see how Newton's theory of gravity, as expressed by the inverse square law, can be reformulated as a **field theory**, based on the idea of a *gravitational field* that obeys *differential equations* similar to those satisfied by the electric and magnetic fields of electromagnetism.

To this end, we first define the Newtonian **gravitational field** $g(r)$ to be a function of position $r = (x, y, z)$ that specifies the Newtonian gravitational force per unit mass that would act on a test particle at the point r. This means that the gravitational force on a particle of mass m at r would be $m\,g(r)$. It follows from Newton's law of gravitation (Equation 4.1) that in the case of a uniform spherical body of total mass M centred on the origin of coordinates ($r = 0$), the gravitational field is given by

$$g(r) = -G\frac{M}{|r|^2}e_r, \tag{4.14}$$

where e_r is a unit vector in the radial direction, pointing away from the origin. The minus sign in Equation 4.14 means that $g(r)$ is directed towards the origin at every point, as shown in Figure 4.9.

If we suppose that the sphere of mass M is enclosed by a larger sphere of radius R also centred on the origin, we can define the **flux** of the gravitational field leaving the larger sphere by a surface integral:

$$\text{outward gravitational flux} = \int_S g \cdot \widehat{n}\, \mathrm{d}S,$$

where $\widehat{n}$ is an outward-pointing unit vector normal to the spherical surface S at every point. From the spherical symmetry of the situation, it is easy to see that in this case the surface integral will be given by the surface area of the sphere ($4\pi R^2$) multiplied by the constant field strength on the surface of the sphere (GM/R^2), multiplied by -1 because in this case the field points inwards, so $e_r \cdot \widehat{n} = -1$. Thus

$$\int_S g \cdot \widehat{n}\, \mathrm{d}S = -4\pi GM.$$

Now, according to the **divergence theorem** of vector calculus, this kind of surface integral of the field can be rewritten as a volume integral of a quantity known as the **divergence** of the field, $\boldsymbol{\nabla} \cdot g$, throughout the volume V bounded by the surface S, so

$$\int_V \boldsymbol{\nabla} \cdot g\, \mathrm{d}V = -4\pi GM, \tag{4.15}$$

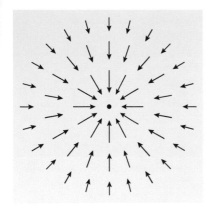

Figure 4.9   The gravitational field due to a uniform sphere of total mass M centred on the origin.

where, in terms of Cartesian components, the vector operator $\boldsymbol{\nabla}$ represents $\left(\frac{\partial}{\partial x}, \frac{\partial}{\partial y}, \frac{\partial}{\partial z}\right)$, so the divergence is defined by

$$\boldsymbol{\nabla} \cdot \boldsymbol{g} = \frac{\partial g_x}{\partial x} + \frac{\partial g_y}{\partial y} + \frac{\partial g_z}{\partial z}. \tag{4.16}$$

If we now write the mass of the sphere as an integral over its density ρ, we have

$$\int_V \boldsymbol{\nabla} \cdot \boldsymbol{g} \, \mathrm{d}V = -4\pi G \int_V \rho \, \mathrm{d}V. \tag{4.17}$$

Though not a proof, this last relation at least makes plausible a general relationship that can be proved by more rigorous methods, namely the differential relationship

$$\boldsymbol{\nabla} \cdot \boldsymbol{g} = -4\pi G \rho. \tag{4.18}$$

This is actually one of the fundamental equations of Newtonian gravitation, relating derivatives of the gravitational field to the mass density that is the source of the field. It is not restricted to spherical bodies, nor even to cases where the density is uniform. Nor is it quite the end of our argument.

The gravitational force is *conservative*. That means that the work done against the gravitational force in moving a body from one point to another is independent of the path followed. That's why it is possible to associate the gravitational force with a *gravitational potential energy*. The gravitational field $\boldsymbol{g}(\boldsymbol{r})$ can be similarly related to a **gravitational potential** field $\Phi(\boldsymbol{r})$ that describes the gravitational potential energy per unit mass located at $\boldsymbol{r}$. The precise relationship is usually written in terms of a three-dimensional **gradient** as

$$\boldsymbol{g} = -\boldsymbol{\nabla}\Phi = -\left(\frac{\partial \Phi}{\partial x}, \frac{\partial \Phi}{\partial y}, \frac{\partial \Phi}{\partial z}\right). \tag{4.19}$$

Substituting Equation 4.19 into Equation 4.18 leads to

$$\boldsymbol{\nabla} \cdot \boldsymbol{\nabla}\Phi = 4\pi G \rho. \tag{4.20}$$

The combination $\boldsymbol{\nabla} \cdot \boldsymbol{\nabla}$ occurs so frequently in some areas of mathematics and physics that it is given a name, the **Laplacian operator**, and denoted by the symbol ∇^2. Following this convention we can say that

$$\nabla^2 \Phi = 4\pi G \rho. \tag{4.21}$$

Written out in full, in terms of Cartesian coordinates, this equation says that

$$\frac{\partial^2 \Phi}{\partial x^2} + \frac{\partial^2 \Phi}{\partial y^2} + \frac{\partial^2 \Phi}{\partial z^2} = 4\pi G \rho. \tag{4.22}$$

Equation 4.21 is called **Poisson's equation**. It provides the essential summary of Newtonian gravitation in terms of a differential equation that we have been seeking. It is entirely equivalent to Newton's inverse square law but has the advantage that it is a differential equation for a scalar quantity that may be straightforward to solve. The Newtonian gravitational field (which is a vector) can then be obtained via Equation 4.19, which involves differentiating the scalar field $\Phi(\boldsymbol{r})$. Notice that both the gravitational potential Φ and the mass density ρ are functions of the *same* position variable $\boldsymbol{r}$.

> **Poisson's equation and gravitation**
>
> The essence of Newtonian gravitation as a field theory is expressed in the Poisson equation
>
> $$\nabla^2 \Phi = 4\pi G \rho, \tag{Eqn 4.21}$$
>
> which relates a combination of second derivatives of the Newtonian gravitational potential Φ to the mass density ρ that is the source of the Newtonian gravitational field. The Newtonian gravitational field $\boldsymbol{g}$ is related to Φ by
>
> $$\boldsymbol{g} = -\boldsymbol{\nabla}\Phi. \tag{Eqn 4.19}$$

It will be shown later that general relativity predicts that an equation of this type provides an approximate description of gravitation under appropriate circumstances (usually referred to as the Newtonian limit). It is in this sense that general relativity is consistent with the successful predictions of Newtonian gravitation, even though it makes no use of gravitational forces. General relativity is also consistent with special relativity in the sense that the results of special relativity hold true locally in general relativity.

4.2   The basic ingredients of general relativity

The principles outlined in the previous section led Einstein to formulate general relativity using covariant tensor equations. But what tensor quantities should be involved in those equations? It was obvious that a theory of gravity should involve the distribution of matter, and it was part of Einstein's genius to realize that if gravity was somehow built into the geometric structure of spacetime, then it would act equally on all forms of matter and the universality of free fall would cease to be an unexplained accident. All forms of matter are subject to the same spacetime geometry, even though they may not be subject to identical forces. Such thoughts eventually led Einstein to consider two particular tensors as basic ingredients of general relativity — one describing the properties of matter, the other concerned with aspects of spacetime geometry. This section introduces those two tensor quantities and relates them to other tensors with which you are already familiar.

4.2.1   The energy–momentum tensor

In Newton's theory of gravity, mass is a conserved quantity that is the 'source' of gravitation. (See, for instance Equation 4.21.) In special relativity, the mass m of a particle is no longer conserved, but it is related to the energy and momentum magnitude of the particle by

$$E^2 = p^2 c^2 + m^2 c^4, \tag{Eqn 2.43}$$

and there are conservation laws that relate to energy (including mass–energy) and to momentum. Hence we should expect that in a relativistic theory, the source of gravitation cannot be mass alone but must also involve energy and momentum.

Since these sources of gravitation must somehow appear in a tensor, you will not be surprised to learn that one of the basic ingredients of general relativity is known as the **energy–momentum tensor**. The only issues are: what is it, what is its rank, what are its symmetries, and how is it defined?

The energy–momentum tensor describes the distribution and flow of energy and momentum in a region of spacetime. It is a rank 2 tensor, so at any event (i.e. a 'point' in spacetime) it is specified by sixteen components, usually denoted $T^{\mu\nu}$ ($\mu, \nu = 0, 1, 2, 3$). It is a symmetric tensor, so $T^{\mu\nu} = T^{\nu\mu}$, and that means that only ten of its components are independent (the four components $T^{\mu\mu}$ and six of the twelve components $T^{\mu\nu}$ where $\mu \neq \nu$). Each component can be measured in units of energy density ($\mathrm{J\,m^{-3}}$), though it is sometimes appropriate to use other equivalent units. Each component is a function of the spacetime coordinates, with the following general significance in the neighbourhood of each event in spacetime:

- T^{00} is the local energy density, including any mass–energy contribution.

- $T^{0i} = T^{i0}$ is the rate of flow of energy per unit area at right angles to the i-direction, divided by c, or, equivalently, the density of the i-component of momentum, multiplied by c.

- $T^{ij} = T^{ji}$ is the rate of flow of the i-component of momentum per unit area at right angles to the j-direction.

Figure 4.10 tries to give some feeling for the meaning of these components by considering the special case of a group of identical, non-interacting particles, each of mass m and velocity $\boldsymbol{v} = (v_x, v_y, 0)$, where we identify x, y and z with the 1-, 2- and 3-directions, respectively. Each of these particles will have a relativistic momentum $m\gamma(v)\boldsymbol{v}$ and a total relativistic energy $m\gamma(v)c^2$, where $v = |\boldsymbol{v}|$ represents the common speed of the particles and $\gamma(v) = 1/\sqrt{1 - v^2/c^2}$.

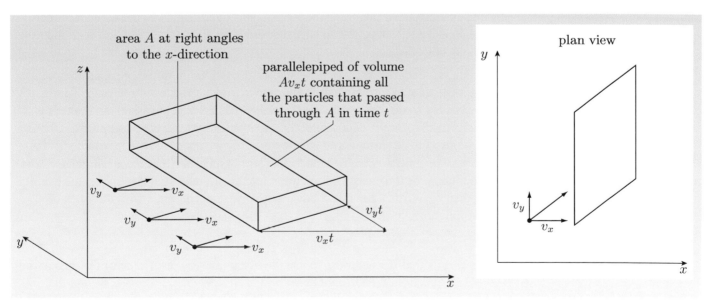

Figure 4.10 The transport of energy and momentum by non-interacting particles with a common velocity $\boldsymbol{v} = (v_x, v_y, 0)$.

If the number of particles per unit volume is n, their energy density will be $T^{00} = nm\gamma(v)c^2$. Because the particles each have a velocity component v_x, the number crossing an area A perpendicular to the x-direction in time t will be $nv_x At$; and since each carries energy $m\gamma(v)c^2$, the rate of flow of energy per unit area through a surface at right angles to the x-direction, divided by c, will be $T^{01} = nv_x At m\gamma(v)c^2/Atc = nmv_x\gamma(v)c$. Since each of the particles has an x-component of momentum given by $m\gamma(v)v_x$, you can see that the density of the x-component of momentum, multiplied by c, is given by the same expression, so $T^{10} = nmv_x\gamma(v)c$. A similar argument shows that $T^{02} = T^{20} = nmv_y\gamma(v)c$, while $T^{03} = T^{30} = 0$ because we have chosen to consider particles with $v_z = 0$. Finally, we note that in a time t, particles with y-component of momentum $m\gamma(v)v_y$ are crossing an area A perpendicular to the x-direction at a rate given by $nv_x At/At = nv_x$, so the rate of flow of the y-component of momentum per unit area through a surface at right angles to the x-direction is $T^{21} = nv_x m\gamma(v)v_y = nmv_y v_x \gamma(v)$, which is also the value of T^{12}. By similar arguments, $T^{11} = nmv_x^2\gamma(v)$ and $T^{22} = nmv_y^2\gamma(v)$, but $T^{13} = T^{31}$, $T^{23} = T^{32}$ and T^{33} are all zero because they involve v_z, which is zero in this particular case.

Putting all these results together gives

$$[T^{\mu\nu}] = \begin{pmatrix} T^{00} & T^{01} & T^{02} & T^{03} \\ T^{10} & T^{11} & T^{12} & T^{13} \\ T^{20} & T^{21} & T^{22} & T^{23} \\ T^{30} & T^{31} & T^{32} & T^{33} \end{pmatrix} = \begin{pmatrix} nm\gamma c^2 & nmv_x\gamma c & nmv_y\gamma c & 0 \\ nmv_x\gamma c & nmv_x^2\gamma & nmv_xv_y\gamma & 0 \\ nmv_y\gamma c & nmv_yv_x\gamma & nmv_y^2\gamma & 0 \\ 0 & 0 & 0 & 0 \end{pmatrix}.$$

In general, the precise form of the energy–momentum tensor will depend on what occupies the region concerned. A particularly simple example to consider is that of a region occupied by a cloud of non-interacting particles, each of mass m. This kind of matter is usually described as **dust**. For present purposes it's best to think of the dust cloud as a continuous body of matter that may contain internal currents — rather like a fluid but without any internal pressure. The nature of the dust cloud at any spacetime event in the region of interest can be characterized by the three-velocity v of the flow at the event, and by the value of the cloud's proper mass density ρ, that is, the density measured by an observer moving with the flow at the event of interest.

Of course, we really want to describe the dust cloud in terms of parameters that have well-known transformation properties under changes of reference frame. This is easy to do: the proper mass density ρ is a scalar invariant, so it already transforms as simply as possible; the three-velocity v is more complicated, but it can be used to determine a four-velocity $[U^\mu] = (c\gamma(v), \gamma(v)v)$ (where $v = |v|$ and $\gamma(v) = 1/\sqrt{1-v^2/c^2}$) that transforms as a rank 1 contravariant tensor. The components of the energy–momentum tensor of the dust at any spacetime event can then be written down in a covariant way, in accordance with the rules of tensor algebra, as

$$T^{\mu\nu} = \rho U^\mu U^\nu. \tag{4.23}$$

This means that if we choose to use the instantaneous rest frame of the dust at the event in question, then *at that event* and *in that frame*, $[U^\mu] = (c, \mathbf{0})$ and the energy–momentum tensor can be represented by the matrix

$$[T^{\mu\nu}] = \begin{pmatrix} \rho c^2 & 0 & 0 & 0 \\ 0 & 0 & 0 & 0 \\ 0 & 0 & 0 & 0 \\ 0 & 0 & 0 & 0 \end{pmatrix}. \tag{4.24}$$

So, in its local instantaneous rest frame, the only non-zero component of the energy–momentum tensor of the dust is T^{00}, which represents the energy density, and that is entirely accounted for by the density of mass–energy in the dust.

Another simple example of an energy–momentum tensor is that of an **ideal fluid**. Such a fluid is slightly more complicated than dust, since its nature at any spacetime event is characterized by a mass density ρ, a four-velocity $[U^\mu]$ and a **pressure** p that acts equally in all directions at that point. At an event where the metric is $g^{\mu\nu}$, the components of the energy–momentum tensor of an ideal fluid are given covariantly by

$$T^{\mu\nu} = (\rho + p/c^2)\, U^\mu\, U^\nu - p\, g^{\mu\nu}. \tag{4.25}$$

If we restrict ourselves to using locally inertial frames with Cartesian coordinates, then at any chosen spacetime event, the metric can be represented by the Minkowski metric, and the components of the energy–momentum tensor will be given by

$$T^{\mu\nu} = (\rho + p/c^2)\, U^\mu\, U^\nu - p\, \eta^{\mu\nu}. \tag{4.26}$$

If we again take the additional step of considering things from the point of view of an observer using the instantaneous rest frame of the fluid at that point, then, in that frame and at that point, the energy–momentum tensor of the ideal fluid is represented by the matrix

$$[T^{\mu\nu}] = \begin{pmatrix} \rho c^2 & 0 & 0 & 0 \\ 0 & p & 0 & 0 \\ 0 & 0 & p & 0 \\ 0 & 0 & 0 & p \end{pmatrix}. \tag{4.27}$$

In this case there will generally be thermal effects leading to flows of energy and momentum. However, because we have chosen to use the instantaneous rest frame, those flows will make no net contribution to the flow of energy, so it will still be the case that $T^{0i} = T^{i0} = 0$, and the lack of interactions between the particles will ensure $T^{ij} = 0$ for $i \neq j$. Consequently, the only non-zero components will be the total energy density $T^{00} = \rho c^2$ (which will include contributions from the random thermal motion of the particles in the fluid) and the three components $T^{ii} = p$ for $i = 1, 2, 3$ (which represent the effect of momentum being transferred with equal magnitude per unit area per unit time in all directions by the thermal motion of the particles).

● Show that for vanishing pressure ($p \to 0$), the energy–momentum tensor of an ideal fluid reduces to that of dust.

○ For $p \to 0$ we get

$$T^{\mu\nu} = \rho\, U^\mu\, U^\nu,$$

which is Equation 4.23 for the energy–momentum tensor for dust.

● Show that the units of pressure (Pa $=$ N m^{-2}) are equivalent to those of energy density (J m^{-3}), and also equivalent to those used to measure the rate of flow of momentum per unit area.

○ In SI units, $1\,\mathrm{J} = 1\,\mathrm{N\,m}$, so the unit of energy density may be written as

$$\mathrm{J\,m^{-3}} = \mathrm{N\,m\,m^{-3}} = \mathrm{N\,m^{-2}} = \mathrm{Pa},$$

which is the unit of pressure. Similarly, the unit of rate of flow of momentum per unit area will be $\mathrm{kg\,m\,s^{-1}\,m^{-2}\,s^{-1}} = \mathrm{kg\,m^{-1}\,s^{-2}}$, but $1\,\mathrm{N} = 1\,\mathrm{kg\,m\,s^{-2}}$, so the unit of rate of momentum flow per unit area per unit time can be written as

$$\mathrm{kg\,m^{-1}\,s^{-2}} = \mathrm{N\,m^{-2}} = \mathrm{Pa}.$$

Exercise 4.6 Verify the matrix in Equation 4.27 by explicitly evaluating T^{00}, T^{0i} and T^{ij} for $i,j = 1,2,3$ from Equation 4.26. ∎

As a final example of an energy–momentum tensor, we note that in the case of a region of space that contains electric and magnetic fields but no matter (a region occupied by electromagnetic radiation, for example), the components of the energy–momentum tensor are

$$T^{\mu\nu} = \frac{1}{\mu_0}\left(\sum_\sigma \mathsf{F}^\mu{}_\sigma\, \mathsf{F}^{\nu\sigma} - \frac{1}{4}\sum_{\rho,\sigma} g^{\mu\nu}\, \mathsf{F}^{\rho\sigma}\, \mathsf{F}_{\rho\sigma}\right), \tag{4.28}$$

where $\mathsf{F}^{\mu\nu}$ is the electromagnetic field tensor that was introduced in Chapter 2. We shall not discuss this energy–momentum tensor in detail, but its existence indicates that in general relativity, electromagnetic radiation alone can be a source of gravitation even though the associated particles (photons) have no mass at all.

At this stage it's useful to recall another result from Chapter 2: in electromagnetism, the conservation of electric charge is represented by the *equation of continuity*

$$\frac{\partial\rho}{\partial t} + \frac{\partial J_x}{\partial x} + \frac{\partial J_y}{\partial y} + \frac{\partial J_z}{\partial z} = 0. \tag{Eqn 2.76}$$

This equation describes how any change in the electric charge density must be balanced by a flow of charge due to electric currents. It is often written more compactly in terms of a three-vector divergence as

$$\frac{\partial\rho}{\partial t} + \boldsymbol{\nabla}\cdot\boldsymbol{J} = 0,$$

or more compactly still, using the current four-vector, by the Lorentz-covariant equation

$$\sum_{\nu=0}^{3} \frac{\partial J^\nu}{\partial x^\nu} = 0. \tag{Eqn 2.77}$$

This suggests that we might expect the conservation of relativistic energy and momentum in a locally inertial frame (where special relativity holds true) to be represented by a relation of the form

$$\sum_\mu \frac{\partial T^{\mu\nu}}{\partial x^\mu} = 0, \tag{4.29}$$

and this is indeed the case. The tensor relationship has a free index ν, so it actually represents four different equations, each of which is similar to the equation of continuity. The first (corresponding to $\nu = 0$) relates the rate of change of the energy density T^{00} to the spatial derivatives of the energy flows T^{0i} in the 1-, 2- and 3-directions. The other three each relate the rate of change of one of the momentum density terms T^{0i} to the spatial derivatives of the corresponding momentum flows T^{ji} for $j = 1, 2, 3$.

It also turns out that in arbitrary coordinates and in a spacetime that may be flat or curved, the energy–momentum tensor has the more general property

$$\sum_{\mu} \nabla_{\mu} T^{\mu\nu} = 0. \tag{4.30}$$

This is sometimes described by saying that the **covariant divergence** of $T^{\mu\nu}$ is zero. In the absence of gravity, in a flat Minkowski spacetime, this result simply allows us to describe the conservation of energy and momentum using general coordinates. However, if the spacetime is curved, then it turns out that Equation 4.30 does not generally describe the conservation of energy and momentum for the contents of spacetime. And that's a good thing, because in the presence of gravitation (i.e. curvature), the conservation of energy is not expected to apply to matter and radiation alone — we also have to take the gravitational energy into account, and that is not included in the energy–momentum tensor. We shall return to the significance of the covariant divergence in curved spacetime later; for the moment we just need to emphasize the following.

The energy–momentum tensor

The energy–momentum tensor $[T^{\mu\nu}]$ describes the distribution and flow of energy and momentum due to the presence and motion of matter and radiation in a region of spacetime. It is a rank 2, symmetric tensor with ten independent components. At any event in the region of interest, its components describe the energy density, the flow of energy in various directions, divided by c (or, equivalently, the density of the corresponding momentum component, multiplied by c), and the flow of the various momentum components in the various directions.

For pressure-free dust, the components of the energy–momentum tensor are given by

$$T^{\mu\nu} = \rho U^{\mu} U^{\nu}; \tag{Eqn 4.23}$$

for an ideal fluid,

$$T^{\mu\nu} = (\rho + p/c^2) U^{\mu} U^{\nu} - p g^{\mu\nu}; \tag{Eqn 4.25}$$

and for electromagnetic fields,

$$T^{\mu\nu} = \frac{1}{\mu_0} \left(\sum_{\sigma} F^{\mu}{}_{\sigma} F^{\nu\sigma} - \frac{1}{4} \sum_{\rho,\sigma} g^{\mu\nu} F^{\rho\sigma} F_{\rho\sigma} \right). \tag{Eqn 4.28}$$

An important general property of the energy–momentum tensor is that its covariant divergence is zero; that is,

$$\sum_{\mu} \nabla_{\mu} T^{\mu\nu} = 0. \tag{Eqn 4.30}$$

4.2.2   The Einstein tensor

The equivalence principle led Einstein to propose that gravity should be regarded not as a force in the conventional sense, but as a manifestation of the curvature of spacetime. Einstein was therefore looking for a *geometric* theory of gravity, so he needed to find a geometric object that could be related to the energy–momentum tensor. Clearly, he needed a rank 2 tensor involving the components of the metric tensor. However, from the example of the electromagnetic field equations, or even from Newtonian gravity formulated as a field theory and based on Poisson's equation, we should expect the final equations to be differential equations, so the metric should enter through its derivatives. We might also expect that the required geometric tensor will be symmetric and will have a vanishing covariant divergence.

Even with so many clues, it took Einstein some time to find the appropriate tensor quantity. What he eventually arrived at involved contractions of the Riemann curvature tensor that was introduced in Chapter 3. Here is the full form of the Riemann tensor for a four-dimensional spacetime:

$$R^{\delta}{}_{\alpha\beta\gamma} \equiv \frac{\partial \Gamma^{\delta}{}_{\alpha\gamma}}{\partial x^{\beta}} - \frac{\partial \Gamma^{\delta}{}_{\alpha\beta}}{\partial x^{\gamma}} + \sum_{\lambda} \Gamma^{\lambda}{}_{\alpha\gamma}\,\Gamma^{\delta}{}_{\lambda\beta} - \sum_{\lambda} \Gamma^{\lambda}{}_{\alpha\beta}\,\Gamma^{\delta}{}_{\lambda\gamma}. \quad \text{(Eqn 3.35)}$$

As you can see, it involves the connection coefficients $\Gamma^{\delta}{}_{\alpha\beta}$, which are defined in terms of the metric and its derivatives by

$$\Gamma^{\lambda}{}_{\alpha\beta} = \frac{1}{2} \sum_{\sigma} g^{\lambda\sigma} \left(\frac{\partial g_{\sigma\beta}}{\partial x^{\alpha}} + \frac{\partial g_{\alpha\sigma}}{\partial x^{\beta}} - \frac{\partial g_{\alpha\beta}}{\partial x^{\sigma}} \right). \quad \text{(Eqn 3.23)}$$

You will recall from Chapter 3 that the vanishing of all components of the Riemann tensor is the necessary and sufficient condition for a spacetime to be flat.

The Riemann tensor has four indices, each of which can take four values (in four-dimensional spacetime), so it has $4^4 = 256$ components. However, the tensor has various symmetries, so there are just 20 independent components.

Although the Riemann tensor is fundamental to the study of curved spaces, there are two other tensors that have been found to be very useful. If we contract the first and last indices on the Riemann tensor, then we get a new rank 2 tensor with components

$$R_{\alpha\beta} \equiv \sum_{\gamma} R^{\gamma}{}_{\alpha\beta\gamma}, \quad (4.31)$$

which is known as the **Ricci tensor**. It follows from the definition of the Riemann tensor that the Ricci tensor is symmetric with respect to interchanging its indices, i.e. $R_{\alpha\beta} = R_{\beta\alpha}$. Further, contracting the indices on the Ricci tensor gives

$$R \equiv \sum_{\alpha,\beta} g^{\alpha\beta} R_{\alpha\beta}, \quad (4.32)$$

which is known as the **curvature scalar** (or sometimes the **Ricci scalar**). Note that all of these curvature-related quantities are ultimately expressed in terms of the components of the metric tensor $[g^{\mu\nu}]$ and their derivatives.

The quantity that Einstein found to be a basic ingredient of general relativity is defined in terms of the Ricci tensor and the curvature scalar. It is called the **Einstein tensor** and its components are given by the following equation.

The Einstein tensor

$$G^{\mu\nu} \equiv R^{\mu\nu} - \tfrac{1}{2} g^{\mu\nu} R. \tag{4.33}$$

Since both $R^{\mu\nu}$ and $g^{\mu\nu}$ are symmetric, it follows that $G^{\mu\nu}$ must also be symmetric. This means that only 10 of its 16 components will be independent, just like the energy–momentum tensor. Moreover, it can be shown that the covariant divergence of the Einstein tensor vanishes $\left(\sum_\mu \nabla_\mu G^{\mu\nu} = 0 \right)$, again just like the energy–momentum tensor.

We are now in a position to introduce Einstein's field equations, the mathematical relations that are at the core of general relativity.

4.3 Einstein's field equations and geodesic motion

The central ideas of general relativity were famously summed up by the American physicist John Wheeler:

> Matter tells space how to curve.
> Space tells matter how to move.

This is very memorable (and worth remembering!), though not completely accurate. (You should already be asking yourself: 'Doesn't he mean spacetime rather than space, and doesn't he mean matter and radiation rather than matter?') Unpacking Wheeler's quote somewhat, to be more accurate, we can say that the central physical ideas of general relativity are that the energy and momentum in a region of spacetime determine the geometry of spacetime in that region. The spacetime geometry then determines a special class of spacetime pathways — the geodesics. Moving under the influence of gravity alone, massive particles travel along time-like geodesics (where $ds^2 > 0$), while light rays follow null geodesics (with $ds^2 = 0$). Thus the distribution of energy and momentum in a region determines the motion of freely falling matter and radiation in that region.

Another helpful but overly simple view is that in Newtonian gravitation, matter tells matter how to move, with the gravitational force playing the role of intermediary. This can be contrasted with general relativity where energy and momentum tell matter and radiation how to move, with spacetime geometry playing the role of intermediary.

Non-gravitational effects, such as nuclear and electromagnetic forces, cause departures from geodesic motion and are usually beyond the scope of general relativity.

The rest of this section is devoted to spelling out these ideas with greater accuracy and improved precision.

4.3.1   The Einstein field equations

As we have seen, Einstein's objective became the formulation of a 'geometric' theory of gravity that would naturally act on all kinds of matter in the same way. He identified the energy–momentum tensor as an important quantity for describing the 'sources' of gravitation, and found another symmetric rank 2 tensor, the Einstein tensor, containing derivatives of the metric coefficients $g^{\mu\nu}$, that he could relate to it. Both the energy–momentum tensor $[T^{\mu\nu}]$ and the Einstein tensor $[G^{\mu\nu}]$ have zero covariant divergence, so it is natural to suggest that the two tensors are proportional. This led Einstein to propose what are now called the **Einstein field equations**, which are usually written as in terms of tensor components as follows.

The Einstein field equations

$$R_{\mu\nu} - \tfrac{1}{2} R\, g_{\mu\nu} = -\kappa\, T_{\mu\nu}. \tag{4.34}$$

Here κ is a constant, sometimes called the **Einstein constant**. We shall show later that requiring the consistency of general relativity and Newtonian gravitation forces us to set $\kappa = 8\pi G/c^4$.

The Einstein field equations are the fundamental field equations of general relativity, analogous to the Poisson equation in Newtonian gravitation. They are the feature of general relativity that Wheeler was referring to when he said (rather loosely) 'matter tells space how to curve'. The Einstein field equations have two free indices, μ and ν, so they actually represent a set of 16 equations, though due to symmetries only 10 of them are independent. They are usually regarded as differential equations for the 10 independent metric tensor components $g^{\mu\nu}$. But they are generally very complicated.

The reason for the complication is not hard to see. The Ricci tensor and the curvature scalar involve combinations of components of the Riemann tensor. Its components $R^{\mu}{}_{\nu\alpha\beta}$ are defined in terms of the connection coefficients $\Gamma^{\mu}{}_{\alpha\beta}$, which are in turn defined in terms of the metric tensor components $g_{\mu\nu}$ and the components of its inverse $g^{\mu\nu}$. The way in which the connection coefficients appear in $R^{\mu}{}_{\nu\alpha\beta}$ means that the Riemann tensor involves second-order derivatives of the metric coefficients with respect to the spacetime coordinates. However, because the connection coefficients involve both the metric tensor and its inverse, the Einstein field equations are **non-linear** in $g_{\mu\nu}$. (An equation is said to be non-linear in a variable y if replacing y by αy throughout the equation does *not* produce an equation that is equivalent to the original equation multiplied by α.) It is the non-linearity that makes the Einstein field equations particularly difficult to solve.

Solving the Einstein field equations means finding the metric tensor $[g_{\mu\nu}]$ that corresponds to a given energy–momentum tensor $[T_{\mu\nu}]$. As you saw in Chapter 3, the metric tensor, once it is known, will determine the connection coefficients, the curvature tensor, the geodesic pathways and all the other geometric features of the spacetime that it describes. Given that gravitation is 'built in' to the geometry of spacetime in general relativity, the metric tensor that corresponds to a given set of source terms (i.e. a given energy–momentum tensor) *is* the gravitational field, even though it is not the 'force per unit mass' of the Newtonian gravitational field.

The act of solving the Einstein field equations might sound straightforward, but the ten independent field equations form a set of simultaneous, non-linear, second-order partial differential equations and, depending on the energy–momentum tensor, the task of finding a solution varies between difficult and impossible. In fact, it is remarkable that the first (and probably most important) exact non-trivial solution was announced very soon after Einstein first proposed his equations. We shall describe that solution in the next chapter.

In addition to various numerical procedures for finding solutions to the field equations, there are three different ways to approach the search for solutions.

1. As already suggested, we could specify the energy–momentum tensor and then work very hard to solve for the metric components $g_{\mu\nu}$. This approach has actually been very successful for some energy–momentum tensors.

2. We could specify the metric tensor and then work out the energy–momentum tensor. This is generally easier since it is more straightforward to differentiate a function than to solve a non-linear partial differential equation. However, it usually turns out that the resulting energy–momentum tensor is non-physical, so this approach is not as useful as might be hoped.

3. We could try to partly determine both the metric tensor and the energy–momentum tensor directly from the physics of a particular situation and then use the field equations as constraints to complete the determination of $[g_{\mu\nu}]$ and $[T_{\mu\nu}]$. This sometimes yields useful results.

In any case, a significant part of the discovery of any new solution of the Einstein field equations is to check that the solution really is new, and not merely an old solution expressed in a different coordinate system. This is an interesting problem but its consideration would take us well beyond the limits of this book.

● Taking the metric tensor components $g_{\mu\nu}$ to be dimensionless quantities (i.e. pure numbers), show that the connection coefficients $\Gamma^{\lambda}{}_{\mu\nu}$ can be expressed in units of m^{-1}, while the Ricci tensor $[R_{\mu\nu}]$ and the curvature scalar R can both be expressed in units of m^{-2}. Combine this with your knowledge of the appropriate units for $T_{\mu\nu}$ to show that $8\pi G/c^4$ has the right units to be the Einstein constant κ.

○ Since $g_{\mu\nu}$ is dimensionless, it follows from

$$\Gamma^{\lambda}{}_{\alpha\beta} = \frac{1}{2}\sum_{\sigma} g^{\lambda\sigma}\left(\frac{\partial g_{\sigma\beta}}{\partial x^{\alpha}} + \frac{\partial g_{\alpha\sigma}}{\partial x^{\beta}} - \frac{\partial g_{\alpha\beta}}{\partial x^{\sigma}}\right) \qquad \text{(Eqn 3.23)}$$

that $\Gamma^{\lambda}{}_{\mu\nu}$ can be expressed in units of m^{-1}. It then follows from

$$R^{\delta}{}_{\alpha\beta\gamma} \equiv \frac{\partial \Gamma^{\delta}{}_{\alpha\gamma}}{\partial x^{\beta}} - \frac{\partial \Gamma^{\delta}{}_{\alpha\beta}}{\partial x^{\gamma}} + \sum_{\lambda}\Gamma^{\lambda}{}_{\alpha\gamma}\,\Gamma^{\delta}{}_{\lambda\beta} - \sum_{\lambda}\Gamma^{\lambda}{}_{\alpha\beta}\,\Gamma^{\delta}{}_{\lambda\gamma} \quad \text{(Eqn 3.35)}$$

that $R^{\delta}{}_{\alpha\beta\gamma}$ can be expressed in units of m^{-2}, but $[R_{\mu\nu}]$ and R are sums of components of $R^{\delta}{}_{\alpha\beta\gamma}$, so they too can be expressed in units of m^{-2}. With this in mind and recalling that the components of the energy–momentum tensor can be expressed in units of J m^{-3} = kg m^{-1} s^{-2}, it can be seen that suitable units for κ are $(1/\mathrm{m}^2)(1/(\mathrm{kg\,m}^{-1}\,\mathrm{s}^{-2})) = \mathrm{kg}^{-1}\,\mathrm{m}^{-1}\,\mathrm{s}^2$, and the units of $8\pi G/c^4$ are indeed $\mathrm{N\,m}^2\,\mathrm{kg}^{-2}\,\mathrm{s}^4\,\mathrm{m}^{-4} = \mathrm{kg}^{-1}\,\mathrm{m}^{-1}\,\mathrm{s}^2$.

Exercise 4.7   Show that Equation 4.34 can also be written as

$$R_{\mu\nu} = -\kappa\left(T_{\mu\nu} - \tfrac{1}{2}g_{\mu\nu}\,T\right),\qquad\qquad (4.35)$$

where $T \equiv \sum_\mu T^\mu{}_\mu$. (*Hint*: Multiply the Einstein field equations by $g^{\mu\nu}$, and contract.) ∎

In some regions of spacetime, it may be that $T_{\mu\nu} = 0$. In such regions, spacetime is said to be **empty**. Equation 4.35 shows that in such a region, the Einstein field equations may be written as

$$R_{\mu\nu} = 0.\qquad\qquad (4.36)$$

Note that this does not necessarily mean that spacetime in the region is flat. The necessary and sufficient condition for flatness is that the components of the Riemann tensor should vanish at all events in the region, but that tensor has 20 components while the Ricci tensor has only 10. The vanishing of $R_{\mu\nu}$ in some region does not necessarily imply the vanishing of $R^\mu{}_{\nu\alpha\beta}$, nor, therefore, does it imply that $g_{\mu\nu}$ describes a flat spacetime. However, setting $T_{\mu\nu} = 0$ does indicate that there is no matter or radiation in the region concerned, so solutions of Equation 4.36 are said to be **vacuum solutions** of the field equations. The study of vacuum solutions is an important sub-field of general relativity.

4.3.2   Geodesic motion

Einstein completed his long search for the field equations in 1915 and announced the basic principles of general relativity in a talk at the Prussian Academy of Sciences in Berlin in November 1915. The details of the theory were published in 1916. At that time Einstein clearly understood that in addition to using the field equations to find the spacetime metric, the theory also required that the metric should be used to determine the geodesics of the spacetime via the geodesic equations. These were introduced in Chapter 3. For a four-dimensional spacetime with metric tensor $[g_{\mu\nu}]$, they take the form

$$\frac{\mathrm{d}^2 x^\rho}{\mathrm{d}\lambda^2} + \sum_{\alpha,\beta} \Gamma^\rho{}_{\alpha\beta}\,\frac{\mathrm{d}x^\alpha}{\mathrm{d}\lambda}\,\frac{\mathrm{d}x^\beta}{\mathrm{d}\lambda} = 0,\qquad\qquad (\text{Eqn 3.27})$$

where λ is an affine parameter and, as usual,

$$\Gamma^\rho{}_{\alpha\beta} = \frac{1}{2}\sum_\sigma g^{\rho\sigma}\left(\frac{\partial g_{\sigma\beta}}{\partial x^\alpha} + \frac{\partial g_{\alpha\sigma}}{\partial x^\beta} - \frac{\partial g_{\alpha\beta}}{\partial x^\sigma}\right).\qquad\qquad (\text{Eqn 3.23})$$

The functions $x^\rho(\lambda)$ that satisfy the geodesic equation describe parameterized curves through spacetime that represent the most direct routes between events. (A tangent to such a curve, parallel transported along the curve, remains a tangent.) So these geodesic curves are the analogues of straight lines in a curved space.

You will recall from Chapter 3 that given a curve specified by the coordinate functions $x^\rho(\lambda)$, the components of the *tangent vector* to the curve at the point specified by λ are

$$t^\rho(\lambda) = \frac{\mathrm{d}x^\rho(\lambda)}{\mathrm{d}\lambda}. \tag{4.37}$$

We can associate a sort of 'length' with this vector (actually called its **norm**) defined by the quantity $\sum_{\alpha,\beta} g_{\alpha,\beta}\, t^\alpha\, t^\beta$. In the case of an affinely parameterized geodesic, where the tangent vector remains a tangent vector under parallel transport, this norm will be the same at all points. Thus we can separate the geodesics into three distinct classes:

- **time-like geodesics**, where the tangent vector always has positive norm
- **null geodesics**, where the tangent vector always has zero norm
- **space-like geodesics**, where the tangent vector always has negative norm.

In the case of the time-like and space-like geodesics, the line element separating neighbouring points on the geodesic, given by

$$\mathrm{d}s^2 = \sum_{\mu,\nu=0}^{3} g_{\mu\nu}\, \mathrm{d}x^\mu\, \mathrm{d}x^\nu, \tag{Eqn 3.11}$$

will always be non-zero, and we can use the square root of its magnitude $|\mathrm{d}s^2|^{1/2}$ to define a distance element that we can use when parameterizing the geodesic. These geodesics are collectively described as non-null geodesics. In the contrasting case of a null geodesic, the line element separating neighbouring points will always be zero, so there is no possibility of using the 'distance' along the curve as the parameter λ in that case, even though it can still be parameterized in other ways.

What is the significance of all this for general relativity and gravity? It is contained in the following **principle of geodesic motion**.

The principle of geodesic motion

In general relativity, the time-like geodesics of a spacetime represent the possible world-lines of massive particles falling freely under the influence of gravity alone. And, similarly, the null geodesics of a spacetime represent the possible world-lines of massless particles moving under the influence of gravity alone.

This is what Wheeler was referring to when he said (somewhat loosely) 'space tells matter how to move'.

The principle implies that, in the absence of any non-gravitational effects, the path through spacetime followed by a planet as it orbits a star will be a time-like geodesic of the spacetime that surrounds the star. And, similarly, the spacetime pathway of a flash of light leaving the star will be a null geodesic of that spacetime.

In 1915–16, Einstein thought that the principle of geodesic motion was a separate postulate that was needed alongside the field equations to make general relativity a complete theory of gravity. However, later work by Einstein and others eventually showed that the geodesic motion of freely falling matter and radiation is actually predicted by the field equations through the requirement that

$$\sum_{\mu} \nabla_{\mu} T^{\mu\nu} = 0. \tag{Eqn 4.30}$$

It is a remarkable feature of general relativity that it predicts the equations of motion of the matter and radiation that is also the source of gravitation. This is another aspect of the non-linearity of the theory.

4.3.3   The Newtonian limit of Einstein's field equations

One of the guiding principles in Einstein's search for a geometric theory of gravity was what we have called the principle of consistency, so it is important to show that under appropriate circumstances, the Einstein field equations are consistent with Poisson's equation

$$\nabla^2 \Phi = 4\pi G \rho. \tag{Eqn 4.21}$$

The 'appropriate circumstances' that define what is usually referred to as the **Newtonian limit** of general relativity suppose that the gravitational effects are weak and that any motions are sufficiently slow to be considered 'non-relativistic'. Also, remember that Newtonian gravitation concerns the movement of only matter, not radiation.

The assumption that gravitational effects are weak allows us to assume that the metric coefficients are close to those of the Minkowski metric $\eta_{\mu\nu}$, so we can write

$$g_{\mu\nu} \approx \eta_{\mu\nu} + h_{\mu\nu}, \tag{4.38}$$

where $|h_{\mu\nu}| \ll 1$, and we can choose to work to first order in $h_{\mu\nu}$. We can also suppose that the metric is not changing significantly with time, so $h_{\mu\nu}$ is not a function of time.

Now, if we consider the simple case of a region filled with dust, for which $T_{\mu\nu} = \rho\, U_{\mu} U_{\nu}$ and $T = \sum_{\mu} T^{\mu}{}_{\mu} = \rho c^2$, we can see that the Einstein field equations given in Equation 4.35 take the form

$$R_{\mu\nu} = -\kappa \left(\rho\, U_{\mu} U_{\nu} - \tfrac{1}{2} g_{\mu\nu}\, \rho c^2 \right). \tag{4.39}$$

Substituting our simplified form of the metric gives

$$R_{\mu\nu} = -\kappa \left(\rho\, U_{\mu} U_{\nu} - \tfrac{1}{2}(\eta_{\mu\nu} + h_{\mu\nu})\rho c^2 \right). \tag{4.40}$$

Examining the R_{00} term, and remembering that speeds are low, so $U_0 \approx c$, and that $|h_{\mu\nu}| \ll 1$, we see that

$$R_{00} \approx -\kappa \left(\rho c^2 - \tfrac{1}{2}\rho c^2 \right) = -\kappa \tfrac{1}{2}\rho c^2. \tag{4.41}$$

However, in the same limit, it can be shown from the definition of the Ricci tensor that

$$R_{00} \approx -\sum_{i=1}^{3} \frac{\partial \Gamma^{i}{}_{00}}{\partial x^{i}}, \tag{4.42}$$

and from the definition of the connection coefficient that

$$\Gamma^{i}{}_{00} \approx -\frac{1}{2} \sum_{j} \eta^{ij} \frac{\partial h_{00}}{\partial x^{j}},$$

and consequently

$$R_{00} = \frac{1}{2} \sum_{i,j} \eta^{ij} \frac{\partial^2 h_{00}}{\partial x^i \partial x^j} = -\frac{1}{2} \nabla^2 h_{00}. \tag{4.43}$$

From Equations 4.41 and 4.43 we now have two expressions for R_{00} and equating these, we see that in the Newtonian limit,

$$-\tfrac{1}{2}\nabla^2 h_{00} \approx -\kappa \tfrac{1}{2}\rho c^2, \tag{4.44}$$

and so

$$\nabla^2 h_{00} \approx \kappa \rho c^2. \tag{4.45}$$

This result already looks something like Poisson's equation, but to really make the link we need to know how h_{00} is related to the Newtonian gravitational potential Φ. This relationship can be determined from the geodesic equation of motion of a particle. We shall not go through the detailed argument, but it turns out that in the Newtonian limit, $\Phi = h_{00}c^2/2$ (so, $g_{00} = 1 + 2\Phi/c^2$). Using this identification, we see that in the Newtonian limit, general relativity predicts that

$$\nabla^2 \Phi \approx \kappa \frac{\rho c^4}{2}, \tag{4.46}$$

which approximates Poisson's equation

$$\nabla^2 \Phi = 4\pi G \rho,$$

provided that we identify $\kappa = 8\pi G/c^4$.

Thus general relativity agrees with Newtonian gravitation in the limit of low speeds and weak fields, provided that $\kappa = 8\pi G/c^4$.

4.3.4 The cosmological constant

We shall end this discussion of the field equations with a brief introduction to a topic that will be discussed at greater length in the final chapter. It concerns a modification to the field equations that Einstein proposed but later described as 'the greatest blunder of my life', though it is now regarded as a very important aspect of general relativity.

The field equations that have been presented in this chapter are those that Einstein presented in 1916 and on which he based a number of astronomical predictions that were used to test general relativity. (These tests will be discussed later.) However, in 1917 he turned his attention to cosmology — the study of the Universe — and realized that he had omitted a term that was mathematically justified and might be important. Including this additional cosmological term, the modified field equations take the form

$$R_{\mu\nu} - \tfrac{1}{2} R\, g_{\mu\nu} + \Lambda\, g_{\mu\nu} = -\kappa\, T_{\mu\nu}, \tag{4.47}$$

where Λ represents a new universal constant of Nature known as the **cosmological constant**. Einstein's original motivation for introducing this constant was that, at the time, the Universe was thought to be static (i.e. neither expanding nor contracting), and he found that a non-zero value of Λ could lead to static solutions of the field equations (although they later turned out to be unstable). In the

Newtonian limit, a positive value of Λ provides a repulsive effect that can counterbalance the usual gravitational attraction. It was the subsequent discovery that the Universe was in fact expanding that prompted Einstein to make his comment about the cosmological constant being his 'greatest blunder'.

Ironically, observational evidence now favours the view that the Universe is not only expanding, but is doing so at an accelerating rate. The cosmological constant, a new fundamental constant, is one way of explaining this. But there are others.

From a mathematical point of view, we can transfer the cosmological term to the right-hand side of the field equations, giving

$$R_{\mu\nu} - \tfrac{1}{2}R\,g_{\mu\nu} = -\kappa\left(T_{\mu\nu} + \frac{\Lambda}{\kappa}\,g_{\mu\nu}\right). \qquad (4.48)$$

The cosmological term now begins to look like some additional contribution to the energy and momentum. We can further this impression by regarding the $-(\Lambda/\kappa)g_{\mu\nu}$ term as arising from a new part of the energy–momentum tensor that we represent by $\overline{T}_{\mu\nu}$. The modified field equations then take the form

$$R_{\mu\nu} - \tfrac{1}{2}R\,g_{\mu\nu} = -\kappa(T_{\mu\nu} + \overline{T}_{\mu\nu}). \qquad (4.49)$$

If we take the additional step of treating the new contribution as if it comes from an ideal fluid with density ρ_Λ and pressure p_Λ, then we can use Equation 4.25 to write

$$\overline{T}_{\mu\nu} = (\rho_\Lambda + p_\Lambda/c^2)\,U_\mu U_\nu - p_\Lambda\,g_{\mu\nu}, \qquad (4.50)$$

where we say that $\rho_\Lambda\,c^2$ represents the density of **dark energy** and p_Λ is the pressure due to dark energy. We can ensure that

$$\overline{T}_{\mu\nu} = \frac{\Lambda}{\kappa}\,g_{\mu\nu} \qquad (4.51)$$

by requiring that

$$p_\Lambda = -\frac{\Lambda}{\kappa} \quad \text{and} \quad \rho_\Lambda = -\frac{p_\Lambda}{c^2} = \frac{\Lambda}{\kappa c^2}. \qquad (4.52)$$

However, this shows that the fluid is a very strange one, since a positive density of dark energy implies a negative pressure that will have the effect of driving things apart on the cosmic scale rather than drawing them together.

The modified field equations are then

$$R_{\mu\nu} - \tfrac{1}{2}R\,g_{\mu\nu} = -\kappa(T_{\mu\nu} + \rho_\Lambda\,c^2\,g_{\mu\nu}),$$

We shall have more to say about dark energy and its cosmological effect in the final chapter.

Summary of Chapter 4

1. A freely falling frame in a gravitational field is a locally inertial frame.

2. The weak equivalence principle states that: 'Within a sufficiently localized region of spacetime adjacent to a concentration of mass, the motion of bodies subject to gravitational effects alone cannot be distinguished by any experiment from the motion of bodies within a region of appropriate uniform acceleration.'

3. The strong equivalence principle states that: 'Within a sufficiently localized region of spacetime adjacent to a concentration of mass, the physical behaviour of bodies cannot be distinguished by *any* experiment from the physical behaviour of bodies within a region of appropriate uniform acceleration.'

4. A general coordinate transformation takes the form $x'^{\mu} = x'^{\mu}(x^{\nu})$, where the four x'^{μ} terms are functions of the four variables x^{ν}. This is more general than the Lorentz transformation, which takes the form

$$x'^{\mu} = \sum_{\nu=0}^{3} \Lambda^{\mu}{}_{\nu}\, x^{\nu}, \qquad \text{(Eqn 2.61)}$$

where the sixteen $\Lambda^{\mu}{}_{\nu}$ terms are constants.

5. Tensors are multi-component mathematical objects that transform in well-defined ways under general coordinate transformations, indicated by the position (up or down) of their indices.

6. A contravariant tensor of rank 1 has the index up and transforms like

$$A'^{\alpha} = \sum_{\beta=0}^{3} \frac{\partial x'^{\alpha}}{\partial x^{\beta}} A^{\beta}, \qquad \text{(Eqn 4.10)}$$

while a covariant tensor of rank 1 has the index down and transforms like

$$A'_{\alpha} = \sum_{\beta=0}^{3} \frac{\partial x^{\beta}}{\partial x'^{\alpha}} A_{\beta}. \qquad \text{(Eqn 4.9)}$$

7. The rank of a tensor is the number of indices, e.g. $R_{\mu\nu}$ is a rank 2 tensor. The type of the indices can be mixed, as in $R^{\mu}{}_{\nu}$.

8. According to the principle of general covariance, the laws of physics should take the same form in all frames of reference. In practice this means that they should be expressed as balanced tensor relationships that will be covariant under general coordinate transformations.

9. Legitimate algebraic operations involving tensors include scaling, addition and subtraction (provided that the types are identical), multiplication and contraction. The partial differentiation of a tensor does not generally produce another tensor, but the process of covariant differentiation does. This may be applied to a tensor of any rank and is exemplified by

$$\nabla_{\lambda} T^{\mu}{}_{\nu} = \frac{\partial T^{\mu}{}_{\nu}}{\partial x^{\lambda}} + \sum_{\rho} \Gamma^{\mu}{}_{\rho\lambda}\, T^{\rho}{}_{\nu} - \sum_{\rho} \Gamma^{\rho}{}_{\nu\lambda}\, T^{\mu}{}_{\rho}. \qquad \text{(Eqn 4.13)}$$

10. According to the principle of consistency, the predictions of general relativity should be consistent with the successful predictions of Newtonian gravitation.

11. The essence of Newtonian gravitation as a field theory is expressed in the Poisson equation

$$\nabla^2 \Phi = 4\pi G \rho, \qquad \text{(Eqn 4.21)}$$

which relates a combination of second derivatives of the Newtonian gravitational potential Φ to the mass density ρ that is the source of the

Newtonian gravitational field. The Newtonian gravitational field g and the gravitational potential Φ are related by

$$g = -\nabla\Phi. \qquad \text{(Eqn 4.19)}$$

12. The energy–momentum tensor (usually denoted $T^{\mu\nu}$) is a symmetric, rank 2 tensor with vanishing divergence $\sum_{\mu}\nabla_{\mu}T^{\mu\nu} = 0$ whose components can be interpreted in terms of the energy density, energy flow, momentum density and momentum flow. The exact form of the energy–momentum tensor depends on the details of the physical system being considered.

13. The components of the energy–momentum tensor for a collection of non-interacting particles (knows as 'dust') with proper mass density ρ and four-velocity U^{μ} are given by

$$T^{\mu\nu} = \rho\, U^{\mu}\, U^{\nu}. \qquad \text{(Eqn 4.23)}$$

The components of the energy–momentum tensor for an ideal fluid of density ρ and pressure p are given by

$$T^{\mu\nu} = (\rho + p/c^2)\, U^{\mu}\, U^{\nu} - p\, g^{\mu\nu}. \qquad \text{(Eqn 4.25)}$$

14. The geometry of spacetime is determined by the metric tensor $g_{\mu\nu}$ through the line element given by

$$\mathrm{d}s^2 = \sum_{\mu,\nu} g_{\mu\nu}\, \mathrm{d}x^{\mu}\, \mathrm{d}x^{\nu}. \qquad \text{(Eqn 3.11)}$$

15. The connection coefficients $\Gamma^{\alpha}{}_{\beta\gamma}$ are given by

$$\Gamma^{\alpha}{}_{\beta\gamma} = \frac{1}{2}\sum_{\delta} g^{\alpha\delta}\left(\frac{\partial g_{\delta\gamma}}{\partial x^{\beta}} + \frac{\partial g_{\beta\delta}}{\partial x^{\gamma}} - \frac{\partial g_{\beta\gamma}}{\partial x^{\delta}}\right). \qquad \text{(Eqn 3.23)}$$

They do *not* transform as the components of a tensor.

16. The components of the Riemann tensor are defined by

$$R^{\delta}{}_{\alpha\beta\gamma} \equiv \frac{\partial \Gamma^{\delta}{}_{\alpha\gamma}}{\partial x^{\beta}} - \frac{\partial \Gamma^{\delta}{}_{\alpha\beta}}{\partial x^{\gamma}} + \sum_{\lambda}\Gamma^{\lambda}{}_{\alpha\gamma}\,\Gamma^{\delta}{}_{\lambda\beta} - \sum_{\lambda}\Gamma^{\lambda}{}_{\alpha\beta}\,\Gamma^{\delta}{}_{\lambda\gamma}. \qquad \text{(Eqn 3.35)}$$

17. The components of the Ricci tensor are defined by

$$R_{\alpha\beta} \equiv \sum_{\gamma} R^{\gamma}{}_{\alpha\beta\gamma}. \qquad \text{(Eqn 4.31)}$$

18. The curvature scalar is defined by

$$R \equiv \sum_{\alpha,\beta} g^{\alpha\beta} R_{\alpha\beta}. \qquad \text{(Eqn 4.32)}$$

19. The components of the Einstein tensor are defined by

$$G^{\mu\nu} \equiv R^{\mu\nu} - \tfrac{1}{2}g^{\mu\nu} R. \qquad \text{(Eqn 4.33)}$$

20. The Einstein field equations are

$$R_{\mu\nu} - \tfrac{1}{2}R\, g_{\mu\nu} = -\kappa\, T_{\mu\nu}, \qquad \text{(Eqn 4.34)}$$

where $\kappa = 8\pi G/c^4$. The equations are second-order in spacetime derivatives and non-linear in $g_{\mu\nu}$.

21. A region of spacetime is empty if $R_{\mu\nu} = 0$.

22. Solving the Einstein field equations implies finding the metric tensor that corresponds to a given energy–momentum tensor. Once this has been done, the geodesic equations can be used to determine the geodesics of the spacetime. These may be time-like, space-like or null.

23. According to the principle of geodesic motion, in general relativity the time-like geodesics of a spacetime represent the possible world-lines of massive particles falling freely under the influence of gravity alone. And, similarly, the null geodesics of a spacetime represent the possible world-lines of massless particles moving under the influence of gravity alone.

24. A non-zero value of the cosmological constant Λ introduces an additional term into the Einstein field equations so that

$$R_{\mu\nu} - \tfrac{1}{2}R\,g_{\mu\nu} + \Lambda\,g_{\mu\nu} = -\kappa\,T_{\mu\nu}. \qquad \text{(Eqn 4.47)}$$

This may be reinterpreted in terms of a dark energy contribution to the energy–momentum tensor, in which case we write the modified field equations as

$$R_{\mu\nu} - \tfrac{1}{2}R\,g_{\mu\nu} = -\kappa(T_{\mu\nu} + \rho_\Lambda\,c^2\,g_{\mu\nu}),$$

where the dark energy density is $\rho_\Lambda\,c^2 = \Lambda/\kappa$, and the associated pressure due to dark energy has the negative value $p_\Lambda = -\rho_\Lambda\,c^2$, leading to an effective gravitational repulsion on the cosmic scale.

Chapter 5 Schwarzschild spacetime

Introduction

The previous chapter introduced Einstein's field equations of general relativity. These equations assert the direct proportionality of the geometric Einstein tensor $[G_{\mu\nu}]$ that represents the gravitational 'field', and the energy–momentum tensor $[T_{\mu\nu}]$ that represents the 'sources' of the gravitational field. However, at a deeper level, once the Einstein tensor has been expanded in terms of the Ricci tensor $[R_{\mu\nu}]$, the Ricci tensor expressed in terms of components of the Riemann tensor $[R^{\rho}{}_{\sigma\mu\nu}]$, and the Riemann tensor related to the connection coefficients and hence to components of the metric tensor $[g_{\mu\nu}]$, it is seen that the Einstein field equations are actually a set of complicated non-linear differential equations that relate the metric coefficients $g_{\mu\nu}$ of some region of spacetime to quantities that describe the density and flow of energy and momentum in that region. Solving the Einstein field equations for some specified region (if that can be done) provides all the information needed to determine the four-dimensional line element $(\mathrm{d}s)^2$ in that region along with all the other geometric properties that follow from it. This includes the set of time-like and null geodesic pathways through an event that represent the possible world-lines of massive and massless particles present at that event.

In four-dimensional spacetime the Einstein field equations can have non-trivial solutions even in regions where there are no sources, i.e. in regions of spacetime that are devoid of matter and radiation (in this chapter we shall ignore dark energy). In the absence of sources $[T_{\mu\nu}] = 0$, and the field equations require that the Ricci tensor must vanish, but the relationship between the Ricci and Riemann tensors is such that the vanishing of the Ricci tensor does not necessarily imply that the Riemann tensor should be zero. If the Riemann tensor is not zero, then the spacetime must be curved and the metric tensor $[g_{\mu\nu}]$ that satisfies the Einstein field equations must differ from the 'trivial' Minkowski metric $[\eta_{\mu\nu}]$ that describes a flat spacetime. In this sense the Einstein field equations can describe gravitational fields in empty space, just as Maxwell's equations can describe non-trivial electric and magnetic fields in a vacuum. As we noted in the previous chapter, the solutions that arise when $[T_{\mu\nu}] = 0$ are called *vacuum solutions*.

This chapter is mainly concerned with one of these vacuum solutions — the *Schwarzschild solution*, the first and arguably the most important non-trivial solution of the Einstein field equations. We shall start by simply writing down the Schwarzschild solution so that you can see what a solution looks like and how it is conventionally presented. Next we shall outline how this particular solution can be obtained and then go on to examine its properties and some of its consequences for observations regarding intervals in space and time. These investigations of a particularly simple curved spacetime can be seen as the analogues of those that we carried out in Chapter 1 when investigating time dilation and length contraction in the flat spacetime described by the Minkowski metric of special relativity.

In Section 5.4 we shall use the metric provided by the Schwarzschild solution to determine geodesic pathways in a region described by that solution. This will enable us to study the motion of massive and massless particles in such a region and thus discuss the behaviour of massive bodies and light pulses that move under the influence of gravity alone.

In case all of this sounds like a purely mathematical exploration of some particular solution of the Einstein field equations, it's worth pointing out that many years after its discovery the Schwarzschild solution was recognized as describing the most basic type of *black hole*. The study of the Schwarzschild solution is therefore the natural precursor and preparation for the study of black holes, which have done much to revolutionize thinking in astrophysics. Black holes will be the subject of the next chapter.

5.1 The metric of Schwarzschild spacetime

The Schwarzschild solution takes its name from the German astrophysicist Karl Schwarzschild (Figure 5.1) who published the relevant results in 1916, shortly after Einstein completed his theory of general relativity. Schwarzschild had been a university professor and Director of the Potsdam Observatory outside Berlin but joined the German army at the outbreak of the First World War and was serving on the Eastern front when he made his discovery. He posted his results to Einstein, who was surprised that such a simple solution could be found.

5.1.1 The Schwarzschild metric

The 'exterior' Schwarzschild solution discussed here describes the spacetime geometry in the empty region surrounding a non-rotating, spherically symmetric body of mass M. (You might like to think of that body as a simplified model of a star.) The presentation of the Schwarzschild solution, like that of any solution of the Einstein field equations, involves specifying, as explicit functions of the spacetime coordinates x^0, x^1, x^2, x^3, the sixteen components of the metric tensor $[g_{\mu\nu}]$ that correspond to the energy–momentum tensor $[T_{\mu\nu}]$ in the region of interest. In the case of the Schwarzschild solution, the relevant energy–momentum tensor is $[T_{\mu\nu}] = 0$ since we are dealing with the empty region *outside* the mass distribution. Nonetheless, the symmetry of the region involved suggests that it would be wise to use a system of spherical coordinates originating at the centre of the massive body, and it also seems likely that the solution will involve the mass M in some way. We shall have more to say about the significance of M and the precise meaning of the coordinates later; for the moment we shall simply refer to the coordinates as **Schwarzschild coordinates** and denote them by $x^0 = ct$, $x^1 = r$, $x^2 = \theta$, $x^3 = \phi$.

Due to the symmetry of the metric tensor, only ten of its sixteen components $g_{\mu\nu}$ are independent. Moreover, in the particular case of the Schwarzschild solution, thanks to the spherical symmetry, the lack of time-dependence and the judicious choice of coordinates, only four of the components turn out to be non-zero, and none of them depends on x^0. In fact, the solution can be represented by the diagonal matrix

$$[g_{\mu\nu}] = \begin{pmatrix} 1 - \frac{2GM}{c^2 r} & 0 & 0 & 0 \\ 0 & -\frac{1}{1 - \frac{2GM}{c^2 r}} & 0 & 0 \\ 0 & 0 & -r^2 & 0 \\ 0 & 0 & 0 & -r^2 \sin^2\theta \end{pmatrix}. \tag{5.1}$$

Figure 5.1 Karl Schwarzschild (1873–1916) discovered the first exact solution of the Einstein field equations. He served as an artillery officer in the First World War, but contracted a serious skin disease and was invalided out of the army. He died in May 1916, not long after completing the work for which he is mainly remembered.

Though clear, this is a rather cumbersome way of presenting the metric, so it is actually more common to see the non-zero components presented as the metric coefficients in the four-dimensional line element of the spacetime region being described. This is usually written as follows.

The Schwarzschild metric

$$(\mathrm{d}s)^2 = \left(1 - \frac{2GM}{c^2 r}\right) c^2 (\mathrm{d}t)^2 - \frac{(\mathrm{d}r)^2}{1 - \frac{2GM}{c^2 r}}$$
$$- r^2 (\mathrm{d}\theta)^2 - r^2 \sin^2 \theta \, (\mathrm{d}\phi)^2. \tag{5.2}$$

Although the terminology that we have been using leads us to refer to this expression as a line element, what it really tells us is the functional form of the non-zero components of the metric tensor. Because of this it is often referred to as the **Schwarzschild metric**. You should also be aware that built into it is the choice that we made regarding the use of an x^0 coordinate to represent time (some authors prefer x^4) and some other decisions regarding signs and symbols. The upshot of all this is that although we have adopted a range of common conventions, you should not be surprised to find that other authors may make different decisions and will therefore write the Schwarzschild solution in a related but different form.

5.1.2   Derivation of the Schwarzschild metric

In empty space $T_{\mu\nu} = 0$, so the Einstein field equations become

$$R_{\mu\nu} - \tfrac{1}{2} g_{\mu\nu} R = 0. \tag{5.3}$$

These equations are known as the **vacuum field equations**. Multiplying them by $g^{\mu\nu}$ and contracting over the indices μ and ν gives

$$\sum_{\mu,\nu} g^{\mu\nu} \left(R_{\mu\nu} - \tfrac{1}{2} g_{\mu\nu} R \right) = 0, \tag{5.4}$$

that is,

$$\sum_{\nu} \left(R^\nu{}_\nu - \tfrac{1}{2} \delta^\nu{}_\nu R \right) = 0. \tag{5.5}$$

Summing $R^\nu{}_\nu$ over all values of ν gives the curvature scalar R, while summing $\delta^\nu{}_\nu$ over all possible values of ν gives $\delta^0{}_0 + \delta^1{}_1 + \delta^2{}_2 + \delta^3{}_3 = 4$. Substituting these results into Equation 5.5, we get

$$R - \tfrac{1}{2} 4R = 0,$$

showing that $R = 0$ in this case and hence (from the vacuum field equations) that $R_{\mu\nu} = 0$ for all values of μ and ν. Thus the Ricci tensor and the curvature scalar must both vanish for a vacuum solution, but remember, this is not sufficient to make spacetime flat.

It would be straightforward (though time-consuming) to show that the Schwarzschild metric written down earlier does indeed lead to a vanishing Ricci tensor and therefore *is* a solution of the vacuum field equations. However, that is not the aim of this section. Rather, our approach here is to write down the most general metric that exhibits the symmetries expected of the Schwarzschild solution and then use the additional requirement that the metric satisfies the vacuum field equations to lead us to a specific metric that will turn out to be the Schwarzschild solution. This is closer to the approach actually followed by Schwarzschild.

Note that you are not expected to remember all the steps in this derivation, but you should be able to follow them and they should provide helpful examples of many of the tensor quantities that were introduced earlier. The derivation omits a lot of detailed algebra, simply quoting results in its place. If you really want to get a feel for relativity, you might like to fill in some of the missing steps, but don't try this if you are short of time!

Since the Schwarzschild solution describes the geometry of the empty spacetime region surrounding a spherically symmetric body, it is natural to use a system of spherical coordinates centred on the middle of that spherically symmetric body (see Figure 5.2). In addition we shall assume the following.

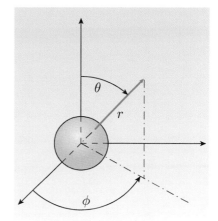

Figure 5.2 The spatial part of the Schwarzschild coordinate system, with origin at the centre of a spherically symmetric body.

1. The spacetime far from the spherically symmetric body is flat. This is described by saying that the metric is **asymptotically flat** and is consistent with the idea that gravitational effects become weaker as the distance from their source increases.

2. The metric coefficients do not depend on time. This is described by saying that the metric is **stationary** and is consistent with the idea that nothing is moving from place to place.

3. The line element is unchanged if t is replaced by $-t$. This is described by saying that the metric is **static** and is consistent with the idea that nothing is rotating.

We shall say more about these assumptions and about the definition and meaning of the Schwarzschild coordinates later. For the moment we shall simply use them.

The most general spacetime line element that meets all of the listed requirements may be written as

$$(\mathrm{d}s)^2 = \sum_{\mu,\nu} g_{\mu\nu}\,\mathrm{d}x^\mu\,\mathrm{d}x^\nu$$
$$= \mathrm{e}^{2A}(c\,\mathrm{d}t)^2 - \mathrm{e}^{2B}(\mathrm{d}r)^2 - r^2(\mathrm{d}\theta)^2 - r^2\sin^2\theta(\mathrm{d}\phi)^2, \qquad (5.6)$$

where A and B are functions of the radial coordinate r alone. You may wonder why we choose to include exponential functions of the form e^{2A} and e^{2B} rather than simply using functions such as $f(r)$ and $g(r)$. The reason is that the use of exponentials ensures that the signs of the metric components will be preserved in the desired $(+,-,-,-)$ pattern. The absence of terms proportional to $\mathrm{d}x^i\,\mathrm{d}t$ (where $i = 1, 2$ or 3) reflects the static property of the spacetime, while the absence of $\mathrm{d}x^i\,\mathrm{d}x^j$ terms reflects the spherical symmetry.

Our aim now is to determine the precise form of the functions $A(r)$ and $B(r)$ using the fact that the metric must satisfy the vacuum field equations. The first step in this process is the determination of the connection coefficients that correspond to the metric given in Equation 5.6. This involves applying the general formula

$$\Gamma^{\sigma}{}_{\mu\nu} = \frac{1}{2} \sum_{\rho} g^{\sigma\rho} \left\{ \frac{\partial g_{\rho\nu}}{\partial x^{\mu}} + \frac{\partial g_{\mu\rho}}{\partial x^{\nu}} - \frac{\partial g_{\mu\nu}}{\partial x^{\rho}} \right\}$$

to the case where $g_{00} = \mathrm{e}^{2A}$, $g_{11} = -\mathrm{e}^{2B}$, $g_{22} = -r^2$ and $g_{33} = -r^2 \sin^2 \theta$. Because the metric is represented by a diagonal matrix in this case, each contravariant component $g^{\mu\nu}$ is simply the reciprocal of the corresponding covariant component $g_{\mu\nu}$, so $g^{00} = \mathrm{e}^{-2A}$, $g^{11} = -\mathrm{e}^{-2B}$, $g^{22} = -1/r^2$ and $g^{33} = -1/r^2 \sin^2 \theta$. Substituting these values into the expression for $\Gamma^{\sigma}{}_{\mu\nu}$ shows that only nine of the forty independent connection coefficients for this metric are non-zero. Using a prime to indicate differentiation with respect to r, so that $A' = \mathrm{d}A(r)/\mathrm{d}r$ and $B' = \mathrm{d}B(r)/\mathrm{d}r$, these nine independent non-zero connection coefficients can be written as

$$\Gamma^0{}_{01} = A' \; (= \Gamma^0{}_{10}),$$
$$\Gamma^1{}_{00} = A'\mathrm{e}^{2(A-B)},$$
$$\Gamma^1{}_{11} = B',$$
$$\Gamma^1{}_{22} = -r\mathrm{e}^{-2B},$$
$$\Gamma^1{}_{33} = -\mathrm{e}^{-2B} r \sin^2 \theta,$$
$$\Gamma^2{}_{12} = \frac{1}{r} \; (= \Gamma^2{}_{21}),$$
$$\Gamma^2{}_{33} = -\sin\theta\cos\theta,$$
$$\Gamma^3{}_{13} = \frac{1}{r} \; (= \Gamma^3{}_{31}),$$
$$\Gamma^3{}_{23} = \cot\theta \; (= \Gamma^3{}_{32}).$$

These non-zero connection coefficients can be used to determine the non-zero components of the Riemann curvature tensor using the general formula

$$R^{\rho}{}_{\sigma\mu\nu} = \frac{\partial \Gamma^{\rho}{}_{\sigma\nu}}{\partial x^{\mu}} - \frac{\partial \Gamma^{\rho}{}_{\sigma\mu}}{\partial x^{\nu}} + \sum_{\lambda} \Gamma^{\lambda}{}_{\sigma\nu} \Gamma^{\rho}{}_{\lambda\mu} - \sum_{\lambda} \Gamma^{\lambda}{}_{\sigma\mu} \Gamma^{\rho}{}_{\lambda\nu}.$$

Again, there are many symmetries so not all the non-zero curvature tensor components are independent, though these are the six that are:

$$R^0{}_{101} = A'B' - A'' - \left(A' \right)^2,$$
$$R^0{}_{202} = -r\mathrm{e}^{-2B} A',$$
$$R^0{}_{303} = -r\mathrm{e}^{-2B} A' \sin^2 \theta,$$
$$R^1{}_{212} = r\mathrm{e}^{-2B} B',$$
$$R^1{}_{313} = r\mathrm{e}^{-2B} B' \sin^2 \theta,$$
$$R^2{}_{323} = \left(1 - \mathrm{e}^{-2B} \right) \sin^2 \theta,$$

where the double prime indicates the second derivative with respect to r. Contraction of the Riemann tensor gives the Ricci tensor with components

$$R_{\mu\nu} = \sum_{\lambda} R^{\lambda}{}_{\mu\nu\lambda},$$

and reveals (after much algebra) that only the four diagonal components of the Ricci tensor are not identically zero:

$$R_{00} = -\mathrm{e}^{2(A-B)}\left(A'' + \left(A'\right)^2 - A'B' + \frac{2A'}{r}\right),$$

$$R_{11} = A'' + \left(A'\right)^2 - A'B' - \frac{2B'}{r},$$

$$R_{22} = \mathrm{e}^{-2B}\left(1 + r\left(A' - B'\right)\right) - 1,$$

$$R_{33} = \sin^2\theta\left(\mathrm{e}^{-2B}\left[1 + r\left(A' - B'\right)\right] - 1\right).$$

Now, we already know that for a vacuum solution all four of these components must be equal to zero. Nonetheless, for the sake of completeness, we shall use the expressions that we have obtained to calculate the curvature scalar

$$R = \sum_{\mu,\nu} g^{\mu\nu} R_{\mu\nu},$$

which in this case becomes

$$R = g^{00} R_{00} + g^{11} R_{11} + g^{22} R_{22} + g^{33} R_{33}$$

and yields

$$R = -2\mathrm{e}^{-2B}\left(A'' + \left(A'\right)^2 - A'B' + \frac{2}{r}\left(A' - B'\right) + \frac{1}{r^2}\right) + \frac{2}{r^2}.$$

When evaluated, this too must vanish for a vacuum solution.

Combining the results for the curvature scalar and the components of the Ricci tensor, we can determine the Einstein tensor components given by

$$G_{\mu\nu} = R_{\mu\nu} - \tfrac{1}{2} g_{\mu\nu}\, R,$$

the only ones that are not identically zero in this case being

$$G_{00} = -\frac{2\mathrm{e}^{2(A-B)}}{r} B' + \frac{\mathrm{e}^{2(A-B)}}{r^2} - \frac{\mathrm{e}^{2A}}{r^2},$$

$$G_{11} = -\frac{2A'}{r} + \frac{\mathrm{e}^{2B}}{r^2} - \frac{1}{r^2},$$

$$G_{22} = -r^2 \mathrm{e}^{-2B}\left(A'' + \left(A'\right)^2 + \frac{A' - B'}{r} - A'B'\right),$$

$$G_{33} = -r^2 \mathrm{e}^{-2B} \sin^2\theta\left(A'' + \left(A'\right)^2 + \frac{A' - B'}{r} - A'B'\right).$$

Now, the vacuum field equations demand that even these Einstein tensor components should each be zero in the space outside the spherically symmetric body. One consequence of this is that $\mathrm{e}^{-2A} G_{00} + \mathrm{e}^{-2B} G_{11} = 0$, but this implies that

$$\frac{2\mathrm{e}^{-2B}}{r}\left(A' + B'\right) = 0,$$

implying that $A' + B' = 0$, which can be integrated to give $A(r) + B(r) = C$, where C is a constant. This constant can be set to zero without loss of generality, since any other choice can be represented by a rescaling of the r-coordinate, which still has an arbitrary scale at this stage. (This is one of the points that we shall return to later.) Making use of this freedom to set $C = 0$, we see that $A(r) = -B(r)$, and the equation $G_{00} = 0$ can be rewritten as

$$\frac{1}{r^2} \frac{\mathrm{d}\left(r[1 - \mathrm{e}^{-2B}]\right)}{\mathrm{d}r} = 0,$$

which, after ignoring $1/r^2$, can also be integrated, to yield $\mathrm{e}^{-2B} = 1 - R_{\mathrm{S}}/r$, where the integration constant, R_{S}, has the units of distance. The constant R_{S} is called the **Schwarzschild radius**.

Since $\mathrm{e}^{2A} = \mathrm{e}^{-2B}$, we can now identify the explicit form that must be taken by the two exponential functions in the line element of Equation 5.6 if the corresponding metric is to satisfy the vacuum field equations. Explicitly,

$$\mathrm{e}^{2A} = 1 - \frac{R_{\mathrm{S}}}{r}, \quad \mathrm{e}^{2B} = \frac{1}{1 - \frac{R_{\mathrm{S}}}{r}}.$$

This shows that the line element of the Schwarzschild solution can be written as

$$(\mathrm{d}s)^2 = \left(1 - \frac{R_{\mathrm{S}}}{r}\right) c^2 (\mathrm{d}t)^2 - \frac{1}{1 - \frac{R_{\mathrm{S}}}{r}} (\mathrm{d}r)^2$$
$$- r^2 \left((\mathrm{d}\theta)^2 + \sin^2\theta\, (\mathrm{d}\phi)^2\right). \tag{5.7}$$

The final step in our modern derivation is to use the principle of consistency and the Newtonian limit to relate the Schwarzschild radius to the mass M of the spherically symmetric body centred on the origin. We saw in Section 4.3.3 that for weak fields, in the Newtonian limit $g_{00} = 1 + h_{00} = 1 + 2\Phi/c^2$, where Φ is the Newtonian gravitational potential (i.e. the potential energy per unit mass). In the case of a spherically symmetric body of mass M centred on the origin, the Newtonian gravitational potential outside the body, at a distance r from the origin, is $\Phi = -GM/r$. It follows that in the Newtonian limit $g_{00} = 1 - 2GM/rc^2$, and comparing this with the metric coefficient that occupies the position of g_{00} in Equation 5.7, we see that the two will agree provided that we assign the Schwarzschild radius the value

$$R_{\mathrm{S}} = 2GM/c^2. \tag{5.8}$$

We can now represent the metric tensor of the Schwarzschild solution in the diagonal matrix form

$$[g_{\mu\nu}] = \begin{pmatrix} 1 - \frac{2GM}{c^2 r} & 0 & 0 & 0 \\ 0 & -\frac{1}{1 - \frac{2GM}{c^2 r}} & 0 & 0 \\ 0 & 0 & -r^2 & 0 \\ 0 & 0 & 0 & -r^2 \sin^2\theta \end{pmatrix} \tag{Eqn 5.1}$$

or in its more common form as the line element

$$(\mathrm{d}s)^2 = \left(1 - \frac{2GM}{c^2 r}\right) c^2 (\mathrm{d}t)^2 - \frac{(\mathrm{d}r)^2}{1 - \frac{2GM}{c^2 r}}$$
$$- r^2 (\mathrm{d}\theta)^2 - r^2 \sin^2\theta\, (\mathrm{d}\phi)^2, \tag{Eqn 5.2}$$

which relates incremental changes in the spacetime interval $\mathrm{d}s$ to incremental changes in intervals of Schwarzschild coordinate time t and the Schwarzschild spatial coordinates r, θ, ϕ between neighbouring events.

There are shortcuts that could have been taken in this section; for instance, we could have used the condition that the components of the Ricci tensor must vanish in the case of a vacuum solution rather than working out the Einstein tensor components and applying the full field equations. The approach we have taken has the advantage of showing you explicit examples of each of the major tensor quantities. Now that we know what they look like, we can investigate their meaning and significance in this particular case.

Exercise 5.1   Confirm the value for G_{00} given above.   ■

5.2   Properties of Schwarzschild spacetime

Several properties of the Schwarzschild metric were mentioned early in the previous section, where they were used to determine the general line element given in Equation 5.6. One of the most basic was spherical symmetry. We shall start by considering that property in more detail.

5.2.1   Spherical symmetry

At any particular value of t, call it T, fixing the value of r to have some particular value R ensures that $\mathrm{d}t = 0$ and $\mathrm{d}r = 0$, and reduces the Schwarzschild line element to

$$(\mathrm{d}s)^2 = -R^2(\mathrm{d}\theta)^2 - R^2\sin^2\theta\,(\mathrm{d}\phi)^2, \tag{5.9}$$

which describes the two-dimensional geometry on the surface of a sphere of radius R. Now, from a physical point of view, no point on this spherical surface is any more 'special' than any other point. The fact that no value of ϕ is picked out is clear from the fact that ϕ does not appear in any of the metric coefficients. However, the same is not true of θ — that does appear in the metric coefficient that multiplies $(\mathrm{d}\phi)^2$. This makes it seem that there might be something special about certain values of θ even though we have already said that there can't be. The reason why θ is picked out in this way has nothing to do with the gravitation of a spherically symmetric body; it is entirely due to the way in which we define spherical coordinates. When we use such coordinates we have to choose some radial direction to be the 'north polar axis'. That direction is assigned the special coordinate value $\theta = 0$ even though in the case of a non-rotating spherically symmetric body there is nothing physically 'special' about the direction chosen to play that role. Any other direction from the origin could just as easily have been chosen as the north polar axis.

This illustrates an important point in general relativity that we shall come back to later. Locations that appear to be 'special' in metrics and line elements may be physically special in some way, or they may only appear to be special because of some particular feature of the coordinate system being used. It is always important to distinguish between real physical effects and non-physical effects produced by the coordinate system alone. The need for this distinction is clear, but as you will soon see it is not always easy to tell whether a particular feature is the result of coordinates or gravitation.

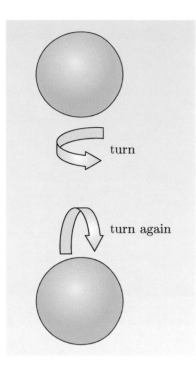

Figure 5.3 A sphere (spherical shell) exhibits spherical symmetry; the sphere is invariant under arbitrary rotations about the origin.

The Schwarzschild solution is *spherically symmetric*: at any given value of t, all points with the same value of r are physically equivalent. The spacetime has the same symmetries as a sphere (by which mathematicians mean it has the symmetries of the surface of a ball), so it is said to be 'invariant under rotations about the origin' (see Figure 5.3).

Of course, this does not mean that points with *different* values of r are physically equivalent. Indeed, we have already seen that in the Newtonian limit, points at different values of r will correspond to different values of the gravitational potential. Also, one of the main outcomes of the derivation was that the metric coefficients in the Schwarzschild line element contain terms of the form $1 - 2GM/c^2r$ that are functions of r.

Exercise 5.2 Suppose that the Schwarzschild coordinate system ct, r, θ, ϕ used to describe the spacetime outside a non-rotating spherically symmetric body is replaced by a different system that uses the coordinates ct, r, θ, ϕ', where $\phi' = \phi + \phi_0$.

(a) Show that the Schwarzschild metric is form-invariant when the new coordinates are substituted for the old ones.

(b) Give a physical justification for the mathematical fact stated in part (a). ∎

5.2.2 Asymptotic flatness

In the Schwarzschild line element, the factor $1 - 2GM/c^2r$ appears in the metric coefficients of the $c^2(\mathrm{d}t)^2$ term and the $(\mathrm{d}r)^2$ term. The factor is independent of direction and approaches 1 as r becomes large. The meaning of 'large' in this context depends on the value of M; what is meant is that r is sufficiently large to make the term $2GM/c^2r$ very much smaller than 1. Where that condition is satisfied, $1 - 2GM/c^2r \to 1$ and the Schwarzschild line element

$$(\mathrm{d}s)^2 = \left(1 - \frac{2GM}{c^2r}\right)c^2(\mathrm{d}t)^2 - \frac{(\mathrm{d}r)^2}{1 - \frac{2GM}{c^2r}} - r^2(\mathrm{d}\theta)^2 - r^2\sin^2\theta\,(\mathrm{d}\phi)^2 \qquad \text{(Eqn 5.2)}$$

takes the form of the Minkowski line element

$$(\mathrm{d}s)^2 = c^2(\mathrm{d}t)^2 - (\mathrm{d}r)^2 - r^2(\mathrm{d}\theta)^2 - r^2\sin^2\theta\,(\mathrm{d}\phi)^2 \qquad (5.10)$$

that describes the flat spacetime of special relativity in spherical coordinates. This is the form that we should expect the Schwarzschild line element to take 'far' from the origin where gravitational effects due to the mass of the spherically symmetric body will be negligible.

Remembering that this 'flatness' only applies at sufficiently large values of r, we say that the Schwarzschild metric has the property of *asymptotic flatness*.

5.2.3 Time-independence

Two other properties of the Schwarzschild metric that were briefly mentioned earlier related to its time-independence. The first of these is the property of being *stationary*, implying that none of the metric coefficients depends on t. So, if t_1 and t_2 represent the time coordinates of neighbouring events, then $dt = t_2 - t_1 = (t_2 + t_0) - (t_1 + t_0) = t'_2 - t'_1 = dt'$, and the metric is invariant under a coordinate transformation of the form $t \rightarrow t' = t + t_0$, where t_0 is a constant. This specific aspect of time-independence is described as 'invariance under translation in time' and is another symmetry of the solution.

The second feature relating to time-independence introduced earlier was the property of being *static*. This concerns invariance under transformations that reverse time, such as $t \rightarrow -t$. The fact that the Schwarzschild metric is stationary ensures that time reversal will have no effect on any of the metric coefficients since they do not depend on t at all. However, in order that the metric should be static, it is also important that the line element should not contain any terms of the form $dr\, dt$, $d\theta\, dt$ or $d\phi\, dt$. Such terms are often referred to as 'cross terms' or 'mixed terms' and are typical of situations involving rotation.

The Schwarzschild metric is both stationary and static.

5.2.4 Singularity

A striking feature of the Schwarzschild metric is its odd behaviour as r approaches the Schwarzschild radius $R_S = 2GM/c^2$. As $r \rightarrow R_S$, the factor $1 - R_S/r$ causes the metric coefficient $g_{00} \rightarrow 0$ while the factor $(1 - R_S/r)^{-1}$ causes $g_{11} \rightarrow \infty$. The unlimited growth of the latter factor is described by saying that there is a **singularity** in the Schwarzschild metric. This particular singularity is in fact a consequence of the coordinates that we are using to describe the Schwarzschild solution. That is, it is a **coordinate singularity**, not a physically meaningful **gravitational singularity**. As a coordinate singularity it can be removed by an appropriate transformation of coordinates in a way that would not be possible for a true gravitational singularity. Nonetheless it is a feature of the solution as described by Schwarzschild coordinates and an indicator of the significance of R_S.

When considering this coordinate singularity it is important to remember that the exterior Schwarzschild solution that we are discussing describes the spacetime *outside* a spherically symmetric body of mass M. It is therefore interesting to ask if R_S is likely to be larger or smaller than the radius of such a body. If R_S is smaller than the body's radius, the coordinate singularity will be outside the domain in which the Schwarzschild solution is applicable, and the solution itself will be non-singular throughout the region that it actually describes.

For a body with the mass of the Sun (about 2.0×10^{30} kg), the Schwarzschild radius is 3.0 km. This compares with a solar radius of about 0.7 *million* km. So in the case of a normal star-like body, the Schwarzschild radius is deep inside the body. Of course, not all bodies of astronomical interest are 'normal' or 'star-like'. As you will see later, the Schwarzschild radius is of great importance in the study of black holes. A body can become a black hole if its surface shrinks within its Schwarzschild radius.

A final point to note is that the Schwarzschild metric also has a singularity at $r = 0$. This is a gravitational singularity, marked by the unlimited growth of invariants related to the curvature, and cannot be removed by any change of coordinates. This singularity is of little relevance to the exterior solution that we have been discussing in this section, but it will be significant when we come to discuss black holes.

5.2.5 Generality

According to the Schwarzschild solution, the spacetime geometry outside a static spherically symmetric body is characterized by a single quantity M, which represents the total mass of that body.

In 1923 the American mathematician George Birkhoff proved that even if the source of gravitation is not static (and therefore not necessarily stationary), and as long as its effect is **isotropic** (i.e. the same in all directions), the vacuum solution of the Einstein field equations in the region exterior to the source is still stationary and is still the Schwarzschild solution.

This result is known as **Birkhoff's theorem**. One of its implications is that a spherically symmetric body that is expanding or contracting in a purely radial way, or even one that is pulsating radially, cannot produce any gravitational signs of that radial motion beyond the spherical region that contains the material of the body itself. So, if a fixed mass M were contained within a sphere of radius r_1, then the Schwarzschild metric would apply throughout the region $r > r_1$, but if the mass distribution were to shrink in an isotropic way to a smaller radius r_2, then the spacetime would be unaffected in the region $r > r_1$ but now the Schwarzschild metric would apply throughout the larger region $r > r_2$.

This is a surprising result. It indicates the special nature of vacuum solutions as well as the generality of the Schwarzschild solution. As you will see later when we discuss gravitational radiation, it also indicates that sources that only pulsate radially cannot produce gravitational waves.

To summarize, we have the following.

> **Properties of the Schwarzschild solution**
>
> The Schwarzschild metric is a static (and therefore stationary), spherically symmetric solution of the Einstein field equations in the empty region exterior to any distribution of energy and momentum characterized by mass M that produces purely isotropic effects in that region. The solution is asymptotically flat, approaching the Minkowski metric in spherical coordinates for sufficiently large values of r. The solution has a coordinate singularity at the Schwarzschild radius $r = R_S = 2GM/c^2$ and a gravitational singularity at $r = 0$, though neither of these singularities is within the region described by the solution for normal 'star-like' bodies.

5.3 Coordinates and measurements in Schwarzschild spacetime

We now need to deal with an issue that has been present since we first introduced the Schwarzschild coordinates ct, r, θ, ϕ near the start of this chapter. The issue concerns the relationship between coordinate values and physically meaningful intervals of time and distance.

When confronted by a system of coordinates that includes a t-coordinate and an r-coordinate, it is tempting to assume that the t must represent time and the r radial distance from the origin. However, such an assumption is always dangerous and often wrong.

The simple fact is that in general relativity, coordinates are essentially arbitrary systems of markers chosen to distinguish one event from another. This gives us great freedom in how we define coordinates, a freedom that we exploited in the derivation of the Schwarzschild metric. The relationship between the coordinate differences separating events and the corresponding intervals of time or distance that would be measured by a specified observer must be worked out using the metric of the spacetime. It cannot be assumed that the 'physical' times and distances that would be measured by clocks or measuring sticks are directly specified by the coordinates. This situation is described by saying that:

In general relativity, coordinates do not have immediate metrical significance.

Einstein found this a perplexing feature of general relativity. In his own account of how the general theory developed after 1908 he says:

> Why were another seven years required for the construction of the general theory of relativity? The main reason lies in the fact that it is not easy to free oneself from the idea that coordinates must have an immediate metrical meaning.

> Quoted in Schilpp, P. A. (ed.) (1969) *Albert Einstein — Philosopher Scientist*, 3rd edn, Illinois, Open Court.

Intervals of time and distance must be measured by an observer who must make use of a frame of reference, so we start with a discussion of the observers and frames that will be relevant to our discussion of Schwarzschild spacetime.

5.3.1 Frames and observers

We saw in the discussion of special relativity that the phenomena of time dilation and length contraction made it important to be clear about who was performing measurements of time and distance, and to be especially careful when relating time and distance measurements made by different inertial observers. In special relativity, inertial fames are 'global', in principle stretching out to infinity. We needed to be clear about the frame that an observer was using but we emphasized the distinction between 'seeing' and 'observing', and stressed that observers were concerned with the latter, which made their location irrelevant for most purposes.

In general relativity, the situation is very different. There is no 'special' class of frames, and the frames that are used are generally 'local' so an observer's location is important. In this chapter we shall be particularly concerned with observations made in three 'local' frames: the frame used by a freely falling observer, a frame that is at rest at some specified location, and the frame of a 'distant' observer located far from the spherically symmetric body at the origin of Schwarzschild coordinates. The frame of the freely falling observer is locally inertial; gravity has effectively been 'turned off' and special relativity applies locally. The observer at a fixed location will need to take steps to avoid falling freely; they might need to locate themselves in a rocket, for example. For such an observer special relativity will work locally but only if the observer supposes that every body is subject to a 'gravitational force' that is proportional to the inertial mass of the body. This is really a 'fictitious force', introduced to account for the fact that the observer's frame is not freely falling and is therefore not really locally inertial. To this extent the observer maintaining a fixed position is in a similar situation to a passenger in a bus turning a corner who 'feels' the effect of a (fictitious) centrifugal force. The distant observer will be in a region of spacetime that is effectively flat, so special relativity will again apply locally and there will not be any local effects of gravitation to take into account. Such an observer can remain at rest without needing the support of a rocket and can even be regarded as falling freely while remaining at rest!

5.3.2   Proper time and gravitational time dilation

Consider two events involving the emission of light, that happen in the Schwarzschild spacetime surrounding a static spherically symmetric body. Suppose that the two emission events are described by the Schwarzschild coordinates $(t_{em}, r_{em}, \theta_{em}, \phi_{em})$ and $(t_{em} + dt_{em}, r_{em}, \theta_{em}, \phi_{em})$, so they are separated by a difference in coordinate time dt_{em}, while their other coordinate separations are all zero: $dr_{em} = d\theta_{em} = d\phi_{em} = 0$.

According to the Schwarzschild metric, the infinitesimal spacetime separation of these time-like events is given by

$$(ds_{em})^2 = \left(1 - \frac{2GM}{c^2 r_{em}}\right) c^2 (dt_{em})^2, \tag{5.11}$$

and the **proper time** between the events, as would be measured by a clock at rest at the location of the events, is $d\tau_{em} = ds_{em}/c$, so

$$d\tau_{em} = ds_{em}/c = \left(1 - \frac{2GM}{c^2 r_{em}}\right)^{1/2} dt_{em}. \tag{5.12}$$

Note that the proper time separating the events, according to a stationary clock at the location of the events, is less than the coordinate time separating the events.

Now consider what will be seen by an observer at rest at some other location with the same angular coordinates θ and ϕ but a different radial coordinate $r = r_{ob}$. As will be shown in Chapter 6, in static Schwarzschild spacetime, such an observer will find that the coordinate time separating the signals when they arrive at $r = r_{ob}$ will be the same as the coordinate time between the emission of those signals. We can indicate this by writing $dt_{ob} = dt_{em}$. All the other coordinate differences dr, $d\theta$ and $d\phi$ will still be zero. It follows that the spacetime separation between the observations of the two signals at $r = r_{ob}$ will be

$$(ds_{ob})^2 = \left(1 - \frac{2GM}{c^2 r_{ob}}\right) c^2 (dt_{ob})^2 = \left(1 - \frac{2GM}{c^2 r_{ob}}\right) c^2 (dt_{em})^2, \tag{5.13}$$

and the proper time between the observations of the two signals will be

$$d\tau_{ob} = ds_{ob}/c = \left(1 - \frac{2GM}{c^2 r_{ob}}\right)^{1/2} dt_{em}. \tag{5.14}$$

There are two important consequences that follow from these relationships.

First, for a distant observer fixed at a sufficiently large value of r, effectively at $r_{ob} = \infty$, it follows from Equation 5.14 that

$$d\tau_\infty = dt_{em}. \tag{5.15}$$

Integrating both sides of this equation shows that even for two emission events separated by a finite coordinate time difference Δt_{em}, that difference will still equal $\Delta\tau_\infty$, the difference in the proper time between observations of those events made by a stationary observer at infinity. This establishes that the Schwarzschild coordinate time separating two events at a fixed location can actually be determined by measuring the proper time between observations of those two events using a stationary clock at infinity. This gives us a way, in principle at least, of assigning Schwarzschild coordinate times to events.

● Should we be worried by the fact that this argument involves an observer at infinity? Does that invalidate the process?

○ No. All it means is that the observer should be far enough away to be in the asymptotically flat region of spacetime where $2GM/c^2 r_{ob}$ is negligible compared with 1.

Second, it follows from Equation 5.15 and the relation between $d\tau_{em}$ and dt_{em} in Equation 5.12 that

$$d\tau_\infty = \frac{d\tau_{em}}{\left(1 - \frac{2GM}{c^2 r_{em}}\right)^{1/2}}. \tag{5.16}$$

This shows that the proper time between the observation of the two light signals at infinity, $d\tau_\infty$, is greater than the proper time between their emission as measured at the site of the emission, $d\tau_{em}$.

If we suppose that the two events that we have been discussing represent the beginning and the end of a single tick of a clock fixed at $r = r_{em}$, then our second result shows that the duration of that tick as seen by a distant observer will be increased by a factor $1/\left(1 - \frac{2GM}{c^2 r_{em}}\right)^{1/2}$. This shows that the distant observer will find that the clock at $r = r_{em}$ is running slow.

● If the stationary clock emitting the light signals was moved closer to the surface of the spherically symmetric body, how would the observations of its rate of ticking by a distant fixed observer be affected?

○ The distant observer would find that the clock ticked even more slowly. Moving the clock closer to the surface reduces the value of r_{em}, which has the effect of increasing the factor $1/\left(1 - \frac{2GM}{c^2 r_{em}}\right)^{1/2}$.

This effect, the slowing of the rate of ticking of a clock in a gravitational field, as seen by a distant observer, is sometimes referred to as **gravitational time dilation**. Note, however, that there is a significant difference between this effect and the time dilation in special relativity that we studied in Chapter 1. In that earlier case we were careful to ignore the effects of signal travel time and only considered the time intervals between the events themselves as measured by different inertial observers, irrespective of the observer's location. In the general

Figure 5.4 A schematic representation of the redshift of radiation as it escapes from a massive body.

relativistic case there is no relative motion; both the clock and the distant observer are at rest, and we are very deliberately considering the proper time between the arrival of light signals at that distant observer's location. The distant observer is still making observations, but the observations are of local events — the arrival of the light signals, not their emission.

The general relativistic effect can be given another interpretation. Suppose that the two 'emission' events represent the emission of successive peaks of an electromagnetic wave (a light wave), so that $d\tau_{em}$ represents the period of that wave at its point of emission. Then $d\tau_{ob}$ will represent the period of that same radiation as measured by a distant observer. The periods will still be related by

$$d\tau_\infty = \frac{d\tau_{em}}{\left(1 - \frac{2GM}{c^2 r_{em}}\right)^{1/2}}, \qquad \text{(Eqn 5.16)}$$

but now we can say that the reciprocal of the period represents the frequency of the radiation, so the frequency observed by the distant observer will be

$$f_\infty = f_{em}\left(1 - \frac{2GM}{c^2 r_{em}}\right)^{1/2}. \qquad (5.17)$$

This shows that the observed (proper) frequency is less than the emitted (proper) frequency. It follows that light rising through a gravitational field will be redshifted. This phenomenon is known as **gravitational redshift** (see Figure 5.4). You saw in Section 4.1.1 that a local version of this phenomenon was already predicted as a consequence of the principle of equivalence. Now, with the aid of the Einstein field equations and the Schwarzschild metric, you can see the full effect, not limited to a local frame, but relating quantities that might be measured in two widely separated local frames. This is an effect that might be measured by an astronomer, and we shall discuss such measurements in Chapter 7.

Exercise 5.3 Treating the Sun as a non-rotating, spherically symmetric body, and regarding the surrounding space as well described by the Schwarzschild metric, at what value of the Schwarzschild coordinate r do intervals of proper time $d\tau$ and coordinate time dt differ by no more than 1 part in 10^8? ∎

To summarize, we have the following.

> **Proper time and gravitational time dilation**
>
> The Schwarzschild coordinate time separating two events at a fixed location is equal to the proper time between sightings of those two events by a distant stationary observer.
>
> The rate of ticking of a stationary clock at Schwarzschild radial coordinate r will be seen to be slowed by a factor of $\left(1 - \frac{2GM}{c^2 r_{em}}\right)^{-1/2}$ as measured by a distant stationary observer. This same effect will lead to a gravitational redshift — seen as a reduction in frequency by a factor $\left(1 - \frac{2GM}{c^2 r_{em}}\right)^{1/2}$ — of the radiation from a stationary source as measured by a distant stationary observer.

5.3.3 Proper distance

Just as we related differences in Schwarzschild coordinate time to intervals of
proper time that might be measured by clocks, so we must relate differences in
Schwarzschild coordinate position to proper distances that might be measured
using measuring sticks. Consider two events that happen in Schwarzschild
spacetime at the same coordinate time but at infinitesimally separated positions,
so that their spacetime separation is given by the negative quantity

$$(\mathrm{d}s)^2 = -\frac{(\mathrm{d}r)^2}{\left(1 - \frac{2GM}{c^2r}\right)} - r^2(\mathrm{d}\theta)^2 - r^2\sin^2\theta\,(\mathrm{d}\phi)^2. \tag{5.18}$$

The **proper distance** between those two events will be given by $\mathrm{d}\sigma = \sqrt{-(\mathrm{d}s)^2}$.

We saw earlier, when discussing the spherical symmetry of the Schwarzschild
solution (see Subsection 5.2.1), that the events occurring at fixed values of t and r
form a spherical shell described by the familiar metric of such a shell. To this
extent the Schwarzschild spacetime can be regarded as consisting of a set of
nested spheres surrounding the spherically symmetric body. The proper distance
between neighbouring points on the sphere of coordinate radius r is given by

$$(\mathrm{d}\sigma)^2 = r^2(\mathrm{d}\theta)^2 + r^2\sin^2\theta\,(\mathrm{d}\phi)^2. \tag{5.19}$$

There is nothing unusual about the geometry of any of these spherical surfaces;
the sphere of coordinate radius r has proper circumference $2\pi r$ and proper area
$4\pi r^2$. In principle either of these quantities could be measured using ordinary
measuring rods. This provides a method, in principle at least, of determining
the Schwarzschild radial coordinate r of any event: use measuring sticks to
measure the proper circumference C of a circle centred on the origin that passes
through the location of the event, then divide that circumference by 2π to find the
coordinate radius $r = C/2\pi$.

What is unusual is that the radial coordinate r does not provide a direct measure
of the proper radius of such a sphere, and differences in the radial coordinate r do
not indicate the proper distance between different spherical shells. Consider
two events that occur at the same coordinate time and with the same angular
coordinates θ and ϕ but at different radial coordinates r and $r + \mathrm{d}r$. The proper
distance between those events will be

$$\mathrm{d}\sigma = \frac{\mathrm{d}r}{\left(1 - \frac{2GM}{c^2r}\right)^{1/2}}. \tag{5.20}$$

This equation shows that $\mathrm{d}\sigma$ is generally greater than $\mathrm{d}r$, provided that r is greater
than the Schwarzschild radius. The differences will be particularly large close to
the Schwarzschild radius (see Figure 5.5 overleaf). This result may be integrated
to determine the proper radial distance between any two events on the same radial
coordinate line.

Stretching a point, so to speak, the relation between coordinate distance and
proper distance can be inverted to show that the coordinate distance is contracted
relative to the proper distance. This could be described as 'gravitational length
contraction', but the comparison with the length contraction of special relativity is
very weak since $\mathrm{d}r$ is not really a 'physical' distance at all.

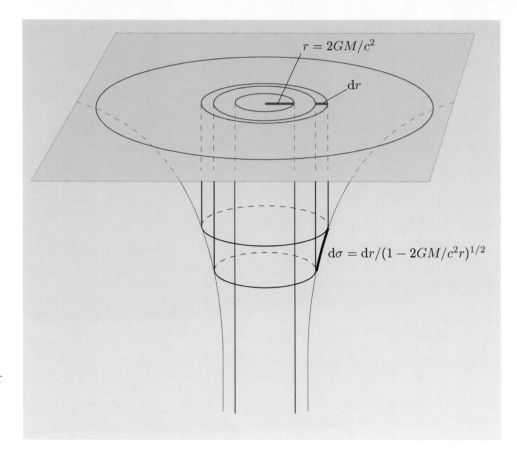

Figure 5.5 A schematic representation of the relation between the Schwarzschild radial coordinate and the proper distance for events close to the Schwarzschild radius $r = R_{\rm S} = 2GM/c^2$.

Exercise 5.4   Confirm that the proper distance around a circle (proper circumference) in the $\theta = \pi/2$ plane centred at $r = 0$ is $C = 2\pi r$, according to the Schwarzschild geometry.   ■

Proper distance

The Schwarzschild metric describes the spacetime around a static, spherically symmetric body as a set of nested spheres. The coordinate radius r of any one of those spheres can be determined by dividing its proper circumference by 2π.

Two events occurring at the same coordinate time and separated only by a radial coordinate distance dr will be separated by a proper radial distance

$$d\sigma = \frac{dr}{\left(1 - \frac{2GM}{c^2 r}\right)^{1/2}}. \qquad \text{(Eqn 5.20)}$$

5.4   Geodesic motion in Schwarzschild spacetime

According to the geodesic principle discussed in Chapter 4, the time-like and null geodesics of a spacetime represent the possible world-lines of massive and massless particles moving under the influence of gravity alone. Remember, a world-line is a pathway through spacetime, not just a trajectory through space. So once we know the world-line of a freely falling particle — i.e. once we know the specific geodesic that it moves along — we know everything about that particular particle's motion. In this section we examine some aspects of geodesic motion in the Schwarzschild spacetime around a static spherically symmetric body. We shall be particularly interested in motions relevant to astrophysics, so we shall be mainly concerned with orbital motion.

5.4.1 The geodesic equations

As you saw in Chapters 3 and 4, the geodesics of a spacetime are usually presented as parameterized curves, represented by four coordinate functions $x^\mu(\lambda)$, where λ is an affine parameter that varies along the geodesic. The choice of parameter is not completely arbitrary. In the case of a massive particle moving along a time-like geodesic, the affine parameter is usually taken to be the proper time τ that would be measured by a clock falling with the particle. It is also possible to use any linearly related parameter such as $a\tau + b$, where a and b are constants, though this would be unusual. These choices are not possible for a null geodesic since $d\tau = ds/c = 0$ for each of its elements, so some other affine parameter must be adopted. In either case the parameter is chosen to be an affine parameter since this ensures that the coordinate functions will satisfy geodesic equations of the relatively simple form

$$\frac{d^2 x^\mu}{d\lambda^2} + \sum_{\nu,\rho} \Gamma^\mu{}_{\nu\rho} \frac{dx^\nu}{d\lambda} \frac{dx^\rho}{d\lambda} = 0,$$

where the $\Gamma^\mu{}_{\nu\rho}$ are the connection coefficients that follow directly from the spacetime metric.

The general form of the non-zero connection coefficients was given in Section 5.1.2 at the start of the derivation of the Schwarzschild metric. Now that we know the explicit form of the Schwarzschild radius and the functions $A(r)$ and $B(r)$, we can write down the explicit form of all the non-zero connection coefficients:

$$\Gamma^0{}_{01} = \frac{GM}{r^2 c^2 \left(1 - \frac{2GM}{c^2 r}\right)} \ (= \Gamma^0{}_{10}),$$

$$\Gamma^1{}_{00} = \frac{GM \left(1 - \frac{2GM}{c^2 r}\right)}{r^2 c^2},$$

$$\Gamma^1{}_{11} = -\frac{GM}{r^2 c^2 \left(1 - \frac{2GM}{c^2 r}\right)},$$

$$\Gamma^1{}_{22} = -r \left(1 - \frac{2GM}{c^2 r}\right),$$

$$\Gamma^1{}_{33} = -r \left(1 - \frac{2GM}{c^2 r}\right) \sin^2 \theta,$$

$$\Gamma^2{}_{12} = \frac{1}{r} \ (= \Gamma^2{}_{21}),$$

$$\Gamma^2{}_{33} = -\sin\theta \cos\theta,$$

$$\Gamma^3{}_{13} = \frac{1}{r} \ (= \Gamma^3{}_{31}),$$

$$\Gamma^3{}_{23} = \cot\theta \ (= \Gamma^3{}_{32}).$$

Using these connection coefficients, the geodesic equations provide the following four differential equations that must be satisfied by the four coordinate functions $x^0 = ct(\lambda)$, $x^1 = r(\lambda)$, $x^2 = \theta(\lambda)$, $x^3 = \phi(\lambda)$ that describe any affinely parameterized geodesic in Schwarzschild spacetime:

$$\frac{\mathrm{d}^2 t}{\mathrm{d}\lambda^2} + \frac{2GM}{c^2 r^2 \left(1 - \frac{2GM}{c^2 r}\right)} \frac{\mathrm{d}r}{\mathrm{d}\lambda} \frac{\mathrm{d}t}{\mathrm{d}\lambda} = 0, \tag{5.21}$$

$$\frac{\mathrm{d}^2 r}{\mathrm{d}\lambda^2} + \frac{GM}{r^2} \left(1 - \frac{2GM}{c^2 r}\right) \left(\frac{\mathrm{d}t}{\mathrm{d}\lambda}\right)^2 - \frac{GM}{c^2 r^2 \left(1 - \frac{2GM}{c^2 r}\right)} \left(\frac{\mathrm{d}r}{\mathrm{d}\lambda}\right)^2$$
$$- r \left(1 - \frac{2GM}{c^2 r}\right) \left[\left(\frac{\mathrm{d}\theta}{\mathrm{d}\lambda}\right)^2 + \sin^2 \theta \left(\frac{\mathrm{d}\phi}{\mathrm{d}\lambda}\right)^2\right] = 0, \tag{5.22}$$

$$\frac{\mathrm{d}^2 \theta}{\mathrm{d}\lambda^2} + \frac{2}{r} \frac{\mathrm{d}r}{\mathrm{d}\lambda} \frac{\mathrm{d}\theta}{\mathrm{d}\lambda} - \sin \theta \cos \theta \left(\frac{\mathrm{d}\phi}{\mathrm{d}\lambda}\right)^2 = 0, \tag{5.23}$$

$$\frac{\mathrm{d}^2 \phi}{\mathrm{d}\lambda^2} + \frac{2}{r} \frac{\mathrm{d}r}{\mathrm{d}\lambda} \frac{\mathrm{d}\phi}{\mathrm{d}\lambda} + 2 \frac{\cos \theta}{\sin \theta} \frac{\mathrm{d}\theta}{\mathrm{d}\lambda} \frac{\mathrm{d}\phi}{\mathrm{d}\lambda} = 0. \tag{5.24}$$

Given the initial location of a particle in Schwarzschild spacetime and the initial values of the four components of its tangent vector $t^\mu = \mathrm{d}x^\mu/\mathrm{d}\lambda$, these four coupled, second-order, ordinary differential equations can be solved (numerically if not analytically) to determine the unique world-line of the particle. If the particle is massless, the magnitude of the initial tangent vector will be zero, showing the particle to be travelling at the speed of light, and the relevant world-line will turn out to be a null geodesic. For a particle with mass, the world-line will be a time-like geodesic.

As far as motion under gravity is concerned, the geodesic equations are the general relativistic analogues of Newton's second law of motion. Both sets of equations may be expressed as differential equations, and their solution allows initial data to be used to predict subsequent motion. However, as you can see, the geodesic equations look formidable and can be very difficult to solve. Because of their difficulty we shall not attempt a direct solution in this case. There are simplifying techniques that can be used based on the Lagrangian approach introduced when we first derived the geodesic equations in Chapter 3, but those methods are beyond the level of this book. Instead, we shall take a lesson from Newtonian mechanics, where problems involving motion are often simplified by making use of constants of the motion such as energy and angular momentum.

Exercise 5.5   Confirm the form of the first of the four geodesic equations given above.   ∎

5.4.2   Constants of the motion in Schwarzschild spacetime

To start, we recall that when geodesics were first introduced we described them as parameterized curves defined by $x^\mu(\lambda)$ with the particular property that the tangent vector $\mathrm{d}x^\mu/\mathrm{d}\lambda$ at any point remained parallel to itself under parallel transport. (This was a property that they shared with straight lines in a flat space.) Choosing the parameter λ to be an affine parameter ensures that as the tangent vector is transported along the geodesic, it not only remains self-parallel but also has a constant magnitude (more properly called a *norm* in this context). The square of that norm at every point on the geodesic is given by

$$n^2 = \sum_{\mu,\nu} g_{\mu\nu} \frac{\mathrm{d}x^\mu}{\mathrm{d}\lambda} \frac{\mathrm{d}x^\nu}{\mathrm{d}\lambda} = \text{constant}, \tag{5.25}$$

and will be zero in the case of a null geodesic.

If we regard the geodesic as the world-line of a massive particle and choose to use the proper time τ (as measured by a clock falling with the particle) as the parameter λ, then the tangent vector components $\mathrm{d}x^\mu/\mathrm{d}\lambda$ become $\mathrm{d}x^\mu/\mathrm{d}\tau$ and are seen to be the components of the particle's four-velocity $[U^\mu]$. Now, for the four-velocity of a massive particle,

$$\sum_{\mu,\nu} g_{\mu\nu} U^\mu U^\nu = c^2. \tag{5.26}$$

So in this case the constant n^2 in Equation 5.25 will be given by $n^2 = c^2$, and we can use our explicit knowledge of the Schwarzschild metric coefficients $g_{\mu\nu}$ to expand Equation 5.25 as

$$c^2 = c^2 \left(1 - \frac{2GM}{c^2 r}\right) \left(\frac{\mathrm{d}t}{\mathrm{d}\tau}\right)^2 - \left(1 - \frac{2GM}{c^2 r}\right)^{-1} \left(\frac{\mathrm{d}r}{\mathrm{d}\tau}\right)^2$$
$$- r^2 \left(\frac{\mathrm{d}\theta}{\mathrm{d}\tau}\right)^2 - r^2 \sin^2\theta \left(\frac{\mathrm{d}\phi}{\mathrm{d}\tau}\right)^2. \tag{5.27}$$

This still looks complicated, but apart from $n^2 = c^2$ there are four other **constants of the motion** that can help to simplify Equation 5.27. There are many ways of deducing these four conserved quantities, most of them drawing on the symmetry of the Schwarzschild solution. There are deep connections between symmetries and conservation laws throughout physics, so it is not surprising that the many symmetries of the Schwarzschild solution should give rise to conserved quantities in this case. In particular, we noted earlier that the static nature of the Schwarzschild solution indicates a symmetry associated with invariance under translation in time. This kind of symmetry is generally associated with the conservation of energy. Similarly, the solution's invariance under rotations about the origin indicates spherical symmetry, and is associated with the conservation of angular momentum.

In the specific context of a freely falling body of non-zero mass m, moving along a time-like geodesic in Schwarzschild spacetime, the conserved quantity that plays the role of total energy (actually the energy per unit mass energy) is

$$\frac{E}{mc^2} = \left(1 - \frac{2GM}{c^2 r}\right) \frac{\mathrm{d}t}{\mathrm{d}\tau}. \tag{5.28}$$

When dealing with the analogue of angular momentum, which is a vector, there are three conserved scalar quantities. These are most conveniently regarded as the magnitude of the angular momentum per unit mass, J/m, and two angles that determine the direction of the angular momentum vector. In practice, rather than dealing with whatever direction the angular momentum actually has, it is usually easier to transform the coordinates so that the angular momentum points along the polar axis, with the consequence that the motion is confined to the plane in which

$\theta = \pi/2$ and consequently $\mathrm{d}\theta/\mathrm{d}t = 0$. So, without any real loss of generality, two of the three constants of the motion associated with angular momentum are represented by the single condition

$$\theta = \pi/2, \tag{5.29}$$

while the third turns out to be

$$\frac{J}{m} = r^2 \sin^2\theta \, \frac{\mathrm{d}\phi}{\mathrm{d}\tau}. \tag{5.30}$$

Take care to note that the quantities E/mc^2 and J/m are specific to the Schwarzschild metric; they do not represent general definitions that can automatically be applied to other cases. If we now use Equations 5.28, 5.29 and 5.30 to simplify Equation 5.27, we see that

$$c^2 = \frac{E^2}{m^2 c^2}\left(1 - \frac{2GM}{c^2 r}\right)^{-1} - \left(1 - \frac{2GM}{c^2 r}\right)^{-1}\left(\frac{\mathrm{d}r}{\mathrm{d}\tau}\right)^2 - \frac{J^2}{m^2 r^2}. \tag{5.31}$$

Rearranging this gives

$$\left(\frac{\mathrm{d}r}{\mathrm{d}\tau}\right)^2 + \frac{J^2}{m^2 r^2}\left(1 - \frac{2GM}{c^2 r}\right) - \frac{2GM}{r} = c^2\left[\left(\frac{E}{mc^2}\right)^2 - 1\right]. \tag{5.32}$$

This equation, which already incorporates the general relativistic analogues of energy conservation and angular momentum conservation, describes the changes in the radial position coordinate with proper time for a freely falling particle of non-zero mass moving in the equatorial plane $\theta = \pi/2$. The phrase 'freely falling' can give the impression that the particle is plummeting radially inwards towards the central body. That is a possible form of freely falling motion, but not the only one. All 'freely falling' really means is that the motion is determined by gravity alone. In this sense the Moon is (very nearly) freely falling around the Earth and the Earth is (very nearly) freely falling around the Sun. So Equation 5.32 holds the key to describing orbital motion about the central massive body in Schwarzschild spacetime, and that is how we shall use it in the next subsection. Before doing that, however, let's see how Equation 5.32 together with the definitions contained in Equations 5.28 and 5.30 *can* be used to solve a problem involving purely radial motion.

Worked Example 5.1

Show that in Schwarzschild spacetime, the motion of a test particle in radial free fall (i.e. directly towards $r = 0$) satisfies the relation

$$\frac{\mathrm{d}^2 r}{\mathrm{d}\tau^2} = -\frac{GM}{r^2}.$$

Solution

To determine the equation of motion for a freely falling body travelling along a radial geodesic, we can use Equation 5.32, together with the supplementary Equations 5.28 and 5.30 that define E and J. In the case of purely radial motion ϕ is constant, so $\mathrm{d}\phi/\mathrm{d}\tau = 0$, and Equation 5.30 shows that $J = 0$. Equation 5.32 therefore reduces to

$$\left(\frac{\mathrm{d}r}{\mathrm{d}\tau}\right)^2 = c^2\left[\left(\frac{E}{mc^2}\right)^2 - 1\right] + \frac{2GM}{r}.$$

Differentiating with respect to τ gives

$$2\left(\frac{\mathrm{d}r}{\mathrm{d}\tau}\right)\frac{\mathrm{d}^2r}{\mathrm{d}\tau^2} = -\frac{2GM}{r^2}\frac{\mathrm{d}r}{\mathrm{d}\tau},$$

and dividing through by $\mathrm{d}r/\mathrm{d}\tau$ gives

$$\frac{\mathrm{d}^2r}{\mathrm{d}\tau^2} = -\frac{GM}{r^2},$$

as required.

The result that has just been derived in this worked example looks very much like the corresponding Newtonian result for free fall under the gravitational pull of a spherically symmetric mass in Euclidean space. Note, however, the several differences between the general relativistic result and its Newtonian counterpart. In the first place, talking about free fall under gravity is fine in general relativity, but talking of the 'pull' of gravity or gravitational 'attraction' would be quite wrong since there is no gravitational 'force' in general relativity, and even the term gravitational 'field' only retains a meaning when interpreted in terms of the metric coefficients, which can vary from place to place. Similarly, the Newtonian result directly relates the second derivative of the radial distance with respect to time to the inverse square of the radial distance, but in the general relativistic result the second derivative is with respect to proper time τ, and r is the coordinate distance, not the 'physical' proper distance. In the Newtonian limit, when $\mathrm{d}r/\mathrm{d}\tau \ll c$ and the particle is sufficiently far from the spherical mass for the field to be weak, these differences vanish, and the general relativistic result does reduce to the Newtonian result. This shows how Einstein's theory of motion under gravity encompasses Newton's theory and reduces to it under appropriate conditions. Nonetheless, away from the Newtonian limit, especially when close to the Schwarzschild radius, the differences are real and significant.

To summarize, we have the following.

Freely falling motion in Schwarzschild spacetime

The motion of a particle of mass m falling freely in the $\theta = \pi/2$ plane of a Schwarzschild spacetime is described by the **radial motion equation**

$$\left(\frac{\mathrm{d}r}{\mathrm{d}\tau}\right)^2 + \frac{J^2}{m^2r^2}\left(1 - \frac{2GM}{c^2r}\right) - \frac{2GM}{r} = c^2\left[\left(\frac{E}{mc^2}\right)^2 - 1\right], \quad \text{(Eqn 5.32)}$$

where τ is the proper time as would be measured by a clock falling with the particle, and the constants of the motion, E/mc^2 and J/m, the Schwarzschild analogues of energy per unit mass energy and angular momentum magnitude per unit mass, are determined by

$$\frac{E}{mc^2} = \left(1 - \frac{2GM}{c^2r}\right)\frac{\mathrm{d}t}{\mathrm{d}\tau}, \qquad \text{(Eqn 5.28)}$$

$$\frac{J}{m} = r^2\sin^2\theta\,\frac{\mathrm{d}\phi}{\mathrm{d}\tau}. \qquad \text{(Eqn 5.30)}$$

5.4.3 Orbital motion in Schwarzschild spacetime

The shape of an orbit in the $\theta = \pi/2$ plane of Schwarzschild spacetime is described by expressing r as a function of ϕ. In the previous subsection we developed a differential equation relating r to τ; we now need to convert that into a tractable relation between r and ϕ, and then investigate its solution. We start by noting that

$$\frac{\mathrm{d}r}{\mathrm{d}\tau} = \frac{\mathrm{d}\phi}{\mathrm{d}\tau}\frac{\mathrm{d}r}{\mathrm{d}\phi}, \tag{5.33}$$

and then use the fact that $J/m = r^2\,\mathrm{d}\phi/\mathrm{d}\tau$, in the plane $\theta = \pi/2$, to eliminate $\mathrm{d}\phi/\mathrm{d}\tau$, giving

$$\frac{\mathrm{d}r}{\mathrm{d}\tau} = \frac{J}{r^2 m}\frac{\mathrm{d}r}{\mathrm{d}\phi}. \tag{5.34}$$

Substituting this result into Equation 5.32 gives

$$\left(\frac{\mathrm{d}r}{\mathrm{d}\phi}\right)^2 + r^2\left(1 - \frac{2GM}{c^2 r}\right) - m^2 r^3 \frac{2GM}{J^2}$$
$$= \left(\frac{r^2 mc}{J}\right)^2\left[\left(\frac{E}{mc^2}\right)^2 - 1\right]. \tag{5.35}$$

Now we apply a standard 'trick' of orbital analysis by introducing the reciprocal variable $u = 1/r$ (so that $\mathrm{d}r/\mathrm{d}\phi = -(1/u^2)\mathrm{d}u/\mathrm{d}\phi$), and rewrite this equation as

$$\left(\frac{\mathrm{d}u}{\mathrm{d}\phi}\right)^2 + u^2 = \left(\frac{mc}{J}\right)^2\left[\left(\frac{E}{mc^2}\right)^2 - 1\right] + \frac{2GMum^2}{J^2} + \frac{2GMu^3}{c^2}.$$

Differentiating with respect to ϕ and dividing the resulting equation by $\mathrm{d}u/\mathrm{d}\phi$ gives the **orbital shape equation** that we need.

Orbital shape equation

$$\frac{\mathrm{d}^2 u}{\mathrm{d}\phi^2} + u = \frac{GMm^2}{J^2} + \frac{3GMu^2}{c^2}. \tag{5.36}$$

It is informative to compare this result with the analogous result from Newtonian mechanics for orbits around a massive spherically symmetric body. In the Newtonian case the result is

$$\frac{\mathrm{d}^2 u}{\mathrm{d}\phi^2} + u = \frac{GMm^2}{J^2}. \tag{5.37}$$

This is the same as the Schwarzschild expression, apart from the absence of the final relativistic term $3GMu^2/c^2$. That additional term will vanish in the limit as u approaches zero, showing that as long as r is sufficiently large, the Newtonian orbits will be recovered from the relativistic orbit equation, as they should be. Of course, for 'small' values of r (meaning close to $2GM/c^2$), the value of u will be large and the additional term will not be negligible. There will then be significant differences between the Newtonian and relativistic behaviours.

Additional insight into the behaviour of orbits comes from a study of energy, so it is useful here to rewrite the radial motion equation (Equation 5.32) that we developed in the previous subsection in a way that emphasizes the role of energy:

$$\frac{c^2}{2}\left[\left(\frac{E}{mc^2}\right)^2 - 1\right] = \frac{1}{2}\left(\frac{dr}{d\tau}\right)^2 + \frac{J^2}{2m^2r^2}\left(1 - \frac{2GM}{c^2r}\right) - \frac{GM}{r}. \quad (5.38)$$

The quantity on the left is not an energy, but for a particle of given mass it is determined by the orbital energy. The expression on the right consists of a 'kinetic' term (proportional to $(dr/d\tau)^2$) added to a sum of terms that depend only on r for given values of J and m. This is sufficient to earn the sum of those r-dependent terms the name 'effective potential' and the symbol V_{eff}. Thus we can write

$$\frac{c^2}{2}\left[\left(\frac{E}{mc^2}\right)^2 - 1\right] = \frac{1}{2}\left(\frac{dr}{d\tau}\right)^2 + V_{\text{eff}}, \quad (5.39)$$

where

$$V_{\text{eff}} = \frac{J^2}{2m^2r^2}\left(1 - \frac{2GM}{c^2r}\right) - \frac{GM}{r}. \quad (5.40)$$

Now, a very similar equation arises in Newtonian orbital analysis, where the constant orbital energy E^{Newton} is given by

$$\frac{E^{\text{Newton}}}{m} = \frac{1}{2}\left(\frac{dr}{dt}\right)^2 + V_{\text{eff}}^{\text{Newton}}, \quad (5.41)$$

with

$$V_{\text{eff}}^{\text{Newton}} = \frac{J^2}{2m^2r^2} - \frac{GM}{r}. \quad (5.42)$$

The Newtonian and Schwarzschild effective potentials for a positive value of J are shown in Figure 5.6. In the Newtonian case the angular momentum magnitude J is the source of an infinite 'effective potential barrier' that prevents particles with non-zero angular momentum magnitude from reaching $r = 0$. In the Schwarzschild case the behaviour at small values of r is quite different. Indeed, for sufficiently small values of J there is no barrier at all.

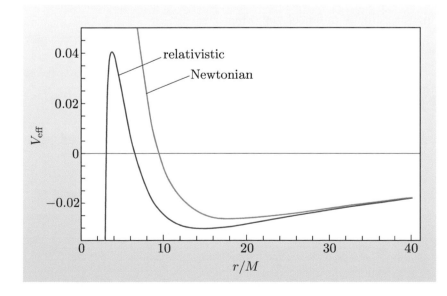

Figure 5.6 Effective potentials for orbital motion with fixed angular momentum magnitude J in Newtonian gravity and general relativity.

The difference between the Newtonian and Schwarzschild effective potentials comes from the extra term $-GMJ^2/m^2c^2r^3$ in the Schwarzschild case. One of its effects is to cause the orbits of particles to rotate in the $\theta = \pi/2$ plane. This effect is negligible at large values of r but significant for small values, preventing elliptical orbits from closing and causing them to follow the kind of rosette pattern shown in Figure 5.7. This is another effect with astronomically observable consequences to which we shall return in Chapter 7.

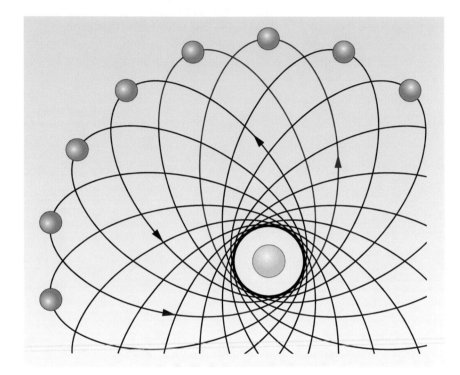

Figure 5.7   The rosette orbit created by rotating a nearly elliptical orbit in its own plane. Part of the path is coloured to clarify the motion.

Exercise 5.6   Both Newtonian and Schwarzschild orbital dynamics allow stable circular orbits to exist at large values of r, but in the Schwarzschild case there is a lower limit to the radius of a stable circular orbit that corresponds to $J/m = 2\sqrt{3}GM/c$.

(a) What is the (coordinate) radius of that orbit?

(b) What is the corresponding value of the parameter E?   ∎

Summary of Chapter 5

1. The Schwarzschild metric tensor is

$$[g_{\mu\nu}] = \begin{pmatrix} 1 - \frac{2GM}{c^2r} & 0 & 0 & 0 \\ 0 & -\frac{1}{1 - \frac{2GM}{c^2r}} & 0 & 0 \\ 0 & 0 & -r^2 & 0 \\ 0 & 0 & 0 & -r^2\sin^2\theta \end{pmatrix}, \quad \text{(Eqn 5.1)}$$

though the term 'Schwarzschild metric' is more often applied to the
corresponding line element

$$(\mathrm{d}s)^2 = \left(1 - \frac{2GM}{c^2r}\right)c^2(\mathrm{d}t)^2 - \frac{(\mathrm{d}r)^2}{1 - \frac{2GM}{c^2r}}$$
$$- r^2(\mathrm{d}\theta)^2 - r^2\sin^2\theta\,(\mathrm{d}\phi)^2. \qquad \text{(Eqn 5.2)}$$

2. The Schwarzschild metric coefficients provide a solution of the Einstein vacuum field equations $R_{\mu\nu} - g_{\mu\nu}R/2 = 0$ in the empty region of spacetime surrounding a non-rotating spherically symmetric body of fixed mass M.

3. The solution is spherically symmetric (having the invariance of a spherical shell), asymptotically flat (approaching the Minkowski metric in spherical polar coordinates at large r), stationary (having metric coefficients that are time-independent) and static (having a line element that is invariant under time reversal).

4. The solution is singular, approaching infinity as $r \to R_S = 2GM/c^2$, the Schwarzschild radius, and as $r \to 0$. The first of these is a coordinate singularity that can be transformed away by an appropriate choice of coordinates; the second is a gravitational singularity that is present in curvature-related invariants and cannot be transformed away. Neither singularity is within the region described by the solution for normal 'star-like' bodies.

5. The solution has great generality, Birkhoff's theorem showing that it applies to the exterior region of any distribution of energy and momentum characterized by mass M that produces purely isotropic effects in that region.

6. The Schwarzschild coordinates t, r, θ, ϕ lack immediate metrical significance. Infinitesimal differences in coordinate time $(\mathrm{d}t)$ and coordinate radial distance $(\mathrm{d}r)$ may be related to infinitesimal differences in measurable proper time $(\mathrm{d}\tau)$ and measurable proper distance $(\mathrm{d}\sigma)$ using the Schwarzschild metric. Finite intervals of proper time and proper distance may be determined by performing appropriate integrals involving the infinitesimal intervals.

7. When considering observations of events in general relativity, the location of the observer is significant as well as the observer's state of motion. When considering events in Schwarzschild spacetime, three observers are commonly mentioned; a local stationary observer at fixed Schwarzschild coordinates, a local freely falling observer, and a distant observer (at $r = \infty$), who may be regarded as freely falling while stationary and whose own 'local' observations concern sightings of the events.

8. Physical meaning can be associated with Schwarzschild coordinates based on the observations that (a) the difference in coordinate time between two events at the same coordinate position is equal to the measurable proper time between sightings of those events by a stationary observer at infinity, and (b) a circle centred on the origin with fixed coordinate radius r has the measurable proper circumference $C = 2\pi r$.

9. Two events that occur at the same coordinate time and with the same angular coordinates, but separated by a coordinate radial distance $\mathrm{d}r$ will, according

to a local stationary observer, be separated by a proper distance

$$d\sigma = \frac{dr}{\left(1 - \frac{2GM}{c^2 r}\right)^{1/2}}. \qquad \text{(Eqn 5.20)}$$

Similarly, two events that occur at the same coordinate position but separated by coordinate time interval dt will, according to a local stationary observer, be separated by a proper time

$$d\tau = \left(1 - \frac{2GM}{c^2 r}\right)^{1/2} dt.$$

10. Due to gravitational time dilation, a clock at rest at radial coordinate r, with ticks of proper duration $d\tau_r$, will be seen to have ticks of longer duration $d\tau_\infty = d\tau_r/(1 - 2GM/rc^2)^{1/2}$ by a stationary distant observer. This implies the existence of an observable gravitational redshift in which a source emitting radiation of proper frequency f_{em} located at fixed radial coordinate r_{em} is seen by a stationary distant observer to have frequency

$$f_\infty = f_{em} \left(1 - \frac{2GM}{c^2 r_{em}}\right)^{1/2}. \qquad \text{(Eqn 5.17)}$$

11. Equations describing the possible world-lines of freely falling massive and massless particles as time-like and null geodesics may be deduced from the geodesic equations applied to Schwarzschild spacetime. The world-line of a specific particle will be determined by the initial position and velocity of that particle. However, for the study of orbital motion it is simpler to consider the quantities that represent constants of the motion, including the norm of the tangent vector, the (generalized) orbital energy and the (generalized) orbital angular momentum.

12. For a freely falling particle of mass m following a geodesic parameterized by the proper time τ (as measured by a co-moving freely falling clock), the conserved total orbital energy per unit mass energy is $E/mc^2 = (1 - 2GM/c^2 r)(dt/d\tau)$ and the conserved orbital angular momentum magnitude per unit mass is $J/m = r^2 \sin^2 \theta \,(d\phi/d\tau)$. In the case of motion in the equatorial plane ($\theta = \pi/2$), the radial motion is described by

$$\left(\frac{dr}{d\tau}\right)^2 + \frac{J^2}{m^2 r^2}\left(1 - \frac{2GM}{c^2 r}\right) - \frac{2GM}{r} = c^2\left[\left(\frac{E}{mc^2}\right)^2 - 1\right] \quad \text{(Eqn 5.32)}$$

while the orbital shape is described using the reciprocal variable $u = 1/r$ by

$$\frac{d^2 u}{d\phi^2} + u = \frac{GMm^2}{J^2} + \frac{3GMu^2}{c^2}. \qquad \text{(Eqn 5.36)}$$

13. At large values of r, far from the central body, the orbits of massive particles approach their Newtonian analogues. At smaller values of r, differences from Newtonian behaviour include the absence of an 'angular momentum barrier' preventing particles with non-zero angular momentum magnitude from reaching $r = 0$, the absence of stable circular orbits with $r < 6GM/c^2$, and the failure of 'elliptical' orbits to close due to a rotation of the ellipse in the orbital plane. These differences can be associated with the action of an additional term in the Schwarzschild 'effective potential' that governs the radial motion in the relativistic case.

Chapter 6 Black holes

Introduction

Black holes are believed to be among the most exotic objects in the Universe. They are regions of spacetime distorted by the gravitational effects of bodies such as collapsed stars to such an extent that light itself is unable to escape.

The study of black holes and their associated astrophysical properties has become an enormous subject. In this chapter we shall address only some of the key points. We start with a wide-ranging section that contains some basic definitions, a brief history of the subject, and a classification of the various types of black hole. We then devote one section to non-rotating black holes and another to rotating black holes. Finally, in Section 6.4, we go beyond the 'classical' black holes of general relativity to discuss some possible implications of quantum physics for black holes, particularly the proposal that quantum physics allows black holes to be sources of radiation. Throughout the discussion there will be references to possible astronomical evidence of black holes, but that subject will be further discussed in Chapter 7, which concerns the testing of general relativity by experiment and observation.

6.1 Introducing black holes

The term 'black hole' was not introduced until the 1960s, though the basic concept can be traced back much further and has its roots in the Schwarzschild solution that was introduced in the previous chapter. We shall begin with some informal definitions and a brief historical survey that will trace the tangled history and even the pre-history of black holes.

6.1.1 A black hole and its event horizon

In general relativity, a **black hole** is a region of spacetime that matter and radiation may enter but from which they cannot escape. It's a 'hole' because matter and radiation can fall into it. It's 'black' because light is unable to escape from it.

Note that a black hole is essentially a spacetime structure, not a material one. This makes it very different from more familiar astronomical bodies, such as stars and planets, which are primarily composed of matter. Also note that our characterization of a black hole implies that it must be bounded by some kind of closed surface that will allow light to enter, but not to leave again. This light-trapping 'one-way' surface is called an **event horizon** and will feature prominently in the discussions that follow.

In the case of the simplest kind of black hole, which is described by the Schwarzschild metric, the event horizon is located at the Schwarzschild radius $r = R_S = 2GM/c^2$ and may be thought of as a sphere, though it follows from what was said about coordinates and distances in the previous chapter that $2GM/c^2$ is its coordinate radius.

6.1.2 A brief history of black holes

Although the term 'black hole' took a long time to emerge, the story of black holes begins with the birth of general relativity and the Schwarzschild solution, both of which were published in 1916. However, long before that, in the context of Newtonian gravitation, there had already been speculations about the possibility of 'dark stars' — material bodies so dense that light would be unable to escape from them. The thinking behind this proposal was simple. If a projectile of mass m is launched from the surface of a spherical body of mass M and radius R, then in order to escape from the gravitational influence of that body the projectile must gain gravitational potential energy GMm/R. If this energy is to come from the projectile's initial kinetic energy at the time of launch, then the required launch speed, sometimes referred to as the **escape speed** v_{es}, is given by

$$\tfrac{1}{2}mv_{es}^2 = \frac{GMm}{R}. \tag{6.1}$$

The projectile mass m cancels, so the escape speed, independent of projectile mass, is

$$v_{es} = \sqrt{\frac{2GM}{R}}. \tag{6.2}$$

It follows from this that the escape speed v_{es} will be greater than the speed of light c if the radius and mass of the body are related by

$$R < \frac{2GM}{c^2}. \tag{6.3}$$

Such a body, it was speculated, would trap light and would therefore be dark.

These ideas, introduced independently by John Michell (1724–1793) and Pierre-Simon Laplace (1749–1827) in the eighteenth century, have very little to do with the black holes of general relativity, but they do show that the physical concept of gravitational light trapping is not new.

That idea was implicit in Schwarzschild's solution when it was developed in 1915, though that was not properly appreciated at the time. In fact, the familiar form of the Schwarzschild solution,

$$(\mathrm{d}s)^2 = \left(1 - \frac{2GM}{c^2r}\right)c^2(\mathrm{d}t)^2 - \frac{(\mathrm{d}r)^2}{1 - \frac{2GM}{c^2r}}$$
$$- r^2(\mathrm{d}\theta)^2 - r^2\sin^2\theta\,(\mathrm{d}\phi)^2, \tag{Eqn 5.2}$$

was introduced about a year later by the mathematician David Hilbert (1862–1943), but even this did not make clear the physical behaviour associated with events at $r = 2GM/c^2$. Additionally, the Schwarzschild radius of real bodies (3 km for a body with the mass of the Sun) was thought to be too small to be of any physical significance, so its physical nature did not receive much attention.

● Regarding the Earth (total mass 5.97×10^{24} kg) as a spherically symmetric body, what is its Schwarzschild radius?

○ For the Earth,

$$R_S = 2GM/c^2 = [2 \times 6.67 \times 10^{-11} \times 5.97 \times 10^{24}/(9.00 \times 10^{16})]\text{ m}$$
$$= 8.84 \times 10^{-3}\text{ m},$$

or about 9 mm.

It was pointed out during the 1920s that not all singularities in the metric $g_{\mu\nu}$ are physically significant; they could be a consequence of the coordinates being used rather than the physics being described. This opened up the possibility that bodies might be able to undergo a complete **gravitational collapse**, shrinking to a point of infinite density irrespective of any singular surfaces that got in their way, provided that those singularities were entirely due to the choice of coordinates. In the case of a spherically symmetric body, surrounded by empty space described by the Schwarzschild metric, the singularity associated with $r = R_S$ was eventually recognized as being a coordinate singularity, but this knowledge was slow to spread and the belief that the singularity was physical remained common at least until the late 1930s. In any case, planets were not sufficiently dense to undergo a complete gravitational collapse; the electrical repulsion between the atoms that they contained was sufficient to balance the gravitational tendency to collapse. Normal stars, such as the Sun, were also resistant to gravitational collapse. The plasma at the centre of the Sun is believed to be roughly ten times denser than lead, but even at these densities the thermal pressure resulting from energy releasing nuclear reactions (together with a contribution from radiation pressure) is sufficient to guarantee a star's equilibrium with a radius in the order of a million kilometres.

The astrophysics of highly evolved stellar bodies, in which nuclear reactions have ceased due to a lack of fuel, became a major topic in the 1930s. It had been suggested in the mid-1920s that the small dense stars known as **white dwarf stars** were supported against gravitational collapse by a **degeneracy pressure** arising from the quantum physics of the electrons that they contained. This idea was taken up by Subrahmanyan Chandrasekhar (Figure 6.1), an Indian theorist studying at the University of Cambridge. In 1931 he proposed that there was an upper limit (about 1.4 times the mass of the Sun) to the mass of any white dwarf supported by electron degeneracy pressure. If the star's mass exceeded that limit, gravity would overwhelm the degeneracy pressure and a gravitational collapse would ensue. Some were doubtful about Chandrasekhar's ideas, most notably the Cambridge-based astrophysicist Sir Arthur Eddington (1882–1944), who had been responsible for much of the foundational work on the internal constitution of stars. Working in the same university, Chandrasekhar came to know Eddington well and admired his work; Eddington's opposition was a professional and personal blow that caused Chandrasekhar to abandon his work on white dwarfs and move to the USA, though his ideas are now an accepted part of astrophysical theory and his insight was eventually rewarded with a Nobel prize for physics.

Another development came in 1932, the year in which the neutron was discovered. Very soon after hearing of the discovery, the Russian theoretical physicist Lev Landau (1908–1968) suggested the possibility of **neutron stars**, the outer parts of which would contain many neutron-rich nuclei while the inner parts (apart, perhaps, from an exotic core) would consist of a quantum fluid largely composed of neutrons. According to Landau, such a 'star' would be stabilized against gravitational collapse by the quantum degeneracy pressure of the neutron fluid. The quantum physics involved was similar to that at work in a white dwarf, but the greater mass of the neutron altered the details allowing neutron stars to be even denser — comparable to the density of an atomic nucleus. A white dwarf with the mass of the Sun was expected to have about a millionth of the Sun's volume, making it about the size of the Earth, with a radius of about $5000\,\text{km}$. A

Figure 6.1   Subrahmanyan Chandrasekhar (1910–1995) recognized the interplay of quantum physics and gravitation in limiting the mass of white dwarf stars. Spending most of his career at the University of Chicago, he worked on many aspects of astrophysics and wrote several books, including *The Mathematical Theory of Black Holes* (1983).

Figure 6.2   J. Robert Oppenheimer (1904–1967) was a leader of American theoretical physics in the 1930s. In 1942 he was appointed scientific director of the Manhattan Project and eventually became known as the father of the atomic bomb. He never resumed his research in relativistic astrophysics.

neutron star of similar mass should be much smaller, more like the size of a city, about 20 km across.

In 1939, J. Robert Oppenheimer (Figure 6.2) and collaborators showed that neutron stars, like white dwarfs, have a maximum mass (now estimated to be about 3 times the mass of the Sun). Above that limit they found nothing to prevent a star that has exhausted its nuclear fuel from undergoing a complete gravitational collapse. Using general relativity they showed that according to a distant observer, such a collapse would take an infinitely long time, the process appearing to slow and freeze as the shrinking surface approached the Schwarzschild radius, though the image would soon become dim and reddened. However, they also found that according to an observer falling with the collapsing stellar surface, there would be no such slowing, only a finite time being required to reach the central singularity. Passing within the Schwarzschild radius would be a natural part of such a fall — relatively uneventful for the falling observer, though actually marking a point of no return. Many regard this work, with its acceptance of complete gravitational collapse and recognition of the coordinate nature of the singularity at $r = R_S$, as the true birth of the black hole concept.

● What general relativistic effect should be expected to cause a distant observer's view of a collapsing star's surface to be reddened compared with the view of an observer falling with the surface?

○ Gravitational redshift will cause radiation emitted from the surface to have a smaller frequency (i.e. to be redder) according to a distant observer than according to an observer moving with the surface.

The 1940s and 1950s are generally regarded as a sterile time for general relativity. There were real achievements but the field faced difficult problems that some thought to be insurmountable, and there was a lack of relevant experimental information to check or challenge the existing theory. However, things began to change at the end of that period, setting the scene for a renaissance of general relativity in the 1960s that would revitalize the field and bring black holes into prominence.

In 1958, rediscovering a coordinate system first used by Eddington in the 1920s, the American mathematical physicist David Finkelstein (1929–) showed how the Schwarzschild metric could be partly freed of its coordinate singularity and used to discuss separately the inward and outward motion of photons in the neighbourhood of the Schwarzschild radius. Then, in 1960, Martin Kruskal (1925–2006) in the USA and George Szekeres (1911–2005) in Australia independently found a coordinate system that allowed a unified description of the Schwarzschild solution, free of coordinate singularities. Soon after came the first observations of peculiar star-like astronomical bodies that would later be given the name **quasars** (short for quasi-stellar objects) and would eventually be recognized as the highly active nuclei of remote but luminous galaxies. So prodigious was the outpouring of energy from quasars that many felt that they had to involve some kind of energy-generating mechanism that was quite different from the nuclear reactions that powered normal stars.

Over a relatively short period during the 1960s, the ideas of gravitational collapse and black holes underwent a rapid development that took them from the fringes to the centre of astrophysical thinking. In 1963 New Zealander Roy Kerr (1934–)

discovered the solution of the vacuum field equations that would later be used to describe realistic rotating black holes, just as the Schwarzschild metric would be used for non-rotating black holes. Roger Penrose (1931–) introduced the first of a number of singularity theorems showing that gravitational singularities were an inevitable consequence of complete gravitational collapse. A number of investigators suggested that the release of gravitational potential energy by matter (about 3 solar masses per year) falling into a compact object with a mass of about 10^8 solar masses could account for the energy emitted by quasars. It was in this fervid atmosphere that John Archibald Wheeler (Figure 6.3), who had been urging the field forward since the late 1950s, introduced the term 'black hole' in 1967. In 1969 the term 'event horizon' (which had been introduced some years earlier in a different context) was applied to the surface surrounding a gravitationally collapsed object that separated the events that might be seen by a distant observer from those that were forever cut off from such an observer. The black hole with its central singularity and surrounding event horizon had arrived.

Figure 6.3 John Archibald Wheeler (1911–2008) was a major contributor to the 1960s renaissance of general relativity. He was well known for coining and popularizing new terms (including black hole) and for providing memorable slogans that summarized complex issues.

Of course, many subsequent developments followed, but to the extent that we discuss them at all we shall treat them as they arise in the discussion below. Let us end this section with some words from Wheeler.

> The black hole epitomizes the revolution wrought by general relativity. It pushes to an extreme — and therefore tests to the limit — the features of general relativity (the dynamics of curved spacetime) that set it apart from special relativity (the physics of static, 'flat' spacetime) and the earlier mechanics of Newton.
>
> J.A. Wheeler (1998) *Geons, Black Holes & Quantum Foam*, Norton

6.1.3 The classification of black holes

The basis of the most common classification scheme for black holes is John Wheeler's pronouncement that 'a black hole has no hair'. What Wheeler meant by this was that a black hole has very few independent, externally measurable properties; namely, its mass, its angular momentum and its electric charge. All black holes must have mass, so there are only four basic types of black hole. An essentially unique metric is now known for each of those types, including the Schwarzschild metric for those with no charge and no angular momentum. The full four-fold classification scheme looks like this.

PROPERTIES	METRIC
Mass only	Schwarzschild
Mass and angular momentum	Kerr
Mass and electric charge	Reissner–Nordström
Mass, angular momentum and electric charge	Kerr–Newman

It is expected that real black holes will have angular momentum, but may well not be charged since atoms tend to be neutral. Because of this we shall discuss rotating and non-rotating black holes but we shall mainly ignore charged black holes.

Another widely used classification scheme for black holes is perhaps more relevant to astrophysics. It is based on the mass of the black hole. The mass limits of the various classes are not precisely defined and several authors have proposed new classes. Here is a version of the scheme.

CLASS	MASS RANGE
Mini black holes	0 to $0.1\,\mathrm{M}_\odot$
Stellar mass black holes	0.1 to $100\,\mathrm{M}_\odot$
Intermediate mass black holes	100 to $10^5\,\mathrm{M}_\odot$
Supermassive black holes	10^5 to $10^{10}\,\mathrm{M}_\odot$

Many authors who discuss mini black holes suppose them to have masses very much less than the mass of the Sun — less, say, than the mass of the Moon — and some have even discussed subdivisions such as micro black holes or nano black holes. However, given the rather imprecise nature of this classification scheme, we shall simply make do with the broad category of mini black holes.

● If the accretion of matter by a black hole, at the rate of a few solar masses per year, explains the luminosity of quasars, what kind of black hole would you expect to be responsible?

○ Real black holes are expected to be rotating and uncharged, so a Kerr black hole is most likely. Also, if the suggested rate of fuelling is to account for the observed energy release from quasars, the black hole would need to have a mass of order 10^8 solar masses, so it would be in the supermassive class.

To summarize, here are the main results of this section.

Black holes

A black hole is a region of spacetime that matter and radiation may enter but from which they may not escape. The region is bounded by an event horizon that separates events that can be seen by an external observer from those that cannot be seen. At the heart of a black hole is a singularity that may arise from the complete gravitational collapse of a star or some other body. The limiting masses of white dwarfs and neutron stars indicate the possibility of gravitational collapse, but the consequences were first investigated in detail by Oppenheimer and his collaborators. The term black hole was introduced by Wheeler in the 1960s when there was a renaissance in the study of general relativity, partly inspired by the need to account for the prodigious energy output from quasars. Black holes are commonly classified according to their mass or according to the solution of the vacuum field equations that describes them. The only independent externally measurable properties of a black hole are its mass, charge and angular momentum.

6.2   Non-rotating black holes

As pointed out in Chapter 5, Birkhoff's theorem establishes the uniqueness of the Schwarzschild solution in describing the spacetime external to a source that has spherically symmetric effects. So, whether discussing the spherically symmetric collapse of a non-rotating star or the spherically symmetric black hole that might be expected to result from such a collapse, the Schwarzschild solution will play a central role.

In this section we shall return to a number of the topics that were introduced in Chapter 5 but our concern will be mainly with events at or around the Schwarzschild radius, which will turn out to be the location of the event horizon of a non-rotating black hole. Since it is described by the Schwarzschild metric, we shall sometimes refer to a non-rotating black hole as a **Schwarzschild black hole**. We shall see some further consequences of the lack of immediate metrical significance of the Schwarzschild coordinates, ct and r, and give further thought to the implications of geodesic motion, including the motion of photons, which we largely ignored earlier.

To start with we shall follow in the footsteps of Oppenheimer and his collaborators by considering the proper time taken for a freely falling observer to reach the central singularity of a Schwarzschild spacetime.

6.2.1 Falling into a non-rotating black hole

In Worked Example 5.1 we showed that in Schwarzschild spacetime the radial motion of a freely falling body with non-zero mass agreed with Newtonian expectations provided that (i) the speed of the body is much less than c, and (ii) the gravitational field is weak (i.e. there is negligible spacetime curvature). Let us now consider the behaviour of a radially falling body that violates these conditions by passing though the event horizon and travelling on towards $r = 0$. As in the worked example, our starting point is the radial motion equation (Equation 5.32) with $J = 0$, but we can use $R_\mathrm{S} = 2GM/c^2$ to write it in the form

$$\left(\frac{\mathrm{d}r}{\mathrm{d}\tau}\right)^2 = c^2 \left[\left(\frac{E}{mc^2}\right)^2 - 1 + \frac{R_\mathrm{S}}{r}\right].$$

The constant E represents the energy, the value of which is determined by the initial conditions. On this occasion we shall suppose that the fall starts from rest at some large value of r which we shall denote r_0, so $\mathrm{d}r/\mathrm{d}\tau = 0$ when $r = r_0$ and

$$\left(\frac{E}{mc^2}\right)^2 = 1 - \frac{R_\mathrm{S}}{r_0}. \tag{6.4}$$

It follows that

$$\left(\frac{\mathrm{d}r}{\mathrm{d}\tau}\right)^2 = c^2 R_\mathrm{S} \left[\frac{1}{r} - \frac{1}{r_0}\right].$$

Taking the negative square root to describe inward motion (r decreasing as τ increases),

$$\frac{\mathrm{d}r}{\mathrm{d}\tau} = -c\sqrt{R_\mathrm{S}}\sqrt{\frac{1}{r} - \frac{1}{r_0}} = -c\sqrt{R_\mathrm{S}}\sqrt{\frac{r_0 - r}{rr_0}}. \tag{6.5}$$

Taking the reciprocal, we can rewrite this as

$$\frac{\mathrm{d}\tau}{\mathrm{d}r} = -\frac{1}{c}\sqrt{\frac{r_0}{R_\mathrm{S}}}\sqrt{\frac{r}{r_0 - r}}. \tag{6.6}$$

Integrating both sides with respect to r, from the starting point r_0 to some general point r', gives the proper duration of the fall as

$$\tau(r') - \tau(r_0) = -\frac{1}{c}\sqrt{\frac{r_0}{R_\mathrm{S}}} \int_{r_0}^{r'} \sqrt{\frac{r}{r_0 - r}}\,\mathrm{d}r.$$

The integral can be found in tables of standard integrals or (with appropriate caution) using an algebraic computing package. It turns out that

$$\tau(r') - \tau(r_0)$$

$$= \frac{r_0}{c}\sqrt{\frac{r_0}{R_S}}\left[\sqrt{\frac{r}{r_0}\left(1 - \frac{r}{r_0}\right)} + \arctan\left(-\sqrt{\frac{r}{r_0 - r}}\right)\right]_{r_0}^{r'}.$$

Substituting the appropriate limits we see that

$$\tau(r') - \tau(r_0)$$

$$= \frac{r_0}{c}\sqrt{\frac{r_0}{R_S}}\left[\frac{\pi}{2} + \sqrt{\frac{r'}{r_0}\left(1 - \frac{r'}{r_0}\right)} + \arctan\left(-\sqrt{\frac{r'}{r_0 - r'}}\right)\right].$$

For the case we are interested in, when $r_0 \gg r'$, expanding the functions on the right in power series leads to the approximation

$$\tau(r') - \tau(r_0) \approx \frac{r_0}{c}\left(\frac{r_0}{R_S}\right)^{1/2}\left[\frac{\pi}{2} - \frac{2}{3}\left(\frac{r'}{r_0}\right)^{3/2}\right]. \tag{6.7}$$

If we allow the general point r' to approach the central singularity by considering the limit $r' \to 0$, we find that the total proper time for the fall is finite and has value

$$\tau_{\text{sing}} = \frac{\pi r_0^{3/2}}{2cR_S^{1/2}}. \tag{6.8}$$

Another significant result that also follows from Equation 6.7 is the proper time required to fall from r_0 to the event horizon at $r' = R_S$. The result is

$$\tau_{\text{horiz}} - \frac{r_0^{3/2}}{cR_S^{1/2}}\left[\frac{\pi}{2} - \frac{2}{3}\left(\frac{R_S}{r_0}\right)^{3/2}\right]. \tag{6.9}$$

The difference between these last two results is the proper time required for the freely falling body to travel from the horizon to the singularity, which is just

$$\tau_{\text{sing}} - \tau_{\text{horiz}} = \frac{2}{3}\frac{R_S}{c}. \tag{6.10}$$

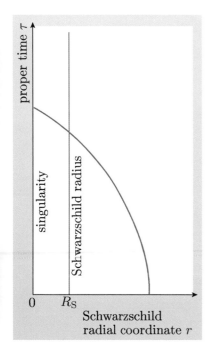

Figure 6.4 The relationship between proper time τ and radial coordinate r for a body falling freely into a black hole of Schwarzschild radius R_S.

The motion of this falling body is indicated in Figure 6.4, where the coordinate position is plotted against proper time as measured by a freely-falling observer travelling with the body. The key points to note are as follows:

Falling into a non-rotating black hole

A body released from rest at a large distance from a non-rotating black hole requires only a finite proper time to reach the central singularity.

Nothing unusual happens at the Schwarzschild radius.

Exercise 6.1   (a) What is the proper time required for a falling body to travel from the Schwarzschild radius to the singularity of a black hole with 3 times the mass of the Sun?

(b) What is the corresponding proper travel time for a fall from the horizon to the singularity of a supermassive black hole of mass $10^9\ \mathrm{M_\odot}$? ∎

6.2.2 Observing a fall from far away

For a distant stationary observer, at rest far from the origin, there is no essential difference between the proper time that would be measured on a clock and the coordinate time t. To avoid confusion with the proper time τ recorded by the freely falling observer, we shall always use t when discussing observations made by the distant observer.

The first thing that we need to know is how long it takes for a light signal emitted by the freely falling body to reach the distant observer. To be specific we shall suppose that the distant observer is located along the same radial line that the falling body is moving along, simply further out. That means we only have to consider photons that travel radially from the falling body to the distant observer. For events along the path of such a photon, $\mathrm{d}\theta = \mathrm{d}\phi = 0$. We already know that the spacetime separation $(\mathrm{d}s)^2$ of events on a photon's world-line is zero, so it follows from the Schwarzschild metric that for two events on the world-line of a photon travelling radially outwards,

$$0 = \left(1 - \frac{R_S}{r}\right) c^2 (\mathrm{d}t)^2 - \frac{(\mathrm{d}r)^2}{1 - R_S/r}. \tag{6.11}$$

Rearranging and taking square roots, we see that for radially moving photons,

$$\frac{\mathrm{d}t}{\mathrm{d}r} = \pm \frac{1}{c} \frac{1}{1 - R_S/r}, \tag{6.12}$$

where the $-$ sign applies to photons travelling radially inwards ($\mathrm{d}r$ deceasing) while the $+$ sign applies to the outward-moving photons that interest us. This relation holds true for neighbouring events all along the world-line of the photon, so for a photon emitted from the falling body at t_1 and r_1 that is observed by the distant observer at t_2 and r_2, the total journey time is given by

$$t_2 - t_1 = \int_{t_1}^{t_2} \mathrm{d}t = \frac{1}{c} \int_{r_1}^{r_2} \frac{\mathrm{d}r}{1 - R_S/r}. \tag{6.13}$$

Evaluating the integral gives

$$t_2 - t_1 = \frac{r_2 - r_1}{c} + \frac{R_S}{c} \ln\left(\frac{r_2 - R_S}{r_1 - R_S}\right). \tag{6.14}$$

There are three important points to note about Equation 6.14.

First, the coordinate time interval is not simply $(r_2 - r_1)/c$. This, of course, is because the coordinates lack immediate metrical significance, especially close to the Schwarzschild radius.

Second, the journey time is always greater than $(r_2 - r_1)/c$ due to the additional logarithmic term. As the point of emission, r_1, gets closer and closer to the Schwarzschild radius, this logarithmic term becomes larger and larger. Indeed, as $r_1 \to R_S$ so $t_2 - t_1 \to \infty$. So, as seen by the distant observer, the falling body will never quite reach the event horizon.

Third, the difference in coordinate time between emission and observation depends only on the coordinate positions of the emitter and observer. As long as the positions remain fixed, signals will always take the same amount of coordinate time to travel from r_1 to r_2, and signals emitted at times separated by Δt will arrive with coordinate time separation Δt. This justifies an assertion that we made in Chapter 5, concerning a stationary emitter and a stationary observer, when we said that the coordinate time interval between the emission of two successive signals was the same as the coordinate time interval between their reception.

We can get a more detailed picture of what the distant observer will see if we determine the position of the freely falling body as a function of coordinate time t. To do this we need to relate the differences in coordinate position dr to differences in coordinate time dt for events on the world-line of the falling body.

● Equation 6.12 already provides a relationship between dr and dt. Why can't we just use that?

○ That equation only applies to events on the world-line of a photon. It was deduced from the metric using the condition $(ds)^2 = 0$. We need a condition that applies to events on the world-line of a freely falling body with non-zero mass.

We considered the motion of a freely falling body in Chapter 5, where one of the results that we introduced (Equation 5.28 after substituting R_S for $2GM/c^2$) was

$$\frac{E}{mc^2} = \left(1 - \frac{R_S}{r}\right) \frac{dt}{d\tau}. \tag{6.15}$$

Now we already know, from Equation 6.4, that for a body starting its fall from rest at a large distance r_0 from the origin, $E/mc^2 = (1 - R_S/r_0)^{1/2}$. Substituting this into Equation 6.15 and rearranging, we see that for events on the world-line of the freely falling body,

$$\frac{dt}{d\tau} = \frac{(1 - R_S/r_0)^{1/2}}{1 - R_S/r}. \tag{6.16}$$

We also considered a freely falling body earlier in this chapter, eventually arriving at

$$\frac{d\tau}{dr} = -\frac{1}{c} \sqrt{\frac{r_0}{R_S}} \sqrt{\frac{r}{r_0 - r}}. \tag{Eqn 6.6}$$

Multiplying these last two results together gives the desired relation between dt and dr for events along the world-line of a freely falling body with non-zero mass:

$$\frac{dt}{dr} = \frac{dt}{d\tau} \frac{d\tau}{dr} = -\frac{1}{cR_S^{1/2}} \frac{(1 - R_S/r_0)^{1/2}}{1 - R_S/r} \sqrt{\frac{rr_0}{r_0 - r}}. \tag{6.17}$$

Analysing this general relationship is possible but complicated, so we shall use the fact that we are mainly interested in effects at or near the event horizon, where r is small compared with r_0, to justify the simplification that

$$\frac{dt}{dr} = -\frac{1}{cR_S^{1/2}} \frac{r^{1/2}}{1 - R_S/r}. \tag{6.18}$$

Integrating both sides with respect to r, from a point at radial coordinate r_* that is much larger than R_S but much less than r_0, to some general point r', gives

$$t(r') - t(r_*) = -\frac{1}{cR_S^{1/2}} \int_{r_*}^{r'} \frac{r^{1/2}}{1 - R_S/r} \, dr.$$

The integral can be found in tables or by using an algebraic computing package:

$$t(r') - t(r_*)$$

$$= -\frac{R_S}{c}\left[\frac{2}{3}\left(\frac{r}{R_S}\right)^{3/2} + 2\left(\frac{r}{R_S}\right)^{1/2} - \ln\left|\frac{(r/R_S)^{1/2}+1}{(r/R_S)^{1/2}-1}\right|\right]_{r_*}^{r'}.$$

Substituting the limits, we get the final answer

$$t(r') - t(r_*) = \frac{R_S}{c}\left(\text{constant} - \frac{2}{3}\left(\frac{r'}{R_S}\right)^{3/2} - 2\left(\frac{r'}{R_S}\right)^{1/2}\right.$$
$$\left. + \ln\left|\frac{(r'/R_S)^{1/2}+1}{(r'/R_S)^{1/2}-1}\right|\right). \qquad (6.19)$$

This relationship is illustrated in Figure 6.5, which also includes a line representing the curve that we obtained earlier when plotting the radial coordinate against proper time. Remembering that we approximated the equation of motion before performing the integral, the constant has been chosen to ensure that the two curves match at $r = r_*$, where intervals of coordinate time t and proper time τ are essentially the same. As r becomes smaller, the two curves separate, with t becoming infinite as $r \to R_S$. So we again see that according to a distant observer it takes an infinite time for a body falling into a black hole to reach the event horizon. Note that this infinity concerns the coordinate time that the falling body requires to reach the horizon; it is quite distinct from the time required for a light signal from the body to reach a distant observer.

As noted earlier, light emitted from a falling body approaching a black hole will exhibit an increasing gravitational redshift according to a distant observer. The formula for gravitational redshift from a stationary source was given in Chapter 5:

$$f_\infty = f_{\text{em}}\left(1 - \frac{2GM}{c^2 r_{\text{em}}}\right)^{1/2}. \qquad \text{(Eqn 5.17)}$$

Using the general relationship $c = f\lambda$, we can express the redshift in terms of wavelength as

$$\lambda_\infty = \frac{\lambda_{\text{em}}}{\left(1 - \frac{2GM}{c^2 r_{\text{em}}}\right)^{1/2}}. \qquad (6.20)$$

The formulae predict that the observed redshift will become greater and greater as the point of emission approaches the event horizon. Indeed, as $r \to R_S$, $\lambda_\infty \to \infty$. For this reason the event horizon is often described as a **surface of infinite redshift**.

Actually, the redshift seen by a distant observer will increase even more rapidly than the formula indicates since our earlier result applied to a *stationary* source while the falling body that we are now considering will be moving away from the distant observer. This motion will cause a Doppler shift that will further increase the observed redshift, though the event horizon will remain a surface of infinite redshift.

Another effect follows from those that we have already mentioned. Suppose that the falling body is emitting light with a constant luminosity L_0 according to an

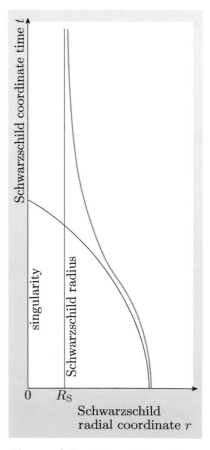

Figure 6.5 The relationship between coordinate time t and radial coordinate r for a body falling freely into a black hole.

observer falling with it. The increasing redshift (which reduces the energy per photon) and the extended time of emission and travel (which reduces the rate at which photons are received) will all tend to decrease the luminosity of the source as seen by a distant observer. During the early part of the fall, the distant observer will see the source becoming dimmer due to its increasing distance from the observer, but the additional dimming due to general relativistic effects will become more pronounced as the falling body is seen to approach the event horizon. Quantitative studies show that if the light is treated as continuous classical radiation (i.e. ignoring the fact that it is actually emitted as photons), then in the final stages of the observed fall, the dimming becomes exponential, measured luminosity halving on a timescale of order R_S/c, so

$$L \to L_0\, \mathrm{e}^{-ct/aR_\mathrm{S}} \quad \text{as} \quad r \to R_\mathrm{S}, \tag{6.21}$$

where a is a constant of order 1. This is such a rapid dimming that, far from the falling body being visible for all eternity, such a body would actually become unobservably dim rather quickly once it gets close to the event horizon.

All this talk of bodies falling into a black hole may sound rather fanciful, but remember that the body concerned might, in principle, be part of the surface of a star undergoing gravitational collapse. In this way the ideas that we have been discussing can form the basis for observational predictions concerning the behaviour of a star as it undergoes gravitational collapse and contracts within its own Schwarzschild radius. The interested reader can pursue this topic elsewhere but we should note again the key points to emerge from our discussion.

Observing a body fall into a non-rotating black hole

A body falling into a black hole takes an infinite amount of coordinate time to reach the event horizon. Light signals emitted from the object also take an increasing amount of (coordinate) time to reach a distant observer. These effects will reduce the rate at which photons from the falling body reach the distant observer. Signals from the falling body are also redshifted according to the distant observer, with the horizon representing a surface of infinite redshift. This reduces the energy per photon received by the distant observer. The combination of all these effects will cause an in-falling body of constant proper luminosity to dim rapidly as it approaches the horizon.

Exercise 6.2   A light pulse is emitted in the outward direction from a source just exterior to the event horizon of a non-rotating black hole. Write down an expression for the radial speed of light according to a stationary local observer and according to a stationary observer at infinity, and show that both are equal to c.

Exercise 6.3   According to a local observer, stationary just outside the event horizon of a non-rotating black hole, what is the speed of a freely falling body, travelling radially inwards, as it nears the event horizon, given that the body was released from rest at a great distance from the black hole?

Exercise 6.4   Imagine watching an astronaut falling freely into a non-rotating black hole, waving goodbye as he or she approaches the event horizon. What might a distant observer expect to see?   ∎

6.2.3 Tidal effects near a non-rotating black hole

It's natural to expect that anyone falling into a stationary black hole will be crushed to death in its central singularity. However, this expectation overlooks **tidal effects**.

Tides are a familiar phenomenon on the Earth. They arise primarily from variations in the gravitational field due to the Moon and the Sun across the diameter of the Earth. The basis of the Newtonian explanation of tides is illustrated for the case of lunar tides in Figure 6.6.

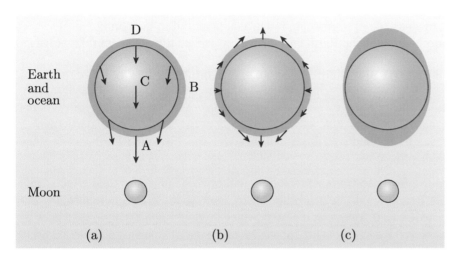

Figure 6.6 Lunar tides result from the variation of the Moon's gravitational field (i.e. the gravitational force per unit mass) across the diameter of the Earth. (a) Gravitational field of the Moon: the gravitational force per unit mass. (b) Tidal field of the Moon: the difference between the local field and the field at the centre of the Earth. (c) Tidal bulges: a gravitational equipotential of the combined Earth–Moon gravitational field.

If the oceans are represented by a uniformly deep layer of water, then at any point on that water surface there is a lunar **tidal field** given by the (vector) difference between the local value of the gravitational field due to the Moon and its value at the centre of the Earth. The effect of this tidal field is to redistribute the oceans in such a way that the water surface forms an equipotential surface of the combined Earth–Moon gravitational field.

If we consider the Earth and the Moon in isolation, the key points to note are as follows.

- If the Earth and the Moon were point particles in an isolated system bound by gravity, each particle would be in free fall about the common centre of mass of the system.

- As they are extended bodies with finite diameters, the individual centres of mass of the Earth and the Moon are in free fall about their common centre of mass (which is actually some way beneath the Earth's surface), but the same is not true of all other points in those bodies.

- The Moon's gravitational field is stronger at point A in Figure 6.6 than at point C, causing material at point A to experience a tidal force *towards* the Moon and therefore *away* from the centre of the Earth.

- The Moon's gravitational field is weaker at point D in Figure 6.6 than at point C, causing material at point D to experience a tidal force *away* from the Moon, but this is also *away* from the centre of the Earth.

- The Moon's gravitational field at point B is inclined at an angle to the gravitational field at point C in Figure 6.6, causing material at point B to experience a tidal force almost perpendicular to the direction towards the Moon and directed towards the centre of the Earth.

- In the case of the solid Earth, the response to the tidal field and the forces that it produces is small. The electrical forces between atoms in a solid are so strong that only a small (but measurable) distortion of the solid Earth is sufficient to produce forces that counterbalance the tidal forces. The same is not true of the oceans. The forces between atoms in a liquid are much weaker than those that act within a solid. In response to the tidal field the oceans rise or fall until the additional weight of the water column at any point counterbalances the tidal force. Put differently, the oceans redistribute themselves in such a way that they form a surface of uniform gravitational potential in the combined gravitational field of the Earth and the Moon. Hence the observed tidal bulges.

Note that this Newtonian argument involves *free fall* and *variations in the gravitational field* across the diameter of the Earth. (Also note that it has nothing to do with 'centrifugal forces' as some sources incorrectly claim.) In reality there are additional effects that arise from the rotation of the Earth and the particular form of ocean basins and coastlines, but these are specific to the Earth, so we shall not pursue them here.

A body falling freely towards a black hole will also be subject to tidal effects. In general relativity it would be inappropriate to describe these effects in terms of the different gravitational forces on the body, since there are no gravitational forces in general relativity. Rather, we should use the language of spacetime curvature and geodesic motion, though we should be able to recover the idea of tidal forces from the relativistic description in the appropriate Newtonian limit.

The usual starting point for a relativistic account of tidal effects is the concept of **geodesic deviation**, which will now be described. Consider a region of spacetime, and suppose that C and D are two parameterized curves passing though that region. More specifically, suppose that C and D are neighbouring geodesics, so each curve is the possible world-line of a particle passing though the region. The geodesic C can be represented by a set of four coordinate functions $[x_C^\mu(\lambda)]$, where λ is an affine parameter, and we shall suppose that its neighbouring geodesic D is affinely parameterized in such a way that it can be described by a similar set of coordinate functions $[x_D^\mu(\lambda)]$. Because C and D are neighbouring geodesics parameterized in similar ways, we can suppose that corresponding to each value of λ is a unique pair of points, one on C and the other on D, separated by a four-dimensional separation $[\xi^\mu(\lambda)]$, where

$$\xi^\mu(\lambda) = x_D^\mu(\lambda) - x_C^\mu(\lambda). \qquad (6.22)$$

This arrangement of geodesics and their separation vector $[\xi^\mu(\lambda)]$ is illustrated in Figure 6.7.

In the absence of gravity, in a region where the Riemann curvature is zero and spacetime is flat, it is easy to imagine that the geodesics will be straight lines that particles move along at constant speed. In such circumstances, the separation vector $[\xi^\mu]$ will be constant. However, in the presence of gravity, spacetime will be curved, the Riemann curvature will be non-zero, particles on neighbouring geodesics can have relative accelerations, and the behaviour of the separation vector might be complicated. In fact, a detailed analysis shows that the changes in

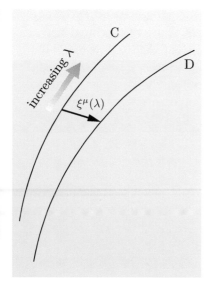

Figure 6.7   Two neighbouring geodesics, C and D, each parameterized by the same affine parameter λ. Points on C and D that correspond to the same value of λ are linked by a separation vector with components $\xi^\mu(\lambda)$. (ξ is the Greek letter xi.)

the separation vector are described by the following **equation of geodesic deviation**.

Equation of geodesic deviation

$$\frac{D^2\xi^\mu}{D\lambda^2} + \sum_{\alpha,\beta,\gamma} R^\mu{}_{\alpha\beta\gamma}\xi^\alpha\frac{dx^\beta}{d\lambda}\frac{dx^\gamma}{d\lambda} = 0. \tag{6.23}$$

This relationship holds at all points along the geodesic C, and the expression $D^2\xi^\mu/D\lambda^2$ represents the second-order **derivative along the curve** C of the separation vector component ξ^μ. This kind of derivative is similar in some respects to the covariant derivative that was introduced in Chapter 4. In the case of the covariant derivative we noted that when differentiating tensor components such as $T^\mu{}_\nu$ with respect to coordinates x^ρ, the partial derivatives $\partial T^\mu{}_\nu/\partial x^\rho$ do not generally transform as the components of a tensor, but we were able to construct a related quantity that we denoted $\nabla_\rho T^\mu{}_\nu$, that was a kind of derivative and produced a result that was a tensor of higher rank. In the present case, when considering changes in ξ^μ as we move from event to event along the geodesic C, we need to differentiate with respect to the affine parameter λ in such a way that the rank 1 tensor nature of ξ^μ will not change. This is what is provided by the derivative along the curve, which is defined by

$$\frac{D\xi^\mu}{D\lambda} = \frac{d\xi^\mu}{d\lambda} + \sum_{\alpha,\beta} \Gamma^\mu{}_{\alpha\beta}\xi^\alpha\frac{dx^\beta}{d\lambda}. \tag{6.24}$$

Taking a second derivative results in a complicated expression that simplifies to Equation 6.23.

In the Newtonian limit, when speeds are low and gravitational fields are weak, the equation of geodesic deviation will provide information about the relative acceleration of freely falling particles as they move along neighbouring geodesics — which is exactly the kind of information needed to work out Newtonian tidal fields. However, the equation of geodesic deviation is not restricted to the Newtonian limit. As a covariant tensor relationship, it provides the essential generalization of Newtonian tidal fields that makes it possible to describe tidal effects throughout curved Schwarzschild spacetime, apart from the central singularity where tidal effects become infinite.

In the case of an astronaut falling feet first towards a non-rotating black hole, the result of geodesic deviation is disastrous. While the astronaut's centre of mass falls into the central singularity in the proper time calculated earlier, the astronaut's head and feet will arrive at significantly different times! During the inward fall, geodesic deviation stretches the astronaut in the radial direction and causes compression in the transverse directions. This process is usually referred to as **spaghettification** and is illustrated schematically in Figure 6.8.

Spaghettification will generally kill an in-falling astronaut before the astronaut reaches the central singularity. Indeed, in the case of a stellar mass black hole, death from spaghettification will usually occur well before the astronaut crosses the event horizon. We can estimate where the effect becomes significant by

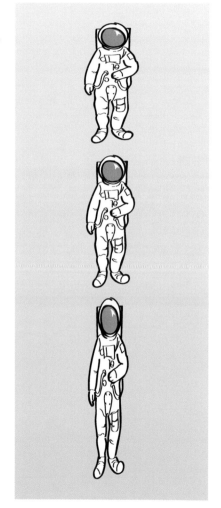

Figure 6.8 An astronaut falling feet first into a black hole will be spaghettified as a result of geodesic deviation.

working in the Newtonian approximation. The magnitude of the Newtonian gravitational field (force per unit test mass) at a distance r from a body of mass M is

$$f(r) = \frac{GM}{r^2}.$$

If δr represents a small change in the radial coordinate, we can use Taylor's theorem to determine the corresponding change in the field. Working to first order,

$$f(r + \delta r) - f(r) = \delta f = \frac{\mathrm{d}f}{\mathrm{d}r}\,\delta r = -\frac{2GM}{r^3}\,\delta r, \tag{6.25}$$

where δf is a measure of the tidal force per unit mass acting along an object of dimension δr. The magnitude of the field gradient $|\mathrm{d}f/\mathrm{d}r| = 2GM/r^3$ provides a useful measure of tidal lethality. This quantity is very large near the event horizon of a stellar mass black hole, partly due to the large mass of the black hole, but more particularly because r is already small near the Schwarzschild radius. A human body is unlikely to survive a gradient of order $10^4\,\mathrm{s}^{-2}$. This is the kind of field gradient that would be encountered at about $1000\,\mathrm{km}$ from a 40 solar mass black hole, far beyond the event horizon, which would be at about $120\,\mathrm{km}$ from the centre. In the case of a supermassive black hole with a mass of 10^7 solar masses, the event horizon would be at $3 \times 10^7\,\mathrm{km}$ and the field gradient at the horizon would be only about $10^{-4}\,\mathrm{s}^{-2}$, too small for a falling astronaut to notice. The falling astronaut who passed through the event horizon would not be able to escape, but would still have a long way to fall before the tidal effects became lethal.

6.2.4   The deflection of light near a non-rotating black hole

When discussing motion in Schwarzschild spacetime in Chapter 5, we started our discussion of the geodesics in a general way that included massless particles such as photons, as well as particles with mass. However, we soon focused on the case of massive particles and essentially ignored the motion of photons. In this chapter we have already used the metric to discuss the radial motion of photons, but we have still not paid any attention to the non-radial motion of photons. We shall now remedy that omission.

Figure 6.9 shows the trajectories of photons (or any other massless particles) moving in a plane that also contains the central singularity of a non-rotating black hole of Schwarzschild radius R_S. The trajectories are initially parallel but each can be identified by its **impact parameter**, that is, the perpendicular (coordinate) distance b from the singularity to the initial direction of motion of the photon. Values of the impact parameter are shown on the vertical axis in the figure, expressed as multiples of the Schwarzschild radius.

As you can see, photons with $b = 3R_\mathrm{S}$ or $b = 4R_\mathrm{S}$ are strongly deflected, though they are not drawn into the black hole. This is an example of the phenomenon of light deflection, mentioned in Chapter 4, that Einstein was able to predict on the basis of the principle of equivalence. The effect becomes weaker as the impact parameter increases but remains detectable even for large multiples of the Schwarzschild radius. We shall have more to say about this phenomenon in the next chapter when we discuss tests of general relativity.

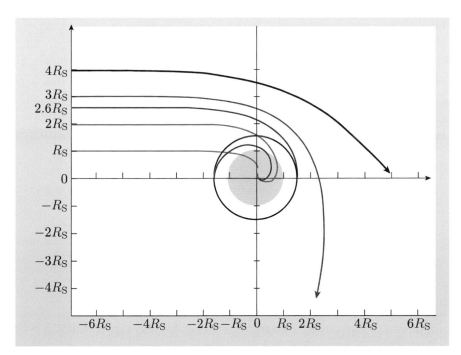

Figure 6.9   The deflection of light by a non-rotating black hole with Schwarzschild radius $R_S = 2GM/c^2$. The region within the event horizon is shaded. The location of the photon sphere is indicated by a black circle at $r = 1.5R_S$. The trajectories are based on computer simulations by H. Cohn, published in the *American Journal of Physics*, vol. 45 (1977) p. 239.

Light with $b = 2.6R_S$ can be captured into a circular orbit of radius $r = 1.5R_S$. An analysis of this orbit, based on an 'effective potential' similar to that used for massive particles in Chapter 5, shows that the orbit is unstable, so light will not linger there for long. Light rays with $b < 2.6R_S$ do not 'orbit' at all but are drawn rapidly to the central singularity.

Since we are dealing with a spherically symmetric black hole, there is nothing physically 'special' about the particular plane that we have chosen to consider in Figure 6.9. Any other plane containing the black hole's central singularity could have been chosen. This shows that any great circle on the sphere of coordinate radius $1.5R_S$ represents a possible unstable circular orbit for a photon. This spherical surface is called the **photon sphere** of the black hole. Any freely falling photon that enters the photon sphere from the outside is certain to be captured by the black hole, but photons emitted from within the photon sphere may escape outwards, and so may photons that are not freely falling such as those reflected by a mirror between the photon sphere and the event horizon. Of course, according to general relativity, any photon that enters the region within the event horizon (shown in grey) is inevitably captured by the central singularity.

6.2.5   The event horizon and beyond

We saw earlier that, as measured by a distant observer, a body falling into a non-rotating black hole takes an infinite amount of coordinate time to reach the event horizon. However, we also saw that such a body, as observed by a freely falling observer travelling with it, requires only a finite proper time to pass through the event horizon and continue on to the central singularity. Interestingly, there are values of the Schwarzschild coordinates that correspond to all events on the inward journey, apart from the coordinate singularity at the Schwarzschild radius. The full journey is shown in Figure 6.10.

As you can see, the extra part of the pathway (shown in orange) from the horizon to the central singularity starts as $t \to \infty$ and leads back in coordinate time to some earlier finite value! It's tempting to interpret this as a sign that the in-falling observer is travelling backwards in time. However, no such fanciful interpretation is needed. It is true that the value of t is decreasing, but you have already learned that in general relativity *coordinates lack immediate metrical significance*. The decreasing value of Schwarzschild coordinate time t for an in-falling observer inside the event horizon simply shows that the Schwarzschild coordinates are especially poorly suited to the task of describing the last stages of the fall.

More evidence of the inappropriateness of Schwarzschild coordinates can be obtained by using them to describe the *lightcones* along the path of an in-falling observer. It was shown in Chapter 1 that lightcones provide a valuable tool for investigating the causal structure of spacetime. In that earlier application we were concerned with the geometrically flat Minkowski spacetime of special relativity, where lightcones could be extended to infinity without any impediment. In contrast, in general relativity, spacetime is generally curved, so lightcones cannot be indefinitely extended. Nonetheless, observers using locally inertial frames (such as freely falling observers) will find that special relativity holds true locally, so any such observer can use lightcones to explore the local structure of spacetime.

The local lightcones in Schwarzschild spacetime can be identified from a spacetime diagram showing incoming and outgoing null geodesics (i.e. possible photon world-lines). Just such a diagram is shown in Figure 6.11. The figure uses Schwarzschild coordinates, the axes being ct and r. The curves are described by Equation 6.12, which was obtained directly from the Schwarzschild metric for the case of radial motion together with the additional requirement that $(\mathrm{d}s)^2 = 0$ for photons. Rearranging that equation slightly, to emphasize the quantity $\mathrm{d}(ct)/\mathrm{d}r$, which describes the gradient of the lightcone's edge, we get

$$\frac{\mathrm{d}(ct)}{\mathrm{d}r} = \pm \frac{1}{1 - R_{\mathrm{S}}/r}. \tag{6.26}$$

Note that far from the horizon, as $r \to \infty$, this equation implies that $\mathrm{d}(ct)/\mathrm{d}r = \pm 1$, so that lightcones take the form that they would have in special relativity. However, when approaching the horizon from outside, $\mathrm{d}(ct)/\mathrm{d}r \to \pm\infty$, causing the lightcones to become very narrow. Just inside the horizon something even more remarkable occurs. The lightcones suddenly become very broad again, and their time-like regions become horizontal, so that the only possible directions of radial motion are towards the singularity. You saw an example of this in Figure 6.10, where the last part of the time-like orange curve was almost horizontal, but Figure 6.11 shows that this is a general phenomenon. The tipping of the lightcones (see Figure 6.12) makes a certain kind of sense since it indicates the inevitability of encountering the singularity once the event horizon has been passed. However, the abrupt switch in direction and the sudden broadening of the lightcones looks very odd and is another sign of inappropriateness of the Schwarzschild coordinates in this region.

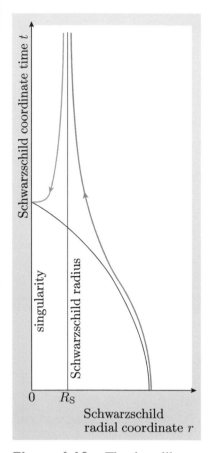

Figure 6.10   The time-like geodesic motion of a body falling freely into a black hole, described in terms of Schwarzschild coordinates.

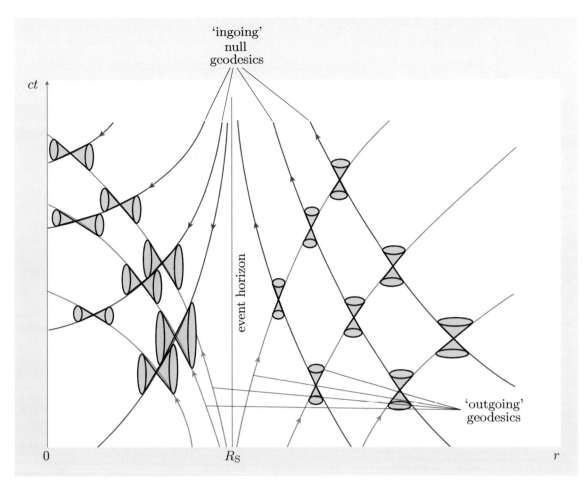

Figure 6.11 Ingoing and outgoing null geodesics in a spacetime diagram drawn in Schwarzschild coordinates. Local lightcones occupy the future and past time-like directions between pairs of null geodesics.

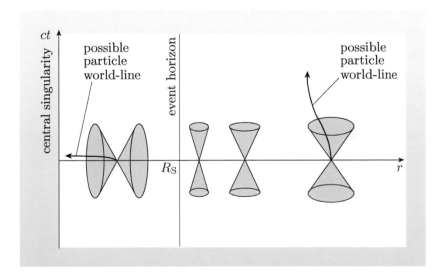

Figure 6.12 In Schwarzschild coordinates, as the event horizon is approached and entered, lightcones show a progressive narrowing followed by an abrupt reopening and reorientation.

Many of the coordinate-related problems associated with non-rotating black holes can be removed by changing the coordinates used to describe them. The necessary transformation was introduced in the late 1950s by Finkelstein, though he was rediscovering coordinates that had been introduced for a different purpose by Eddington in 1924.

In what are known as **advanced Eddington–Finkelstein coordinates**, a new coordinate t' is related to the Schwarzschild t and r coordinates by the equation

$$ct' = ct + R_{\mathrm{S}} \ln\left(\frac{r}{R_{\mathrm{S}}} - 1\right). \tag{6.27}$$

With this modified time coordinate, the line element of Schwarzschild spacetime can be written as

This line element is actually the first step in describing a non-static extension to the exterior Schwarzschild solution, hence the presence of a term involving $\mathrm{d}t'\,\mathrm{d}r$.

$$(\mathrm{d}s)^2 = c^2\left(1 - \frac{R_{\mathrm{S}}}{r}\right)(\mathrm{d}t')^2 - 2\frac{R_{\mathrm{S}}}{r}c\,\mathrm{d}t'\,\mathrm{d}r - \left(1 + \frac{R_{\mathrm{S}}}{r}\right)(\mathrm{d}r)^2$$
$$- r^2\left((\mathrm{d}\theta)^2 + \sin^2\theta\,(\mathrm{d}\phi)^2\right), \tag{6.28}$$

which is non-singular at $r = R_{\mathrm{S}}$. In these coordinates ingoing null geodesics are represented by straight lines while outgoing photons are curves. (Of course, those within the event horizon don't actually go outwards, they just arrive at the central singularity at a later value of t'.) The relevant spacetime diagram for advanced Eddington–Finkelstein coordinates is shown in Figure 6.13, and the corresponding sequence of lightcones is shown in Figure 6.14.

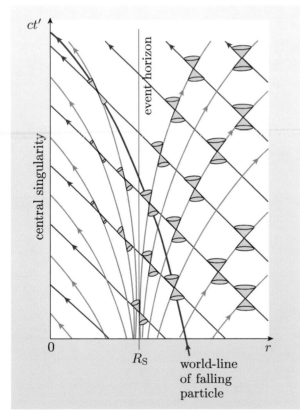

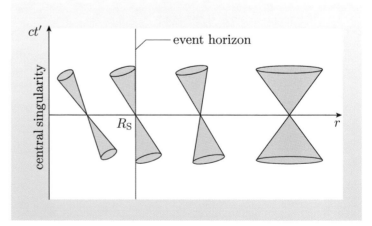

Figure 6.13 Ingoing and outgoing null geodesics in a spacetime diagram drawn in advanced Eddington–Finkelstein coordinates.

Figure 6.14 In advanced Eddington–Finkelstein coordinates, as the event horizon is approached and entered, the lightcones become increasingly tipped and narrowed in a smooth progression.

The 'opening-up' of Schwarzschild spacetime that advanced Eddington–Finkelstein coordinates permit is the start of a new chapter in the investigation of black holes, not the end of one. In Schwarzschild coordinates there is a symmetry between ingoing and outgoing null geodesics, yet in advanced Eddington–Finkelstein coordinates an asymmetry is introduced: the ingoing null geodesics are straight, the outgoing ones are not. This suggests the existence of another coordinate system that would in some sense reverse the asymmetry. Such a coordinate system does exist, and the two types of Eddington–Finkelstein coordinates together were a step towards a further development. In 1960, Martin Kruskal introduced a single set of coordinates that were non-singular everywhere outside the physical singularity. In these coordinates it is natural to extend the domain covered by the usual Schwarzschild solution. Indeed, in this context the Schwarzschild solution is seen to be just one half of a broader domain referred to as its **maximal analytic extension** (see Figure 6.15). The existence of this mathematically extended domain has given rise to many speculations about 'other universes', 'spacetime wormholes', and 'white holes' from which matter and radiation might be expelled with the same kind of inevitability that they are drawn into a black hole. We shall not discuss these aspects of the extended Schwarzschild solution, though you may like to follow them up in other sources. However, it is appropriate to end with two final points. The first is to note that some physicists take the view that the extended domain is physically inaccessible and therefore of little interest and no scientific relevance. The second is to note that in a field as complicated as general relativity it has often taken a long time for the physical significance of mathematical results to be fully appreciated; humility in the face of complexity is sometimes an appropriate response.

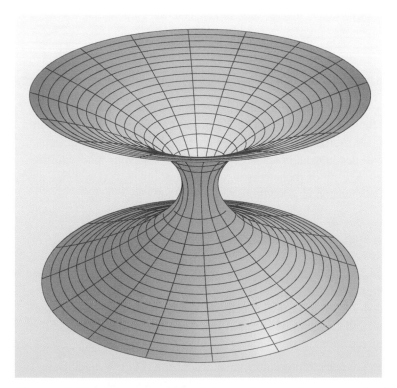

Figure 6.15 The use of Kruskal coordinates shows that the familiar Schwarzschild solution represents only half of its maximal analytic extension, in which two asymptotically flat regions are linked by a throat.

Lightcones, spacetime diagrams and event horizons

Lightcones and spacetime diagrams are valuable tools for investigating local spacetime structure in general relativity, but the behaviour of lightcones will depend on the particular coordinates being used. In Schwarzschild coordinates lightcones show abrupt changes at the Schwarzschild radius, which is the location of a coordinate singularity. Advanced Eddington–Finkelstein coordinates remove the coordinate singularity and produce lightcones that change in a regular way, tipping and narrowing as they approach the Schwarzschild radius. The behaviour of the lightcones at and within the Schwarzschild radius indicates the inevitability of encountering the central singularity, though more powerful methods must be used to prove that inevitability.

Exercise 6.5 When working in advanced Eddington–Finkelstein coordinates, which feature(s) of the lightcones suggest the impossibility of escaping from within the event horizon of a non-rotating black hole?

Exercise 6.6 Using (a) Schwarzschild coordinates and (b) advanced Eddington–Finkelstein coordinates, sketch spacetime diagrams showing the time-like geodesic of a radially in-falling body. In each case add to the geodesic future lightcones representing the development of flashes of light emitted by that body during its fall. Include the region inside the event horizon as well as the region outside the horizon. ■

6.3 Rotating black holes

Real astrophysical systems, such as stars and galaxies, generally possess angular momentum. A body that undergoes a gravitational collapse is expected to retain a good deal of the angular momentum that it has immediately prior to the collapse. In addition, as you will see later, a black hole may acquire angular momentum from in-falling bodies. For all of these reasons, real black holes, if they exist, are expected to rotate. This section is devoted to rotating black holes.

6.3.1 The Kerr solution and rotating black holes

Our starting point for the description of a non-rotating black hole was the Schwarzschild solution, which describes the spacetime outside a spherically symmetric body. The solution has the properties of being stationary (so that the metric coefficients are independent of t), spherically symmetric, asymptotically flat, singular and (loosely speaking) unique.

We cannot expect the Schwarzschild solution to describe a rotating black hole because the black hole's angular momentum will pick out some particular direction in space and that will destroy the spherical symmetry. We might, though, expect there to be some sort of analogue of the Schwarzschild solution with the properties of being stationary, axially symmetric (i.e. having the invariance of a cylinder), asymptotically flat and singular. We might also hope that some kind of extension or generalization of Birkhoff's theorem will again establish the essentially unique character of the solution. Just such a solution was discovered by Roy Kerr in 1963, though it took some time for its uniqueness to be established.

The line element of the **Kerr solution** can be written as follows.

Kerr line element

$$(ds)^2 = \left(1 - \frac{R_S r}{\rho^2}\right) c^2 (dt)^2 + \frac{2 R_S r a c \sin^2 \theta}{\rho^2} \, dt \, d\phi - \frac{\rho^2 (dr)^2}{\Delta}$$
$$- \rho^2 (d\theta)^2 - \left(\left(r^2 + a^2\right) \sin^2 \theta + \frac{R_S r a^2 \sin^4 \theta}{\rho^2}\right) (d\phi)^2. \quad (6.29)$$

This looks (and is) rather complicated, but there are some key points to note.

- The Kerr metric depends on just two parameters, $R_S = 2GM/c^2$ and $a = J/(Mc)$, which in turn depend on the mass M and angular momentum magnitude J. The metric describes a black hole only when $a \le R_S/2$, i.e. when $J \le GM^2/c$, and the important limiting case when $a = R_S/2$ is said to describe an **extreme Kerr black hole**.

- The coordinates used to describe the metric, ct, r, θ, ϕ, are called **Boyer–Lindquist coordinates**. ϕ is a standard spherical coordinate, but θ and r are not. They are related to standard Cartesian coordinates x and y by

$$x = \sqrt{r^2 + a^2} \sin \theta \cos \phi, \qquad (6.30)$$
$$y = \sqrt{r^2 + a^2} \sin \theta \sin \phi. \qquad (6.31)$$

r is still a kind of radial coordinate, but increasing values of r do not correspond to spheres of increasing proper circumference, nor does $r = 0$ identify a unique point. At a fixed value of t, a surface of constant r is an ellipsoid.

- Two functions, Δ and ρ, are introduced to simplify the line element, but they are just useful combinations of the coordinates and parameters — they do not introduce anything new. These two functions are defined by $\Delta = r^2 - R_S r + a^2$ and $\rho^2 = r^2 + a^2 \cos^2 \theta$.

- The metric coefficients $g_{\mu\nu}$ do not depend on the coordinate ϕ. This property ensures the axial symmetry of the solution.

- As $r \to \infty$ it can be seen that $\rho^2 \to r^2$ and $\Delta \to r^2$, with the consequence that $(ds)^2 \to c^2 (dt)^2 - (dr)^2 - r^2 \left((d\theta)^2 + \sin^2 \theta \, (d\phi)^2\right)$. This property ensures the asymptotic flatness of the solution.

- The metric is singular when $\rho = 0$ and when $\Delta = 0$. The first of these is a physical singularity; the second turns out to be a coordinate singularity. Due to the particular character of the Boyer–Lindquist coordinates, the physical singularity corresponding to $\rho = 0$ takes the form of a ring of coordinate radius a in the equatorial plane. The coordinate singularity corresponding to

$\Delta = 0$ is represented by two closed surfaces,

$$r = r_+ \equiv \frac{R_S}{2} + \left[\left(\frac{R_S}{2} \right)^2 - a^2 \right]^{1/2}, \tag{6.32}$$

$$r = r_- \equiv \frac{R_S}{2} - \left[\left(\frac{R_S}{2} \right)^2 - a^2 \right]^{1/2}. \tag{6.33}$$

These surfaces both behave as event horizons. In the case of an extreme Kerr black hole, the two surfaces coincide at $r_+ = r_- = R_S/2$, but in non-extreme cases the surface corresponding to r_- is enclosed within the surface corresponding to r_+, giving the Kerr black hole a complicated internal structure.

• As seen by a distant stationary observer, there is a surface of infinite redshift at

$$r = s_+ \equiv \frac{R_S}{2} + \left[\left(\frac{R_S}{2} \right)^2 - a^2 \cos^2 \theta \right]^{1/2}. \tag{6.34}$$

This ellipsoidal surface (s_+) encloses the outer event horizon (r_+) except at the poles, where the two surfaces meet. For reasons that will be explained in the next section, the surface s_+ is called the static limit, and the region between the **static limit** and the outer event horizon (r_+) is called the **ergosphere**.

• In the limit that $a \to 0$, as the angular momentum goes to zero, the ring singularity shrinks to become a central point-like singularity. The inner event horizon at r_- shrinks to coincide with that central singularity, while the outer event horizon grows to become a sphere of coordinate radius R_S that coincides with the surface of infinite gravitational redshift (s_+) at all points. In short, in the limit $a \to 0$ the Kerr solution approaches the Schwarzschild solution.

● (a) Which property of the Kerr line element shows that it represents a stationary solution of the vacuum field equations?
(b) Which property shows that it is not a static solution?

○ (a) The metric coefficients do not depend on the time coordinate; more formally, $\partial g_{\mu\nu}/\partial t = 0$. This shows that the line element has the property of being stationary.
(b) The presence of a cross-term proportional to $dt \, d\phi$ shows that the line element is not invariant under the transformation $t \to t' = -t$. This shows that it does not have the property of being static.

The main structural features of the Kerr solution are shown in Figure 6.16.

Exercise 6.7 Verify the claims made about the location of the event horizons when (a) J has its maximum value, and (b) J is zero. ■

6.3.2 Motion near a rotating black hole

The Kerr spacetime around a rotating body exhibits a phenomenon known as the **dragging of inertial frames**. This describes the effect of the cross-term proportional to $dt \, d\phi$ in the Kerr line element in dragging the exterior spacetime along with the rotating body, so that time and space are effectively 'skewed' in the ϕ-direction. The effect can be seen by examining the lightcones in the equatorial plane of a rotating black hole, as indicated in Figure 6.17. (The lightcones have been drawn using a modified form of advanced Eddington–Finkelstein coordinates, so they are comparable with those shown in Figures 6.13 and 6.14 for the case of a non-rotating black hole.) In the present case of a rotating black hole, the lightcones are not only tilted towards the centre of the black hole, but also tipped in the direction of increasing ϕ — the direction of rotation of the black hole.

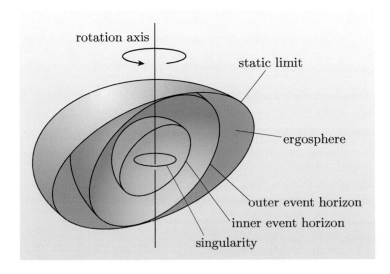

Figure 6.16 The structure of a Kerr black hole, drawn based on Boyer–Lindquist coordinates.

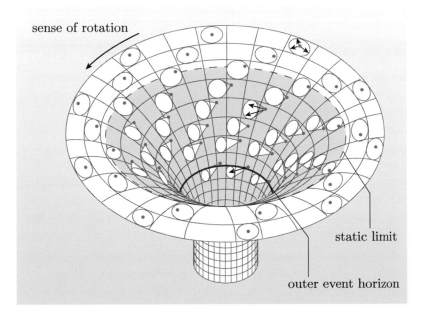

Figure 6.17 Lightcones in the equatorial plane ($\theta = \pi/2$) of a Kerr black hole.

Far from the black hole, light travels with equal ease in all directions. In this asymptotically flat region, lightcones have the usual symmetric form familiar from Minkowski space. Closer to the static limit, the lightcones become increasingly distorted, being tipped towards the origin and tipped in the direction of rotation of the black hole. The static limit marks a particular critical case: imagine a radial line extending from the origin to some point on the static limit and then extending outwards towards the asymptotically flat region. (Any of the radial lines in Figure 6.17 will do.) Now imagine placing a light source on that radial line at the point where it crosses the static limit. As Figure 6.17 indicates, light emitted from that source can travel in directions that take it closer to or further from the origin; it can also travel in directions that take it more-or-less in the direction of rotation of the black hole. What it cannot do is travel in any direction that opposes the direction of rotation of the black hole. At and within the static limit, the skewing of spacetime in the direction of rotation is so strong that motion in the direction of rotation cannot be resisted. Light itself is dragged in that direction, and so, by implication, is anything that travels slower than light. Note that the static limit is not an event horizon; it is quite possible for signals to escape through the static limit, but they must do so by travelling in the direction of rotation. The inability of objects entering the static limit to remain at rest explains why this surface of infinite redshift is called the static limit.

The dragging of inertial frames by a rotating black hole has many consequences. For example, material that starts falling towards the black hole from rest at a great distance will initially move along a radial pathway. However, as it nears the black hole, the effect of frame dragging will increase so, unless it happens to be travelling along the axis of rotation, the in-falling matter will also tend to move in the direction of the black hole's rotation. Once within the static limit it must move in that direction, irrespective of any action taken to move in the opposite direction.

Similarly, photons or other massless particles travelling in the equatorial plane of a rotating black hole will not only be deflected towards the black hole but will also be skewed around the black hole, as indicated for a clockwise-rotating black hole in Figure 6.19.

Another interesting consequence is the extraction of energy from a rotating black hole through what is known as the **Penrose process**, originally proposed by Roger Penrose (Figure 6.18) in the 1960s. The process involves some kind of unstable particle that enters the region between the static limit and the outer event horizon, and while there decays to form two other particles. Penrose showed that under appropriate circumstances, including the requirement that one of the particles produced in the decay passes through the outer horizon and enters the black hole, it is possible for the other decay product to pass out through the static limit and carry away more energy from the black hole than the original particle carried in. As a result of the process, the energy and angular momentum of the black hole are reduced, so the process provides a mechanism for extracting rotational energy from the black hole. It is because of this link with energy that the region between the static limit and the outer horizon is called the *ergosphere*.

As in the case of the non-rotating black hole, there is much that might be said concerning motion within the outer event horizon. The presence of the inner horizon is a sign of internal complexity, and the introduction of Kruskal-like coordinates leads to a maximal analytic extension that can be interpreted in terms of an infinite sequence of interconnected universes. However, the physical significance of these mathematical features is still unclear so we shall not pursue them here.

Figure 6.18   Sir Roger Penrose (1931–) is renowned for his geometrical imagination. His contributions to the theory of relativity include powerful theorems showing the inevitability of singularity formation under a variety of circumstances, and the invention of the Penrose process.

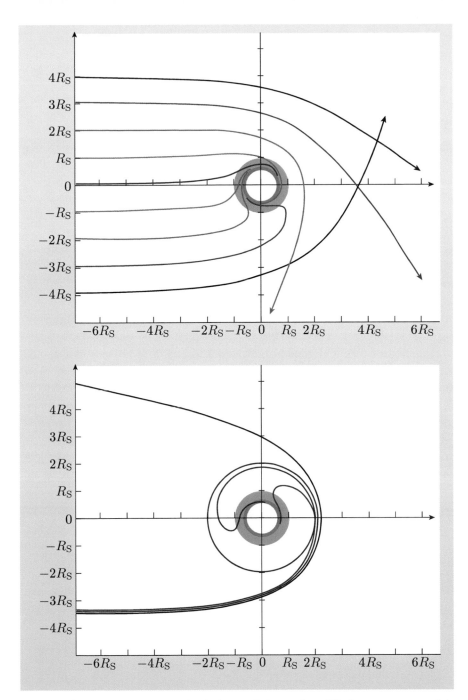

Figure 6.19   Computer calculations of the paths of light rays approaching an extreme Kerr black hole with a range of impact parameters. The light paths shown all lie in the equatorial plane. When a light ray enters the ergosphere, it must move in the direction of rotation of the black hole, even if it was originally circling the black hole in the opposite sense. The lower part of the figure is a zoomed-in detail showing the paths of three light rays with very similar impact parameters.

Exercise 6.8   Consider the representation of a rotating black hole shown in Figure 6.20 overleaf. The path of a spacecraft approaching the static limit is shown as a dashed line.

(a) Explain why this cannot be the path of an observer in free fall.

197

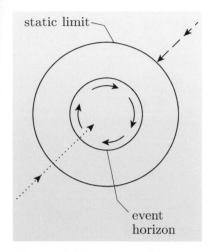

Figure 6.20   A possible trajectory?

Figure 6.21   Stephen Hawking (1942–) collaborated with Roger Penrose on the development of singularity theorems and independently discovered that quantum physics might be expected to allow black holes to act as thermal sources of radiation.

(b) Is it possible for the spacecraft to follow the dashed path? Explain.

(c) Is it possible for a spacecraft to follow the dotted path in Figure 6.20? Explain.   ■

6.4   Quantum physics and black holes

Up to this point, all our discussions of black holes have been based on predictions of the general theory of relativity. There is no doubt that black holes exist as solutions to the equations of general relativity, but the existence of 'real' black holes is a matter that can be settled only by observation. We shall examine some of the relevant evidence in the next chapter, but even if objects that can be described as black holes do exist, it is possible that parts of physics other than general relativity might significantly influence their properties. In particular, scientists are well aware of the wide importance of quantum phenomena in nature and know of many examples where quantum physics has modified or even completely overthrown the predictions of classical theories such as Newtonian mechanics or Maxwellian electromagnetism. Many physicists look forward to an eventual unification of classical general relativity and quantum physics in a yet to be formulated theory of **quantum gravity**. Some think that such a unified theory may already be at hand in the form of **string theory** or the so-called **M theory** that it has spawned; others strongly disagree. Whatever the fate of M theory, there have already been attempts to use general features of quantum physics that seem likely to survive any future unification to gain insight into the modifications that quantum physics might impose on 'classical' black holes. This section is concerned with some of those modifications.

6.4.1   Hawking radiation

In 1975 Stephen Hawking (Figure 6.21) published an influential paper showing that, due to quantum effects, black holes should be *sources* of radiation. In the paper he demonstrated that a black hole would behave as a body with a finite temperature that was inversely proportional to the mass M of the black hole. The relevant temperature is now called the **Hawking temperature**, T_H, and is given by

$$T_\mathrm{H} = \frac{\hbar c^3}{8\pi GkM} = 6.18 \times 10^{-8} \left(\frac{\mathrm{M}_\odot}{M}\right) \mathrm{K}, \qquad (6.35)$$

where $\mathrm{M}_\odot = 2.00 \times 10^{30}$ kg represents the mass of the Sun, $k = 1.381 \times 10^{-23}\,\mathrm{J\,K^{-1}}$ is the Boltzmann constant, and $\hbar = 1.055 \times 10^{-34}\,\mathrm{J\,s}$ is the Planck constant divided by 2π. The effective temperature of a stellar mass black hole is expected to be very small, but the very idea that a real black hole might act as a thermal source that can radiate away its energy is very striking since it is clearly at odds with the classical concept of a *black* hole that only ever absorbs radiation. The radiation that would be emitted by a black hole is now known as **Hawking radiation**.

Hawking's work was originally presented in the highly mathematical context of quantum field theory, but more intuitive interpretations were soon provided. In quantum physics, it was noted, the physical vacuum is subject to **quantum fluctuations** in which particle–antiparticle pairs can enjoy a short-lived existence before undergoing mutual annihilation. This seething quantum vacuum is not the static, featureless void of classical physics; rather, it is a fluctuating sea of transient particles in which quantum physics allows energy conservation to be violated by an amount ΔE for a time interval Δt, provided that, roughly, $\Delta E \, \Delta t \leq \hbar$, as a consequence of Heisenberg's uncertainty principle.

Under normal laboratory circumstances, the effects of the fluctuating quantum vacuum can be measured, but the particles responsible are not directly observed. They are said to be **virtual particles** since their energy and momentum do not generally satisfy the relation $E^2 - p^2 c^2 = m^2 c^4$ that applies to real, directly observable particles. It is possible to imagine a virtual particle pair in which one of the pair has positive energy while the other has the corresponding negative energy; such a zero-energy fluctuation might exist according to quantum uncertainty but would be ruled out by the additional requirement that all real particles have positive energy.

However, in the extreme conditions close to the event horizon of a black hole, particularly a low-mass black hole where the tidal effect would be very strong and particle–antiparticle pairs might quickly separate, the situation is different. Taking the case of a non-rotating black hole for simplicity, the metric coefficients $g_{00} = (1 - R_S/r)$ and $g_{11} = (1 - R_S/r)^{-1}$ change sign at the event horizon, switching the role of space-like and time-like intervals, and allowing particles within the horizon to follow geodesics characterized by negative energy values that would be forbidden outside the horizon. A particle–antiparticle pair, one member of which had a negative energy, might be created just outside the event horizon of a black hole within the limits allowed by quantum uncertainty, and the negative-energy particle might enter the horizon where its negative-energy geodesic is classically allowed. Meanwhile, the positive-energy particle outside the horizon might follow a positive-energy geodesic that would eventually lead to a distant observer. In this way normally short-lived quantum fluctuations might create long-lived observable particles. The positive particle energy measured by a distant observer would be balanced by a negative energy carried into the black hole, so from the point of view of the distant observer there would be no violation of energy conservation. The black hole would emit particles of all kinds and would gradually lose mass as it did so.

Of course, this intuitive argument does not account for details such as the Hawking temperature or the thermal spectrum of Hawking radiation, but it can be extended to make such outcomes plausible. What it does do is indicate the potential interplay of quantum physics and classical general relativity.

In classical physics an ideal thermal source of electromagnetic radiation (a black body) of surface area A and temperature T emits energy at a rate proportional to AT^4. For a Schwarzschild black hole, $A \propto R_S^2 \propto M^2$ and $T = T_H \propto 1/M$, so the rate of energy emission by Hawking radiation is

$$\frac{\mathrm{d}E}{\mathrm{d}t} \propto AT^4 \propto M^2 \times \left(\frac{1}{M}\right)^4 = \frac{1}{M^2}.$$

So, as the mass of the black hole decreases, its rate of energy emission will accelerate, causing a low-mass black hole (if such an object exists) to end its life with an escalating burst of energy emission that would be seen as an explosion! Such explosions are improbable because most black holes are likely to increase their mass by accreting matter from their environment. Nonetheless it is interesting to determine the expected life of an isolated black hole.

To a distant observer, the emission of energy ΔE is compensated by a decrease in the mass of the black hole $\Delta M = -\Delta E/c^2$. Thus

$$\frac{\mathrm{d}M}{\mathrm{d}t} \propto -\frac{\mathrm{d}E}{\mathrm{d}t} \propto -\frac{1}{M^2}.$$

The solution of the corresponding differential equation implies that a black hole of current mass M has a remaining lifetime proportional to M^3. In fact, the approximate total lifetime of an isolated black hole is estimated to be

$$\tau \approx 1.5 \times 10^{66} \left(\frac{M}{\mathrm{M}_\odot}\right)^3 \text{ years.} \tag{6.36}$$

The above takes account of the emission of photons; the production of other particles does not affect the dependence on mass, only the constant of proportionality. The lifetime τ of a black hole of mass $M < 10^{22}$ kg that loses mass by radiating only photons and neutrinos is given by

$$\left(\frac{\tau}{2 \times 10^{10} \text{ years}}\right) \approx \left(\frac{M}{2 \times 10^{11} \text{ kg}}\right)^3. \tag{6.37}$$

Hence an isolated mini black hole of mass 2×10^{11} kg, formed during the Big Bang say, might now be in its death throes.

Exercise 6.9   Why would the discovery of a mini black hole be important for physics?   ■

6.4.2   Singularities and quantum physics

In 1965 Roger Penrose showed that all massive bodies surrounded by an event horizon must contain a gravitational singularity that cannot be eliminated by a clever choice of coordinates. Although the singularity is hidden from outside observers by the event horizon, one identifying feature is that the curvature tensor generates an invariant scalar quantity that diverges and approaches infinity at the singularity. Once anything penetrates the event horizon, its world-line ends up at the singularity with no overshoot. Geodesics come to an end at finite values of their affine parameters in a region of finite mass but zero volume.

Although general relativity implies infinite density, many physicists suspect that quantum physics might somehow prevent such singularities from forming. A number of specific mechanisms have been advanced but there is no general agreement about this at the present time. On very general grounds it is expected that quantum effects and gravitational effects will become comparable at the **Planck scale**, which is characterized by

- Planck energy $E_{\mathrm{Pl}} = (\hbar c^5/G)^{1/2} = 1.22 \times 10^{19}$ GeV
- Planck length $l_{\mathrm{Pl}} = (\hbar G/c^3)^{1/2} = 1.62 \times 10^{-36}$ m
- Planck time $t_{\mathrm{Pl}} = (\hbar G/c^5)^{1/2} = 5.39 \times 10^{-44}$ s.

The Planck units are usually taken to represent the natural domain of quantum gravity, but they are currently far beyond our capacity for direct experimental investigation. If it is only at these extreme scales that the classical view of singularity becomes untenable, then the non-existence of ideal classical singularities might be of little astronomical significance. Supermassive black holes accreting a few solar masses of matter per year could still account for the energy emission from quasars, and lesser amounts of matter being heated to million degree temperatures in a swirling disc around a stellar mass black hole would still account for the intense X-ray sources not explained by neutron stars. Nonetheless an understanding of quantum gravity that included a quantum theory of spacetime singularities could hold many surprises and so it remains one of the main aims of gravitational research.

Summary of Chapter 6

1. According to classical general relativity, a black hole is a region of spacetime that matter and radiation may enter but from which they may not escape. The region is bounded by an event horizon that separates events that can be seen by an external observer from those that cannot be seen. At the heart of a black hole is a gravitational singularity at which invariant quantities related to the curvature of spacetime diverge.

2. Singularities may arise from the complete gravitational collapse of massive bodies such as degenerate stars (white dwarfs and neutron stars) that have exceeded their limiting mass, or even, much more speculatively, from smaller bodies compressed by cosmological processes in the early Universe.

3. Black holes are commonly classified according to their mass or according to the solution of the vacuum field equations that describes them. The only independent externally measurable properties of a black hole are its mass, charge and angular momentum.

4. Supermassive black holes might account for the energy emitted by quasars and other forms of active galaxy. Stellar mass black holes might account for some stellar sources of X-rays, though others can be accounted for by the action of neutron stars.

5. A non-rotating black hole is described by the stationary, spherically symmetric, Schwarzschild solution of the Einstein vacuum field equations. In Schwarzschild coordinates the solution has a gravitational singularity at $r = 0$ and a coordinate singularity at $r = R_\mathrm{S} = 2GM/c^2$, the Schwarzschild radius, which is also the location of the event horizon.

6. A body released from rest at a large distance from a non-rotating black hole only requires a finite proper time to fall freely to the central singularity. Nothing unusual happens to the body as it passes through the event horizon, though this marks a point of no return on the inward motion of the body. Once within the horizon the body will inevitably reach the central singularity.

7. As seen by a distant stationary observer, a body falling into a black hole takes an infinite amount of coordinate time to reach the event horizon. Light signals emitted from the object also take an increasing amount of

(coordinate) time to reach a distant observer. These effects reduce the rate at which photons from the falling body reach the distant observer (for whom coordinate time and proper time agree) and contribute to an observed dimming of the body.

8. Signals from the falling body are redshifted according to the distant observer, with the horizon representing a surface of infinite redshift. This reduces the energy per photon received by the distant observer and further contributes to the observed dimming.

9. Bodies in the neighbourhood of a black hole are subject to tidal effects that arise from the presence of spacetime curvature and are described by the equation of geodesic deviation. These effects can be lethal outside the event horizon of a stellar mass black hole but would be mild at the event horizon of a supermassive black hole.

10. There would be a strong gravitational deflection of light close to a black hole with photons having the possibility of entering an (unstable) circular orbit at the radius of the photon sphere, $1.5R_S$.

11. Lightcones and spacetime diagrams provide valuable tools for investigating local spacetime structure in general relativity, but the behaviour of lightcones will depend on the particular coordinates being used. In Schwarzschild coordinates lightcones show abrupt changes at the Schwarzschild radius, which marks a coordinate singularity. Advanced Eddington–Finkelstein coordinates remove the coordinate singularity and produce lightcones that change in a regular way, tipping and narrowing as they approach the Schwarzschild radius. The behaviour of the lightcones at and within the Schwarzschild radius indicates the inevitability of encountering the central singularity, though more powerful methods must be used to prove that inevitability.

12. A rotating black hole is characterized by a mass M and an angular momentum magnitude $J = Mac$, and is described by the stationary, axi-symmetric Kerr solution of the Einstein vacuum field equations. In Boyer–Lindquist coordinates the solution has a central ring-shaped gravitational singularity of radius a, and coordinate singularities at the ellipsoidal surfaces

$$r = r_+ \equiv \frac{R_S}{2} + \left[\left(\frac{R_S}{2} \right)^2 - a^2 \right]^{1/2}, \qquad \text{(Eqn 6.32)}$$

$$r = r_- \equiv \frac{R_S}{2} - \left[\left(\frac{R_S}{2} \right)^2 - a^2 \right]^{1/2}, \qquad \text{(Eqn 6.33)}$$

which behave as outer and inner event horizons.

13. The ellipsoidal surface

$$r = s_+ \equiv \frac{R_S}{2} + \left[\left(\frac{R_S}{2} \right)^2 - a^2 \cos^2 \theta \right]^{1/2} \qquad \text{(Eqn 6.34)}$$

is a surface of infinite redshift that encloses the outer event horizon, meeting it only at the poles.

14. The surface s_+ also marks the static limit, within which all particles must move in the direction of rotation of the black hole.

15. The motion of massive bodies and light rays in the neighbourhood of a rotating black hole is skewed in the direction of rotation of the black hole as a consequence of the dragging of inertial frames by the black hole.

16. Quantum physics may cause the properties of real black holes to differ significantly from those of black holes in classical general relativity. In particular, Hawking radiation may allow black holes to act as thermal sources of radiation with a Hawking temperature that is inversely proportional to the mass of the black hole. If so, the explosion of isolated (mini) black holes is possible, though unlikely due to the greater probability of the accretion of mass from the surrounding environment. Quantum physics might also prevent the formation of ideal classical singularities, though this will not necessarily affect the ability of black holes to account for the energetic emissions from various galactic and stellar sources.

Chapter 7   Testing general relativity

Introduction

Up to this point, our discussion of general relativity has been mainly theoretical. This chapter concerns the experimental and observational evidence regarding general relativity. We start with the so-called 'classic tests', interpreting that term in its most liberal sense to include some experiments that were not performed until the early 1960s. We draw the dividing line at that point to separate those early tests from a number of more recent satellite-based tests, and astronomical observations of presumed black holes and gravitational lenses. We end with a section on gravitational waves. This last topic might well have been a chapter in its own right, but the theory of gravitational waves is too sophisticated to be treated fully in this book, while the observational aspects are too important to overlook. For that reason the topic is mainly treated as an observational one but is given an unusually detailed theoretical introduction.

There have been many references to tests and observations in earlier chapters. Where appropriate this chapter refers back to the material that inspired them and where necessary builds on it.

7.1   The classic tests of general relativity

7.1.1   Precession of the perihelion of Mercury

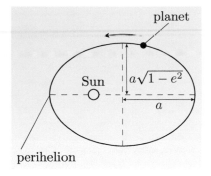

Figure 7.1   The orbit of an isolated planet around the Sun, according to Newtonian mechanics.

A famous prediction of Newtonian mechanics is that the path of an isolated planet moving around the Sun is an ellipse, with the Sun at one focus of the ellipse, as illustrated in Figure 7.1. As well as having a specific size (described by its *semi-major axis*, a) and a specific shape (described by its *eccentricity*, e), an elliptical orbit also has a specific orientation in the orbital plane. This orientation can be specified by the direction of the line joining the Sun to the point of closest approach of the planet; this point is called the **perihelion**. According to Newtonian mechanics, for a spherically symmetric Sun and an isolated planet, this direction should not change — the planet's perihelion should occur at the same point in space, orbit after orbit.

By 1845 it was known that the orbit of the planet Mercury did not behave in this way. With each successive orbit, the orbital orientation changed slightly, as shown in exaggerated form in Figure 7.2. This movement is called **perihelion precession**; a large part of it can be accounted for by using Newtonian mechanics to calculate the gravitational effect on Mercury of the other planets. However, by 1859 the work of Urbain Le Verrier (1811–1877) had shown that there was a small but significant residual movement, amounting to 43 seconds of arc per century, that could not be accounted for by any known Newtonian force. In spite of much effort over many years (including some fairly wild conjectures), no satisfactory reason for the residual precession could be found. Then in 1915, Einstein, using what would later be seen as an approximate form of the Schwarzschild metric, showed that general relativity predicts a perihelion advance of just the right amount. This was an important early triumph for the theory that did much to convince Einstein that he was on the right track.

The changing orientation of orbits in general relativity was mentioned at the end of Chapter 5, in the context of the Schwarzschild solution, where it was associated with an additional non-Newtonian term in the orbital shape equation. It can be shown that for each orbit, the perihelion advances by an angle $\Delta\phi$ given by

$$\Delta\phi = \frac{6\pi GM}{a(1 - e^2)c^2}, \tag{7.1}$$

where M is the total mass of the system (in this case dominated by that of the Sun), a is the semi-major axis, and e is the eccentricity. (A circular orbit has $e = 0$.) Clearly, $\Delta\phi$ becomes larger as a becomes smaller and as e approaches 1. Mercury has an orbit with high eccentricity and a small semi-major axis so it is a good candidate for measuring the advance of the perihelion. The original observations were carried out by means of optical telescopes but now radar ranging is used for greater precision. This enables the effect of general relativity on the precession of the perihelion of other planets (including the minor body Icarus) to be tested, as shown in Table 7.1.

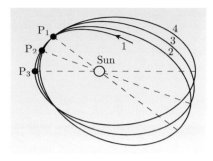

Figure 7.2 The advance of the perihelion of a planet, according to general relativity.

Exercise 7.1 Mercury has a period of 87.969 days, semi-major axis $a = 5.791 \times 10^{10}$ m and eccentricity $e = 0.2067$, and the mass of the Sun is $M_\odot = 1.989 \times 10^{30}$ kg. Calculate the general relativistic contribution to the rate of perihelion precession. Express your answer in seconds of arc per century. ∎

Table 7.1 Predicted and observed rates of residual perihelion advance in seconds of arc per century for various planets and for the minor body Icarus.

Planet	Predicted rate of advance /seconds of arc per century	Observed rate of advance /seconds of arc per century
Mercury	43.0	43.1 ± 0.5
Venus	8.6	8.4 ± 4.8
Earth	3.8	5.0 ± 1.2
Icarus	10.3	9.8 ± 0.8

7.1.2 Deflection of light by the Sun

The second testable prediction of general relativity concerns the deflection of light by a massive body. This was noted by Einstein as a general consequence of the principle of equivalence, and we saw in the previous chapter the extreme case of deflected light paths in the neighbourhood of rotating and non-rotating black holes. In the case of light rays passing close to the limb (i.e. the edge) of the Sun, the effect is small but large enough to be detectable. The effect is illustrated schematically in Figure 7.3.

Using the null geodesics of the Schwarzschild metric to represent the world-lines of light rays that pass close to a spherically symmetric body of mass M, general relativity predicts that the angle of deflection $\Delta\theta$ is given (in radians) by

$$\Delta\theta = \frac{4GM}{c^2 b}, \tag{7.2}$$

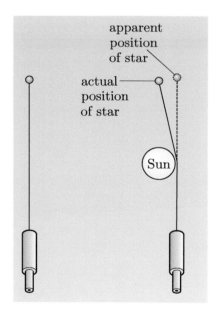

Figure 7.3 The deflection of light due to the curvature of spacetime in the vicinity of the Sun.

where b is the impact parameter (i.e. the perpendicular distance from the initial path of the light ray to the deflecting body). We can see that this effect is largest when b is as small as possible, which occurs for rays just grazing the massive body.

Exercise 7.2   Use Equation 7.2 to calculate the deflection (in seconds of arc) for rays just grazing the limb of the Sun.   ■

The first problem in trying to verify this prediction is that it's not easy to see any stars at all when the Sun is above the horizon, and it is particularly difficult to see stars that appear just beyond the edge of the Sun's disc. Observing such stars is possible during a **total eclipse of the Sun**, when the Moon is directly between the Earth and the Sun, and most of the unwanted sunlight is eliminated. However, a considerable number of experimental difficulties remain, not the least of which is poor weather conditions on the Earth during the $7\frac{1}{2}$ minutes maximum total eclipse time. Table 7.2 lists some attempts at this measurement. In spite of the experimental difficulties, it was the expeditions planned by Sir Arthur Eddington (the first two entries in this table) that gave general relativity its most publicized initial triumph and made Einstein a world-famous figure.

There seems to be little scope for improving these measurements; for example, a measurement in 1975 gave a deflection that was 0.95 ± 0.11 times the prediction of general relativity, which is consistent, but hardly a precision confirmation. Such optical measurements have been superseded by **radio interferometry**. The idea is that by using two radio telescopes, one can measure the very small differences between the times that particular wave crests arrive at the two observatories. The resolution is proportional to the distance between the radio telescopes and this has led to the development of **very long baseline interferometry** (VLBI), involving two or more observatories, often separated by thousands of kilometres, emulating one giant telescope. Using radio transmission from certain quasars (which are so distant as to be almost point sources of radio waves) and measuring the deflection as the source is eclipsed by the Sun, the predicted gravitational deflection has been verified to better than 0.04%.

7.1.3   Gravitational redshift and gravitational time dilation

The third testable prediction of general relativity concerns gravitational time dilation and the related gravitational redshift. This effect was also predicted at an early stage in the development of general relativity, based on the principle of equivalence. A detailed quantitative prediction for a stationary emitter and a stationary observer was given in Chapter 5 using the Schwarzschild metric. The general relationship obtained there was

$$\mathrm{d}\tau_{\mathrm{ob}} = \left(1 - \frac{2GM}{c^2 r_{\mathrm{ob}}}\right)^{1/2} \mathrm{d}t_{\mathrm{em}}, \tag{Eqn 5.14}$$

where $\mathrm{d}t_{\mathrm{em}}$ represents the coordinate time separating two events at the location of the stationary emitter, and $\mathrm{d}\tau_{\mathrm{ob}}$ is the proper time separating sightings of those two events by a stationary observer at radial coordinate position r_{ob}. When the observer is far away, so that $r_{\mathrm{ob}} \to \infty$, we can represent $\mathrm{d}\tau_{\mathrm{ob}}$ by $\mathrm{d}\tau_\infty$ and write

$$\mathrm{d}\tau_\infty = \mathrm{d}t_{\mathrm{em}}. \tag{Eqn 5.15}$$

Table 7.2 History of observations of light bending, 1919–52. (Source: Sciama, D.W. (1972) *The Physical Foundation of General Relativity*, Heinemann Educational Books.)

Observatory (and place of observation)	Eclipse	Number of stars	Minimum distance of star from Sun, in solar radii from centre	Maximum distance of star from Sun, in solar radii from centre	Mean angle of deflection* in seconds of arc	Uncertainty in seconds of arc
Greenwich (Brazil)	29 May 1919	7	2	6	1.98	0.16
		11	2	6	0.93	—
Greenwich (Principe)	29 May 1919	5	2	6	1.61	0.40
Adelaide–Greenwich (Australia)	21 Sept 1922	11–14	2	10	1.77	0.40
Victoria (Australia)	21 Sept 1922	18	2	10	1.75 1.42 2.16	—
Lick I (Australia)	21 Sept 1922	62–85	2.1	14.5	1.72	0.15
Lick II (Australia)	21 Sept 1922	145	2.1	42	1.82	0.20
Potsdam I (Sumatra)	9 May 1929	17–18	1.5	7.5	2.24	0.10
Potsdam II (Sumatra)	9 May 1929	84–135	4	15	—	—
Sternberg (USSR)	19 June 1936	16–29	2	7.2	2.73	0.31
Sendai (Japan)	19 June 1936	8	4	7	2.13 1.28	1.15 2.67
Yerkes I (Brazil)	20 May 1947	51	3.3	10.2	2.01	0.27
Yerkes II (Sudan)	25 Feb 1952	9–11	2.1	8.6	1.70	0.10

* This is the value estimated for a light ray grazing the Sun, obtained by an extrapolation of the shift in apparent position of a number of stars.

At the location of the emitter, where $r = r_{em}$,

$$dt_{em} = \left(1 - \frac{2GM}{c^2 r_{em}}\right)^{-1/2} d\tau_{em},$$

so we get the following relation between the proper time separating events at the emitter and the proper time separating their sighting by the distant observer:

$$d\tau_\infty = \left(1 - \frac{2GM}{c^2 r_{\text{em}}}\right)^{-1/2} d\tau_{\text{em}}. \qquad \text{(Eqn 5.16)}$$

Since frequency is inversely proportional to period, we arrive at the following prediction concerning the gravitational redshift in the radiation from a stationary emitter:

$$f_\infty = \left(1 - \frac{2GM}{c^2 r_{\text{em}}}\right)^{1/2} f_{\text{em}}. \qquad \text{(Eqn 5.17)}$$

It was hoped that this effect would be seen in the spectra of stars, as a reduction in the observed frequency of spectral lines. In fact, in the 1916 paper that contained the first complete formulation of general relativity, Einstein referred to the astronomer Erwin Freundlich, saying:

> According to E Freundlich, spectroscopical observations on fixed stars of certain types indicate the existence of an effect of this kind, but a crucial test of this consequence has not yet been made.

Unfortunately, such a test was very difficult to perform. Early attempts based on normal stars were inconclusive. The spectra were easy to observe, but the anticipated gravitational redshift turned out to be small compared with other effects, such as Doppler shifts due to turbulence in the star's atmosphere. Observing the spectra of dense stars (where M is relatively large and r_{em} is relatively small) provided better prospects of success. The first white dwarf was discovered in 1910 — attention was drawn to it in 1914 — and a second white dwarf, the companion to Sirius, was found by the American astronomer Walter Adams in 1915. Eddington emphasized the exceptional density of these stars in the 1920s and pointed out the large gravitational redshift that they should exhibit. In 1925, careful measurements by Adams confirmed these expectations but the 'test' was not very precise. More precise astronomical measurements were eventually performed but only after gravitational redshift had been used in the first precise laboratory-based test of general relativity.

The Pound–Rebka experiment

In 1960, Robert Pound (1919–) and Glen Rebka (1931–) published the results of a terrestrial measurement of gravitational redshift. Before describing the experiment itself, let's examine the theoretical basis of the test. If we use m to represent the mass of the Earth and f_r to represent the proper frequency of an emitter located at coordinate radius r (measured from the centre of the Earth), the gravitational redshift relationship of Equation 5.17 tells us that

$$f_r = \left(1 - \frac{2Gm}{c^2 r}\right)^{-1/2} f_\infty, \qquad (7.3)$$

and for the values of interest this is well approximated by the relation

$$f_r = \left(1 + \frac{mG}{c^2 r}\right) f_\infty. \qquad (7.4)$$

We now want to relate the frequency of light emitted from the original point at coordinate radius r to the frequency of light received at some different point with radial coordinate $r + h$. The best way to think of this is to imagine a train of waves with period $\Delta \tau_r$ at radius r and period Δt at a point at infinity, i.e. Δt is the coordinate time interval corresponding to $\Delta \tau_r$. At whatever radius the radiation is received, the coordinate time interval (and its reciprocal f_∞) will be the same, so f_{r+h}, the measured frequency at radius $r + h$, must be

$$f_{r+h} = \left(1 + \frac{mG}{c^2(r + h)} \right) f_\infty. \tag{7.5}$$

If h is small, then a first-order Taylor expansion shows that the frequency measured at $r + h$ differs from f_r by

$$\Delta f_r = f_{r+h} - f_r \approx h \times \frac{\mathrm{d}}{\mathrm{d}r} f_r. \tag{7.6}$$

Using Equation 7.4 to evaluate the derivative, we see that

$$\Delta f_r \approx -\frac{mG}{c^2 r^2} f_\infty h \tag{7.7}$$

and therefore, from Equations 7.4 and 7.7, for small $mG/c^2 r$

$$\frac{\Delta f_r}{f_r} \approx -\frac{mGh}{c^2 r^2} \left(1 + \frac{mG}{c^2 r} \right)^{-1} \approx -\frac{mGh}{c^2 r^2}. \tag{7.8}$$

Now suppose that h represents a small difference in height above the Earth's surface. So, with $r = R$, the radius of the Earth, we have

$$\frac{\Delta f_R}{f_R} = -\frac{mG}{c^2 R^2} h. \tag{7.9}$$

But the acceleration due to gravity on the surface of the Earth has magnitude $g = mG/R^2$, so finally

$$\frac{\Delta f_R}{f_R} = -\frac{gh}{c^2}, \tag{7.10}$$

where Δf_R is the difference between the frequency of the emitter in its own rest frame and the frequency that would be measured on receiving its light in a rest frame at a height h above the emitter.

Pound and Rebka were able to measure the gravitational redshift of photons travelling vertically through a distance of just $22.5\,\mathrm{m}$ in a tower at Harvard University's Jefferson Laboratory (Figure 7.4). This was only possible due to the discovery of the **Mössbauer effect** a year or so earlier. Normally, when an atom emits or absorbs a photon, it also recoils a little as required by conservation of momentum. This recoil takes away some energy from the photon, making its frequency a little uncertain. The associated change in photon frequency is typically about five orders of magnitude greater than the expected gravitational redshift for a photon travelling vertically through a distance of $22.5\,\mathrm{m}$. So, normally, recoil effects would ruin any attempt to measure the gravitational redshift. However, in 1958 Rudolf Mössbauer (1929–) showed that in some crystalline solids a significant number of relatively low frequency gamma-ray emissions involve the whole crystal lattice absorbing the recoil momentum. In such cases, the movement of the emitting atom is very small and consequently the frequency of the emitted gamma-ray photon is very well-defined. It turns out that

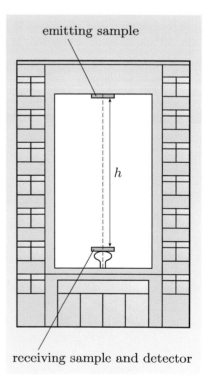

Figure 7.4   A schematic representation of the Pound–Rebka gravitational redshift experiment.

only a few elemental solids satisfy the necessary conditions for observing the Mössbauer effect, and Fe-57 has proved to be by far the most popular.

In the **Pound–Rebka experiment**, a solid sample containing Fe-57, which emits 14 keV gamma rays, was placed in the centre of a loudspeaker cone near the top of the tower. By vibrating the loudspeaker cone, varying Doppler shifts were created in the photons emitted by the gamma-ray source. The Doppler-shifted gamma rays travelled vertically downwards through a Mylar bag filled with helium in order to minimize scattering of the gamma rays. Another sample containing Fe-57 was placed in the basement, and a scintillation counter was placed below this in order to detect the gamma rays that were not absorbed by the receiving sample. When the Doppler shift imparted by the loudspeaker cancelled out the gravitational redshift, the receiving sample selectively absorbed the gamma rays, and the number of gamma rays detected by the scintillation counter dropped significantly. The variation in absorption could be correlated with the vibration frequency of the loudspeaker and hence with the Doppler shift and the gravitational redshift that it cancelled. This experiment by Pound and Rebka confirmed the gravitational redshift predictions of general relativity to about 10%, and this was later improved to better than 1% by Pound and Snyder.

Beyond the Pound–Rebka experiment

In 1976, in an experiment known as **Gravity Probe A**, a hydrogen maser (a stable source of radiation with a very precise frequency) was briefly sent to a height of 10 km above the Earth, while its emissions were monitored from the ground. This experiment confirmed the predictions of gravitational time dilation to about 70 parts per million.

An interesting application of gravitational time dilation is provided by the **Global Positioning System** (GPS). The GPS uses between 24 and 32 satellites that transmit precise microwave signals, enabling GPS receivers on or near the Earth's surface to determine their location, speed, direction and time. Each satellite contains an atomic clock and orbits at about 20 200 km above the Earth's surface. Since a satellite clock is in a weaker gravitational field than a ground-based one, it will tick more rapidly. Corrections are made for this effect by setting the satellite clock frequency to slightly less than the nominal frequency of 10.23 MHz. Because the functioning of the GPS is based on accurate timing, the effect of general relativity is significant, and if appropriate corrections were not made, errors in the positions of GPS receivers would accumulate at the rate of tens of kilometres per day. The continued accurate functioning of the GPS is therefore an experimental verification of general relativity. However, the accuracy of the verification (about 1%) is no better than for other experiments.

Exercise 7.3   (a) Calculate the time dilation due to general relativity for a GPS satellite clock compared to a ground-based clock.

(b) Calculate the time dilation due to special relativity for a GPS satellite clock compared to a ground-based clock. (Ignore the satellite's acceleration.)

(c) Estimate the error that results in a ground-based GPS receiver from the combined effect of (a) and (b). ∎

7.1.4 Time delay of signals passing the Sun

The three tests of general relativity that we have described so far could fairly be called the *classic tests* since they were proposed early in the history of the subject. However, a further classic test of general relativity, exploiting exceptionally high-powered radar, was proposed by Irwin I. Shapiro in 1964. The basic idea of the **Shapiro time delay experiment** is to record the transit times of radar signals from the Earth to a nearby planet (such as Mercury or Venus) and back. If the planet is just slipping around the back of the Sun (see path C–C' in Figure 7.5), then the radar pulse will probe the region close to the Sun where the spacetime metric differs most from that of special relativity. Since the orbit of the planet is well known from other astronomical observations, we can predict the travel times for all pulses going to and returning from the planet at any point in its orbit. If we made predictions assuming that spacetime is flat, we would find that they agree with experiment for all pulses except those that go close to the Sun's edge. These pulses, which are probing the curved spacetime near to the Sun, take a slightly *longer* time than expected to come back.

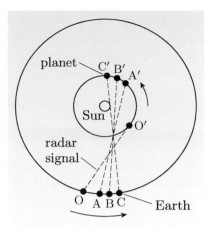

Figure 7.5 A radar time delay experiment between the Earth and a nearby planet.

Using the Schwarzschild metric to represent the spacetime near the Sun, it can be shown that the total round-trip time for a radar pulse that travels from the Earth to the planet and back, with the pulse just grazing the Sun's surface, is approximately given by

$$\Delta T(\text{Earth–planet–Earth}) \approx \frac{2}{c}\left[(R_E^2 - R_\odot^2)^{1/2} + (R_P^2 - R_\odot^2)^{1/2}\right]$$
$$+ \frac{4k}{c}\left\{\ln\left(4\frac{R_E R_P}{R_\odot^2}\right) + 1\right\}, \qquad (7.11)$$

where k is the Schwarzschild metric parameter ($= GM_\odot/c^2$ in this case) and $R_\odot$, R_E and R_P are the radial coordinates of the Sun's surface, the Earth and the planet, respectively, as shown in Figure 7.6. The first thing to notice is what happens to this result if we set k equal to zero. This corresponds to saying that spacetime is everywhere like that of special relativity. The total travel time reduces in this case to

$$\Delta T(k = 0) = \frac{2}{c}\left[(R_E^2 - R_\odot^2)^{1/2} + (R_P^2 - R_\odot^2)^{1/2}\right]. \qquad (7.12)$$

This is just what we would expect; we would obtain precisely this result if we used Euclidean geometry to work out the total distance there and back (contained in the square bracket) and then divided the result by c to get the total travel time of the pulse. It is therefore the last term in curly brackets in Equation 7.11, multiplied by $4k/c$, that represents the effect of curved spacetime on ΔT.

Equation 7.11 allows us to calculate the extra time delay due to the spacetime curvature. We know that light from the Sun takes about 8 minutes to get to the Earth. Thus the first term of Equation 7.11 will be of order 16 to 40 minutes, depending on the planet used. Now $4k/c$ ($= 4GM_\odot/c^3$) is about 20 μs; so unless the term in the curly bracket is very large (which it won't be — typical values are 10 to 15), the extra time delay predicted by general relativity is a tiny fraction of the total travel time. This illustrates the fact that general relativity predicts extremely small departures from Newton's theory everywhere within the Solar System; there are simply no sufficiently large concentrations of mass within the Solar System for it to be otherwise.

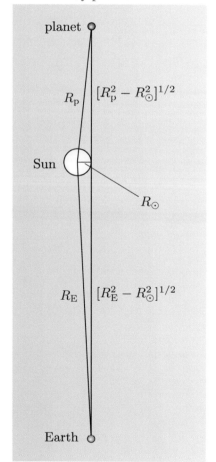

Figure 7.6 A radar pulse from Earth (E) just grazing the Sun on its way to planet P. In Shapiro's experiment, P was Mars, which is more distant from the Sun than is Earth.

We can also see that the effect of the expression in the curly brackets of Equation 7.11 is to *increase* the time of travel of the pulse from that expected for the spacetime of special relativity; general relativity predicts a time *delay*. The quantity whose logarithm is to be taken can be written as

$$4 \left(\frac{R_{\text{E}}}{\text{R}_{\odot}} \right) \left(\frac{R_{\text{P}}}{\text{R}_{\odot}} \right).$$

Since

$$R_{\text{E}} \gg \text{R}_{\odot} \quad \text{and} \quad R_{\text{P}} \gg \text{R}_{\odot},$$

we know that

$$4 \left(\frac{R_{\text{E}}}{\text{R}_{\odot}} \right) \left(\frac{R_{\text{P}}}{\text{R}_{\odot}} \right) \gg 1,$$

and because natural logarithms of numbers greater than unity are positive, it follows that the whole term in curly brackets is positive.

Finally, we can put in some typical values of R_{E} and R_{P}, and the value of $\text{R}_{\odot}$, to get a quantitative estimate of the time delay caused by the effect of the Sun on the spacetime near it. At the outset of this calculation we should mention that the experimental problems involved in measuring radar pulse travel times are considerable, coming from a variety of sources, and we cannot do justice to the experiments here. A variation on Shapiro's suggestion is to measure the time delay experienced by a signal transmitted by an artificial satellite or planetary probe as the signal passes close to the Sun. An example is given by experiments conducted during NASA's Viking mission to Mars. This consisted of two space probes (launched in 1975) that orbited Mars, each equipped with a lander to study the planet from its surface. While one of the landers was on the surface of Mars, the time delay in a signal whose path was close to the Sun was measured. In this case we must interpret R_{P} as the distance of Mars from the Sun: 2.254×10^{11} m. Putting this quantity along with $R_{\text{E}} = 1.496 \times 10^{11}$ m, $\text{R}_{\odot} = 6.960 \times 10^{8}$ m and $4k/c = 4G\text{M}_{\odot}/c^{3} = 1.971 \times 10^{-5}$ s into the expression

$$\frac{4k}{c} \left\{ \ln \left(\frac{4R_{\text{E}}R_{\text{P}}}{\text{R}_{\odot}^{2}} \right) + 1 \right\}$$

gives a predicted maximum time delay of $267 \, \mu$s. The maximum delay observed in the Viking experiment was $250 \, \mu$s; so our general relativistic calculation gives a reasonably accurate prediction of a time-delay effect of the Sun on a radio signal.

Other space probes have subsequently been used in the measurement of the time delay experienced by a signal passing close to the Sun. NASA's Voyager mission consisted of two probes, Voyagers 1 and 2, which were launched in 1977 with the aim of passing close to all the planets in the Solar System. The probes are still functioning and are now in the outer reaches of the Solar System. The time delay obtained using these probes is in agreement with the theoretical predictions with an accuracy of one part in one thousand. The Cassini probe was launched in 1997 with the aim of orbiting Saturn. In 2003, measurements on signals from the Cassini probe confirmed that the time delay agreed with the predictions of general relativity to about 20 parts in a million.

This first section on classic tests of general relativity can be summarized as follows.

Classic tests

The four classic tests of general relativity are as follows.

1. **The precession of the perihelion of Mercury** The observations, which have an uncertainty of about 1%, are consistent with the predictions of general relativity.

2. **Deflection of starlight by the Sun** The observations, which have an experimental uncertainty of about 10% for optical wavelengths, are in agreement with the predictions of general relativity. The agreement is better than 0.04% for VLBI radio telescope observations.

3. **Gravitational redshift** Gravitational redshift has been verified to better than 1% in variants of the Pound–Rebka experiment. Gravity Probe A verified the time dilation due to general relativity to 70 parts per million. The continued functioning of the GPS confirms general relativistic time dilation to about 1% on a daily basis.

4. **Time delay of electromagnetic radiation passing the Sun** The Cassini probe confirmed the effect to about 20 parts per million.

7.2 Satellite-based tests

Soon after the formulation of general relativity, the Dutch astronomer Willem de Sitter (1872–1934) used Einstein's theory to show that there would be a non-Newtonian contribution to the behaviour of the angular momentum of the Earth–Moon system as it orbited the Sun. The de Sitter effect, sometimes called the *solar geodetic effect*, is too small to provide a viable test of general relativity, but its discovery prompted others to consider more generally the way in which spinning bodies would transport angular momentum through curved spacetime. This led to predictions concerning the behaviour of orbiting gyroscopes that have recently been tested. This section first introduces the general relativistic phenomena involved in those tests and then discusses some of the results obtained.

7.2.1 Geodesic gyroscope precession

A **gyroscope** is a device that uses the angular momentum of a spinning body to indicate a particular direction in space. Gyroscope designs vary, but a common sort consists of a heavy rotatable disc mounted in a set of very low friction bearings that allow the disc's axis of rotation to point in any direction (Figure 7.7). The disc is symmetric, so when it is made to spin rapidly, its angular momentum is aligned with the axis of rotation. In a flat spacetime the whole gyroscope can be moved without altering the angular momentum of the disc, so the axis of rotation will indicate a fixed direction in space. This principle is used as the basis of the *gyrocompass*, which has many applications in air and sea navigation.

In a region where spacetime is curved, the situation is rather different. In curved spacetime, the centre of mass of a freely falling gyroscope will move along a geodesic, and the angular momentum of the gyroscope will be transported along that geodesic. We saw earlier that the four-velocity of a freely falling particle is

Figure 7.7 A common form of gyroscope.

parallel transported along the geodesic that the particle follows, and in a similar way the angular momentum associated with the spin of a freely falling gyroscope will also be parallel transported along the geodesic. Even so, the presence of curvature will generally cause the direction of the spin angular momentum to change. (You saw in Chapter 3 that when a vector is parallel transported around a closed loop, the orientation of that vector changes in a way that depends on the spacetime curvature.)

As a comparatively straightforward example, consider a gyroscope moving in free fall around a spherically symmetric body of mass M. Suppose that the gyroscope is in a polar orbit of radius r, and that initially the spin angular momentum vector of the gyroscope points radially away from the centre of the massive body. In a flat spacetime we know that after one complete orbit the angular momentum vector will remain radial and that this will still be true after any number of orbits. However, according to general relativity the spacetime in the vicinity of the gyroscope is not flat but can be described by the Schwarzschild metric. Using this metric, it can be shown that after one orbit the angular momentum vector of the gyroscope is no longer radial but will have precessed by a small angle α in the plane of the orbit, as shown in Figure 7.8. The precession angle α is given by

$$\alpha = 2\pi \left[1 - \left(1 - \frac{3GM}{c^2 r} \right)^{1/2} \right]. \tag{7.13}$$

This effect is sometimes known as **geodesic gyroscope precession**, though it is also often referred to as the **geodetic effect**. It is a very small effect, but since it is cumulative, it can become significant over many orbits.

Exercise 7.4   Confirm that for a gyroscope with angular momentum vector initially radial, in a low Earth orbit, the precession is about $8''$ per year. ∎

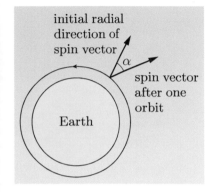

Figure 7.8   Geodesic gyroscope precession. The angle α is exaggerated for clarity.

7.2.2   Frame dragging

In the neighbourhood of a rotating body, such as a rotating black hole, spacetime is more accurately described by the axially symmetric Kerr metric rather than the spherically symmetric Schwarzschild metric. As you saw earlier, the Kerr metric implies the dragging of inertial frames around the rotating body. This too can give rise to gyroscopic precession, though it is quite distinct from the geodesic precession described in the previous section.

The rotational dragging of inertial frames is sometimes referred to as the **Lense–Thirring effect** after Josef Lense (1890–1985) and Hans Thirring (1888–1976), the scientists who deduced the existence of such an effect in 1918, long before the introduction of the Kerr metric. In fact, the rotational dragging of inertial frames is a particular case of a more general phenomenon of *frame dragging* that takes place whenever there is a significant movement of matter (a *mass current*) in the neighbourhood of a locally inertial frame.

For a slowly rotating body, such as the Earth, the Lense–Thirring effect is very small and difficult to observe. One way to understand the consequences of frame dragging is to consider a satellite in a polar orbit about the Earth. If the Earth was isolated, perfectly symmetric, and didn't rotate, then the plane of the satellite's orbit would remain fixed. However, since the Earth does in fact rotate about an axis through the poles, frame dragging predicts that the plane of the satellite's orbit will rotate very slowly in the same direction as the Earth's rotation, as indicated in Figure 7.9. An effect of frame dragging is to induce a very small precession in a gyroscope orbiting the Earth. If the rotation axis of the gyroscope is initially in the equatorial plane of the planet and points radially away from the planet's centre, then the Lense–Thirring effect will cause the spin axis to precess eastward but the rate will be less than 1% of that due to geodesic precession.

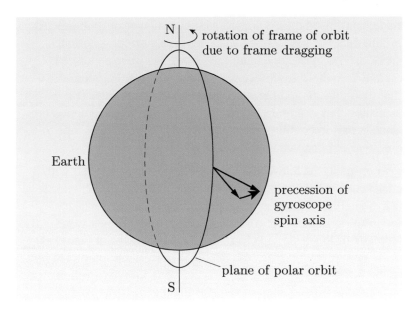

Figure 7.9   Frame dragging for a satellite in a polar orbit.

7.2.3 The LAGEOS satellites

The satellites **LAGEOS I** (launched in 1976) and **LAGEOS II** (launched in 1992) are simply heavy (411 kg) spheres, 60 cm in diameter, that orbit at a height of 5900 km above the Earth's surface. They have no on-board electronics, but are covered in retro-reflectors, which are used for laser ranging from ground tracking stations. One of the satellites is shown in Figure 7.10.

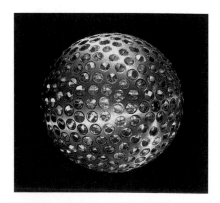

Figure 7.10   A LAGEOS satellite.

The satellites enable very accurate measurements to be made of their positions relative to points on the Earth's surface. Such observations have been used to produce an accurate picture of how the Earth's gravitational field differs from that produced by a uniform sphere, and to make precise measurements of continental drift. One research group claims that the plane of the orbits of the LAGEOS I and II satellites appears to be shifting, confirming the frame dragging prediction of general relativity to better than 10%. However, the result is highly controversial because other estimates of the probable error are very much higher than 10%. The most common view amongst experts in the field is that the LAGEOS results are interesting but inconclusive. They do not call general relativity into question, but nor do they provide any meaningful confirmation of the theory.

7.2.4   Gravity Probe B

Gravity Probe B was an ambitious project using cutting edge technology to test general relativity. It was based on a polar orbiting satellite that was launched in April 2004 to a height of 642 km above the Earth.

To give a greatly simplified description of the experiment, the satellite contained a telescope and a set of four gyroscopes (four were used to increase the sensitivity and provide redundancy). Each gyroscope took the form of an electrically levitated sphere made from fused quartz coated with a thin layer of niobium. At the time of their production, the gyroscopes were the most perfect spherical objects ever constructed. The gyroscopes and their housings were contained within lead shields, and the whole assembly was cooled to a few degrees above absolute zero so that the niobium and the lead were superconducting. The superconductivity ensured that external electromagnetic fields were screened out and played an important part in enabling the rotation axis of each gyroscope to be accurately monitored without disturbing the rotation.

At the start of the experiment, the telescope and gyroscopes were aligned with a guide star and the telescope was kept aligned with that guide star for 50 weeks, during which time the satellite continued in its polar orbit. The idea was to measure the change in the spin axis alignment of each gyroscope over the 50 weeks (a) in the plane of the orbit and (b) in the Earth's equatorial plane, as shown in Figure 7.11. Result (a) indicates the geodesic precession, predicted by general relativity to be 6.606 arcseconds (0.0018°) per year. Gravity Probe B was expected to test this result to an accuracy of 0.01%. Result (b) is the frame dragging precession due to the Lense–Thirring effect and had not previously been measured. Gravity Probe B was expected to test this result to an accuracy of 1%.

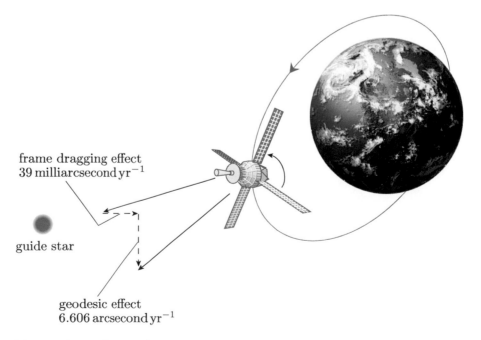

Figure 7.11 Changes in the spin axis alignment of a gyroscope in the Gravity Probe B experiment.

frame dragging effect
39 milliarcsecond yr^{-1}

guide star

geodesic effect
6.606 arcsecond yr^{-1}

The results so far are that (a) the experiment has confirmed the geodesic precession effect to 1.5%, but (b) the expected frame dragging is below the noise level of the data. This noise is due to unexpected torques on the gyroscopes, which the project team is still trying to model at the time of writing.

We summarize the results of this section as follows.

Satellite-based tests

Satellite-based tests aim to detect two effects:

- geodesic gyroscope precession (geodetic effect)
- rotational frame dragging (Lense–Thirring effect).

Two satellite-based tests are:

1. The LAGEOS satellite results, which have been claimed to confirm frame dragging to 10%, but this is disputed.

2. Gravity Probe B results, which confirm geodesic gyroscope precession to 1.5%. The expected frame dragging is below the noise level, though there is still some hope that further analysis might improve the situation.

Exercise 7.5   Calculate the expected geodesic precession per year for a gyroscope in the Gravity Probe B experiment.   ■

7.3   Astronomical observations

This section concerns astronomical observations of *gravitational lenses* and systems believed to contain *black holes*. Neither kind of observation provides a direct test of general relativity, but each concerns non-Newtonian behaviour and contributes to the body of circumstantial evidence that supports general relativity. There is an important additional strand of evidence that comes from observations of *pulsars* (rotating magnetic neutron stars), but this is considered separately in the next section.

7.3.1   Black holes

Black holes were discussed at length in Chapter 6. There, they were mainly treated as idealized classical spacetime structures in which a singularity is contained within an event horizon. It was suggested that such singularities might arise from the catastrophic gravitational collapse of stars that had exhausted their core nuclear fuel and were too massive to exist stably as white dwarfs or neutron stars. It was pointed out that quantum effects might prevent the formation of singularities, but no mechanism for this is currently known, and even if it happened, it would not preclude the existence of bodies that are essentially indistinguishable from black holes. Once a black hole is formed, its mass can increase due to the capture of stars, interstellar matter or other black holes.

Evidence concerning black holes is most easily organized by considering in turn the various mass regimes: *mini*, *stellar*, *intermediate* and *supermassive*.

Mini black holes

Black holes with masses in the range $0 \, \mathrm{M_\odot}$ to $0.1 \, \mathrm{M_\odot}$ (where $\mathrm{M_\odot}$ is the mass of the Sun) have not been observed. Very low mass black holes will be sought in the high-energy proton collisions at the Large Hadron Collider in CERN. Higher mass mini black holes have already been sought astronomically but without success. This is not altogether surprising since there is no obvious route for their production, though they might been formed in the early Universe. As we saw in Chapter 6, evaporating mini black holes are expected to emit Hawking radiation and should end their lives in an explosion. Such explosions could release detectable amounts of gamma radiation. Astronomical sources of gamma-ray bursts have been detected, but their properties are different from those expected of an exploding mini black hole so the two phenomena are currently thought to be unrelated. The Hawking radiation from any mini black holes that do exist will contribute gamma rays and particles such as antiprotons to the *cosmic radiation* that reaches the Earth from space. Studies of the composition of cosmic rays not only fail to give direct evidence of mini black holes, but also impose limits on the abundance of mini black holes in the Universe.

Stellar mass black holes

Black holes with masses in the range of a few $M_{\odot}$ to a few tens of $M_{\odot}$ are such feeble sources of Hawking radiation that, for all practical purposes, they are truly 'black' and therefore not directly observable. Nonetheless, substantial indirect evidence of their existence has been (and continues to be) accumulated. This evidence comes mainly from the study of **binary star systems** in which the supposed black hole is detected via its interaction with a companion star. The components of a binary system can sometimes be sufficiently close together that material from the atmosphere of a star is transferred to the companion body. The transfer is particularly easy if the donor star is a giant or a supergiant with an enormously distended atmosphere and a significant stellar wind, or if the two stars are close enough together for the donor star to fill its Roche lobe. (The *Roche lobe* is the teardrop-shaped region around a star where the gravitational effect of the star is stronger than that due to its binary companion.) Either method of mass transfer can lead to the emission of X-rays if the receiving body is a compact object, such as a black hole, a neutron star or possibly a white dwarf. The transferred material is quite likely to have too much angular momentum to fall directly onto the compact object. If so, it will form a rotating disc around the compact object. The study of these discs has become an important topic in astrophysics and is discussed in detail in this book's companion volume, *Extreme Environment Astrophysics* by Ulrich Kolb.

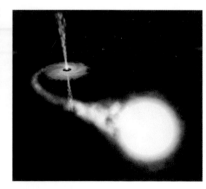

Figure 7.12   An artist's impression of an X-ray emitting binary system that includes an accretion disc. This impression includes two axial jets, which are a feature of some systems. These must originate outside the event horizon and may be magnetically driven.

The material in a rotating disc encircling a black hole is subject to tidal effects and to friction. These will heat the disc material and cause it to spiral inwards to the point where it can be accreted by the compact body. It is for this reason that these discs are usually referred to as **accretion discs**. The heating of the in-falling matter is such that it can emit X-rays, making the system a suitable target for detection by astronomers working at X-ray wavelengths. Many X-ray emitting binary systems are now known, and an artist's impression of such a system is given in Figure 7.12.

The task of the black hole hunter is to distinguish those systems in which the compact object must be a black hole from those in which it might be a neutron star or a white dwarf. This is done on the basis of the compact object's mass. It is known that there is an upper limit to the mass of a white dwarf (the Chandrasekhar limit, about $1.4 \, M_\odot$) and also an upper limit to the mass of a neutron star (the **Oppenheimer–Volkoff limit**, about $2.5 \, M_\odot$). Consequently, an X-ray emitting binary system in which there is a compact partner that can be shown to have a mass that exceeds the Oppenheimer–Volkoff limit is regarded as containing a black hole. The Oppenheimer–Volkoff limit is not particularly well determined so, generally speaking, the greater the mass of the candidate, the better the case for believing it to be a black hole. Unfortunately, the mass determination is rarely straightforward. It is usually based on observations of Doppler shifts in the frequency of the radiation emitted by the system and can be subject to uncertainty arising from the inclination of the compact body's orbit.

One well-known stellar mass black hole candidate is Cygnus X-1, the strongest X-ray source in the constellation of Cygnus. It was first detected in 1964, in the early days of X-ray astronomy, using a rocket-borne detector. Later studies confirmed it as an intense source of X-rays but also showed that it was a highly irregular variable source. Its shortest fluctuations are on timescales of milliseconds, implying that the X-ray emitting region is unlikely to be more than about a millilightsecond across (300 km), which is just what might be expected of a gravitationally collapsed star and the inner part of an accretion disc. In the early 1970s, when the position of Cygnus X-1 was accurately determined for the first time, it was found to be associated with the blue supergiant star HDE 226868. Periodically varying Doppler shifts in the spectral lines of that star indicate that it is part of a binary system with a 5.6-day orbital period. The amplitude of the variations in Doppler shift provides further information about the orbit, and together with the period strongly suggests that the compact companion has a mass that is greater than $4.8 \, M_\odot$. Additional arguments concerning the system's distance and its lack of eclipses suggest that the mass of the compact component is actually well above this minimum, probably in the range 7–$13 \, M_\odot$. All this makes it very likely that Cygnus X-1 consists of a black hole with an accretion disc that is supplied with matter by HDE 226868. About 20 broadly similar stellar mass black hole systems are currently known, with a further 20 or so candidate systems, representing a range of black hole and companion star masses.

The evidence that some X-ray emitting binaries contain a compact object that is too massive to be a neutron star is strong. But the additional step of saying that this object is a black hole is based on the lack of any credible alternative; there is no direct evidence of an *event horizon* or any other feature that might be considered specific to general relativity. However, indirect evidence that an event horizon is present can be obtained from the observed variations in the intensity of X-rays emitted by such binary systems. Much of this variation is attributed to changes in the rate at which matter is being supplied to the central compact object via the accretion disc. When the X-ray intensity is low, it is presumed that the rate of in-fall is small — perhaps little more than a trickle. Under these circumstances material falling onto a neutron star would continue to contribute to the total intensity of the source as long as it was hot, but material falling into a black hole would be lost from sight as it dimmed rapidly when approaching the event horizon. If the observed X-ray emitting binaries are divided into two classes according to whether the compact object has a mass below $2 \, M_\odot$ or above $3 \, M_\odot$, it is found that the former objects have a higher minimum X-ray intensity than the latter. This has been interpreted as evidence that in the latter case, where the compact object has a mass that is above the Oppenheimer–Volkoff limit, an event horizon is indeed present. We shall have more to say about X-ray evidence later.

In addition to the evidence from close binary systems, there is additional evidence for stellar mass black holes from a process known as *gravitational microlensing*. This is sensitive to isolated black holes as well as those in binary systems. It will be mentioned again when we discuss gravitational lensing in the next section.

Intermediate mass black holes

Black holes with masses in the range $100\,M_\odot$ to $10^5\,M_\odot$ have been sought for many years. It is probably fair to say that there is growing evidence that they may exist in various clusters of stars both within the Milky Way and in some external galaxies. However, there are still many astronomers who doubt the existence of black holes in this class, especially because it is not clear how they would form. Since their existence is still in doubt both theoretically and observationally, intermediate black holes cannot currently be said to provide any sort of test of general relativity.

Supermassive black holes

Black holes with masses in excess of $10^5\,M_\odot$ are not only thought to exist, but are believed to be common. The most direct evidence for their existence comes from studying the behaviour of stars and gas clouds close to the centres of galaxies. In the case of our own galaxy, the Milky Way, extensive studies of this kind, based on observations of stellar orbits at infrared wavelengths, have provided strong evidence of a compact central object with a mass of about $2.5 \times 10^6\,M_\odot$, contained within a volume comparable to that of the inner Solar System. This object is associated with Sagittarius A* (pronounced *A-star*), a strong radio source located at the centre of the Milky Way. Another example is at the centre of the galaxy NGC 4258, which has been observed using very long baseline interferometry (VLBI). The results show clear evidence of a compact object with a mass of $4 \times 10^7\,M_\odot$. Many other examples are known, and there is growing evidence that each of these central objects has a mass that is directly related to the mass of the spheroidal component of its host galaxy. This correlation suggests that the formation of galactic centre black holes may be a natural part of the process of galaxy formation rather than something that happens by accident in a few galaxies.

● What are the Schwarzschild radii corresponding to $2.5 \times 10^6\,M_\odot$ and $4 \times 10^7\,M_\odot$?

○ The Schwarzschild radius $R_S = 2GM/c^2$ corresponding to $1\,M_\odot$ is $3\,\text{km}$. The Schwarzschild radius grows in proportion to mass, so $2.5 \times 10^6\,M_\odot$ corresponds to $7.5 \times 10^6\,\text{km}$, and $4 \times 10^7\,M_\odot$ corresponds to $12 \times 10^7\,\text{km}$.

Dynamical studies of stars and gas clouds close to galactic centres give evidence of compact massive bodies but they do not prove that those bodies really are black holes. However, this issue is addressed to some extent by detailed studies of X-ray spectra.

Figure 7.13 shows a distorted spectral line seen in the X-ray spectrum of the galaxy MCG-6-30-15. This feature is believed to be due to ionized iron atoms that travel around the galaxy's central black hole as part of an encircling accretion disc. The atoms involved are thought to be close to the inner edge of the accretion disc and moving at high speed, about a third of the speed of light. The observed shape of the line can be reasonably well explained using a theoretical model that takes account of the rate of rotation of the black hole, the inclination and size of the accretion disc, and a number of special and general relativistic effects, including the gravitational deflection of radiation, gravitational redshift and frame dragging. Spectral studies of this kind have been extended to other systems (including some stellar mass black holes), and are allowing scientists to study behaviour in the 'strong field' region close to the event horizon. As a result there is now evidence that the more rapidly the central object rotates, the smaller the inner radius of the accretion disc. This is exactly what is expected of an accretion disc around a Kerr black hole, where the radius of the event horizon depends on the rate of rotation of the black hole. and the inner edge of the accretion disc is determined by the smallest stable circular orbit that the spacetime allows. This minimum radius varies from about $3R_S$ for a slowly rotating black hole to $0.5R_S$ for a rapidly rotating black hole. Within this radius material cannot orbit; instead, it will simply spiral into the black hole.

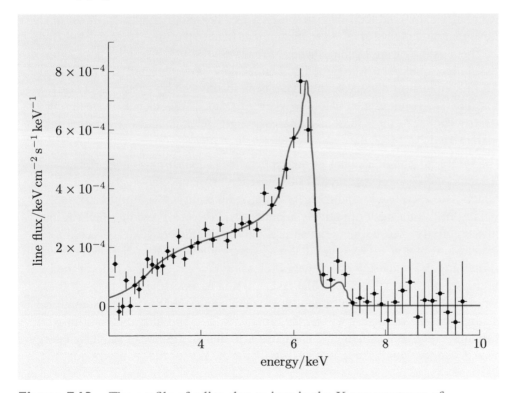

Figure 7.13 The profile of a line due to iron in the X-ray spectrum of MCG-6-30-15.

To many astronomers another strong argument for believing that supermassive black holes are common in galactic centres comes from the observations of *quasars* and other types of active galaxy. You will recall from Chapter 6 that the discovery of quasars in 1963 and the recognition of their very great (and varying) luminosity played an important part in driving the development of relativistic astrophysics throughout the 1960s. Over 100 000 quasars have now been identified, each the result of highly energetic activity in the nucleus of a galaxy. None are nearby and most are at very great distances, though this observation probably tells us more about the evolution of quasars over time than about their distribution in space.

It is believed that quasars were common in all parts of the Universe when it was about a quarter of its present age. Each quasar, it is assumed, was powered by a supermassive black hole swallowing matter from its vicinity via an accretion disc. The black hole might have formed along with the galaxy or as the result of mergers between sub-galactic units. The prodigious amount of energy needed to account for the observed luminosity of a typical quasar is supposed to come from the release of gravitational potential energy by matter falling into the supermasive black hole. The gravitational potential energy would initially be converted to kinetic energy of the in-falling matter itself, but as the matter encountered and passed through the accretion disc, much of its kinetic energy would be converted to radiation. It is estimated that an in-fall rate of a few solar masses per year is enough to account for the luminosity of a typical quasar.

As the Universe aged, the galactic centre black holes responsible for quasar activity would have grown in mass while simultaneously clearing the space around them of consumable matter. In this way most quasars would have eventually exhausted their own fuel supply and ceased their activity. Most of those that we now observe are so distant that (due to the finite speed of light) we see them as they were long ago when still active. As for the smaller population of less remote quasars, it is assumed that either they have managed to remain active throughout cosmic history or they have been reactivated by a new supply of fuel, possibly as a result of a collision between galaxies. If this view is correct, quasar activity should be thought of as a phase through which galaxies pass rather than a characteristic of particular types of galaxy.

The 'youthful phase' account of quasar activity is appealing as a story, but the scientific case for it recognizes two particularly important facts. First, galactic-scale collisions and mergers were common in the youthful Universe, making in-falling matter relatively abundant and thereby providing fuel for the quasar activity. Second, note the surprisingly high efficiency with which the accretion of matter converts gravitational potential energy to radiation. One way of defining the efficiency of an energy releasing process is as the ratio of the rate of energy release to the rate of fuel consumption expressed as the mass of fuel consumed per unit time multiplied by c^2. (This definition of fuel consumption ensures that the efficiency will be the dimensionless ratio of two quantities with the same units, as it should be.) If we use L to denote the rate of radiative energy release (i.e. the *luminosity*), and $c^2 \, \mathrm{d}m/\mathrm{d}t$ for the rate of fuel consumption, the efficiency is

$$\eta = \frac{L}{c^2 \, \mathrm{d}m/\mathrm{d}t}. \tag{7.14}$$

In these terms, the most efficient energy releasing process is **matter–antimatter annihilation**, which has an efficiency of 1, or 100% if you prefer. The efficiency of **gravitational energy release by accretion** onto a black hole depends on the black hole's rate of rotation; it varies from 5.7% for a non-rotating Schwarzschild black hole to 32% for a rapidly rotating Kerr black hole. This should be compared with an efficiency of only 0.7% for the **nuclear fusion** of hydrogen that is largely responsible for starlight. The overall situation as seen by astronomers in 2009 was described in an address by Royal Astronomical Society President, Andrew Fabian:

The visible sky is dominated by objects powered by nuclear fusion such as stars and galaxies. Shifting to shorter wavelengths in the X-ray band reveals an extragalactic sky powered by gravity: gravitational energy released by matter falling into black holes. ... When accretion rates are high, considerable amounts of gravitational energy are released as radiation, and in some circumstances as powerful jets.

In summary, we have the following.

Evidence from black holes

There is good evidence for the existence of both stellar mass black holes and supermassive black holes. This includes indirect evidence of black hole rotation and the presence of an event horizon from analysis of a distorted iron line in the X-ray spectrum. This astronomical evidence gives further support to general relativity but does not provide a precise test.

Gravitational energy release through accretion onto black holes provides a plausible mechanism to account for the luminosity of quasars. The extragalactic X-ray sky is dominated by gravitationally powered sources.

7.3.2 Gravitational lensing

As described earlier, Einstein's prediction of the gravitational deflection of light was first verified using data gathered in the total solar eclipse of 1919. The same physical process underlies the more recent discovery of **gravitational lensing**, the process in which a massive body (such as a galaxy or a cluster of galaxies), located between an observer and a distant source of electromagnetic radiation, causes the observer to see distorted or multiple images of the source.

In 1979, Dennis Walsh (1933–2005) and his colleagues pointed out that two narrowly separated quasars, Q0957+561 A and B (which we shall simply refer to as A and B), have identical optical and radio spectra. They are evidently at the same distance since their spectra are redshifted by the same amount. The most likely interpretation seemed to be that A and B are actually two images of a *single* quasar and that the light from that quasar is reaching the Earth by two different paths due to gravitational lensing (Figure 7.14).

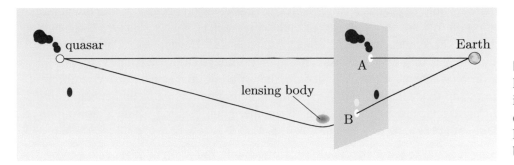

Figure 7.14 Gravitational lensing of a distant quasar by an intermediate body forms a double image as seen from Earth. (The angular scales have been exaggerated.)

The body responsible for the lensing was shown to be a galaxy, faint but detectable, located between the quasar and the Earth. This was the first example of a gravitational lens. It should be understood that a gravitational lens is not a true 'lens' in the optical sense of that term. Figure 7.15 shows the action of a converging optical lens on parallel rays, representing light from a source at an effectively infinite distance. In the case of an optical lens, the deflection of light *increases* with increasing distance from the central axis. Contrast that with the behaviour of parallel light rays passing a massive body, as shown in Figure 7.16.

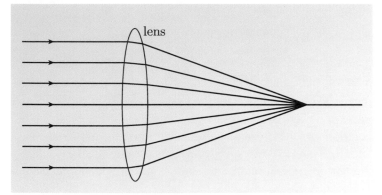

Figure 7.15 In an optical converging lens, the focusing effect relies on a greater deflection of light farther from the axis of the lens.

In the case of a gravitational lens, the deflection *decreases* with increasing distance from the central axis. In fact, for a point-like gravitational lens of mass M, if b represents the impact parameter of a light ray (the perpendicular distance from the initial path of the ray to the lensing body), then the angle of deflection θ is given by

$$\theta = \frac{4GM}{c^2 b}, \tag{7.15}$$

and the distance D from the lens to the point at which the light crosses the axis is given by

$$D \approx \frac{b}{\theta} = \frac{c^2 b^2}{4GM}. \tag{7.16}$$

The theory of gravitational lenses is very different from that of ordinary lenses. Real images of extended objects are never seen. Any intervening body of sufficient mass (such as a black hole) can produce gravitational lensing. If a point source, intervening body and observer all happened to be exactly in line, then the source would appear as a ring. Such circumstances do occur, but it is much more common to see a series of arcs or blobs. Figure 7.17 shows a picture taken by the Hubble space telescope of an object known as the 'Einstein cross' that includes four images of a distant quasar and a central image of the lensing body. An additional effect is that the light from the different images may arrive at different times (up to weeks apart) due to taking different optical paths and encountering different spacetime curvature (this is another manifestation of the Shapiro time delay effect).

Gravitational lensing affects all electromagnetic radiation and has also been observed at radio and X-ray wavelengths. It provides support for general relativity but is not really a stringent test of the theory. Rather it is a useful observational tool with many applications. For example, a gravitational lens may concentrate the light of a faint object to bring it above the threshold of what is detectable. In

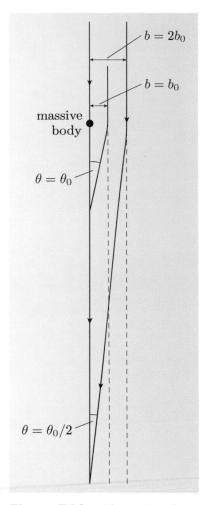

Figure 7.16 The angle of deflection θ of light by an object of mass M is inversely proportional to the impact parameter b.

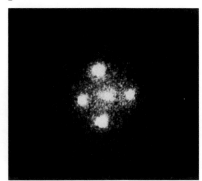

Figure 7.17 The Einstein cross, the result of gravitational lensing of a quasar.

this context, the object known as Abell 2218, a rich cluster of galaxies located about 2 billion light-years away, enables a far more distant object to be detected, as shown in Figure 7.18. The Abell 2218 cluster has produced two images of the distant object (circled in the inset) and amplified the brightness of each by a factor of about 30.

Figure 7.18 Two images of a distant object (inset and circled) due to gravitational lensing by the galaxy cluster Abell 2218.

Exercise 7.6 A gravitational lens does not function in the same way as a converging optical lens. Explain in qualitative terms how, notwithstanding this, the brightness of a very distant object can be amplified by a factor of 30 due to gravitational lensing. ∎

The term gravitational lensing is usually applied to situations in which the lensing body is very massive, typically a galaxy or a cluster of galaxies. However, the process is a general one and there is no reason, in principle, why the lensing body should not be much smaller. In fact, gravitational lensing by bodies of stellar mass or less has been observed since the early 1990s and is generally referred to as **gravitational microlensing**. When dealing with lensing bodies of such low mass it is not practical to detect image distortion, so image brightening is used instead. The technique is straightforward: bright stars in a nearby galaxy are carefully and continuously monitored using equipment capable of recording fluctuations in brightness. If a dense dark body passes across the line of sight from the observing site to any one of the monitored stars, then the brightness of that star will change and its variation with time can be recorded as a light curve. There are many reasons why the brightness of a stellar body might change, but microlensing will produce a characteristic contribution that can be distinguished from other signals and used to model the properties of the lensing body. In this way it is possible to search for isolated stellar mass black hole candidates and to put limits on the abundance of stellar mass black holes in the outer parts of the Milky Way.

Evidence from gravitational lensing

There are many examples of gravitational lenses. These give additional support to general relativity.

7.4   Gravitational waves

In 1993 the Nobel Prize for Physics was awarded to Joseph Taylor (1941–) and his former graduate student Russell Hulse (1950–) for their discovery (in 1974) and subsequent study of a very unusual binary star system that has become a test-bed for general relativity. The Hulse–Taylor system is believed to consist of two neutron stars, one of which is emitting regular pulses of radiation at radio wavelengths and is therefore classified as a **pulsar** and designated PSR B1913+16. Pulsars were first detected in the 1960s by Jocelyn Bell Burnell (1943–) and it was soon proposed that they were actually rapidly rotating neutron stars with a strong magnetic field. Many are now known but PSR B1913+16 was the first **binary pulsar** — a pulsar confirmed as part of a close binary system. In the Hulse–Taylor system, both of the compact stars has a mass of about $1.4\,\mathrm{M}_\odot$, and the pair orbit each other with a period of just 7.75 hours. The star that is a pulsar is thought to turn on its axis 17 times per second, accounting for the observed pulse separation of 59 milliseconds.

According to general relativity, a system of this kind should mainly lose energy through the emission of **gravitational waves**, a form of radiation involving propagating distortions of spacetime that was proposed by Einstein in 1916. As a result of gravitational wave emission, the orbital period of PSR B1913+16 should be decreasing in a predictable way. This prediction has now been tested over more than three decades and has been found to accurately agree with observations to within 0.2% (see Figure 7.19). It is an impressive confirmation of general relativity and also an indirect confirmation of the existence of gravitational waves, which have still not been directly detected here on Earth. (Note that gravitational radiation has nothing to do with electromagnetic waves and is not part of the electromagnetic spectrum. The Hulse–Taylor system is observed using electromagnetic (radio) waves, even though its orbital decay is mainly attributed to the emission of gravitational waves.)

This section is devoted to gravitational waves. It starts by introducing gravitational waves as solutions of the Einstein field equations and then goes on to examine the methods that may be used to detect them and some of the likely sources of such waves.

7.4.1   Gravitational waves and the Einstein field equations

In regions of spacetime where the gravitational field is weak, the curvature will be small and the metric tensor can be written as

$$[g_{\mu\nu}] = [\eta_{\mu\nu}] + [h_{\mu\nu}],$$

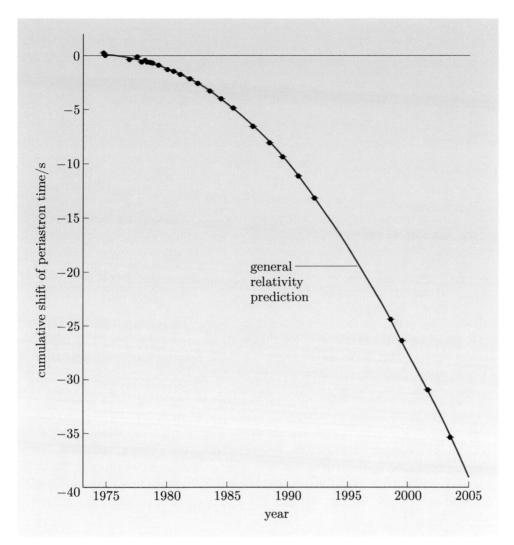

Figure 7.19 The orbital decay of PSR B1913+16. The cumulative shift of periastron time indicates how the time in the orbit at which the two neutron stars are closest together has advanced over time as the orbital period has become shorter.

where $[\eta_{\mu\nu}]$ represents the Minkowski metric of flat spacetime, and $[h_{\mu\nu}]$ describes the small departures from flat geometry. Though the disturbance tensor components $h_{\mu\nu}$ and their partial derivatives will be small, they are significant because they may vary with time. In the context of weak gravitational fields, the problem of finding a non-stationary metric tensor $[g_{\mu\nu}]$ that might represent a gravitational wave is replaced by that of finding the appropriate disturbance tensor $[h_{\mu\nu}]$.

In the case of weak fields, it is possible to show that there are wave-like solutions of the Einstein field equations. The details are not difficult but they are fairly tedious so we only give an outline here. The idea is to start with $\Gamma^{\sigma}{}_{\mu\nu}$ expressed in terms of the metric tensor components $g_{\mu\nu}$, and then write it in terms of $h_{\mu\nu}$. The result is non-linear in $h_{\mu\nu}$ and if carried out exactly would consist of an infinite sum of terms containing products of $h_{\mu\nu}$. However, since each component

$h_{\mu\nu}$ is small, we can make the simplification that we only retain terms linear in $h_{\mu\nu}$. This means that in the case of weak fields, the Einstein field equations

$$R_{\mu\nu} - \tfrac{1}{2}g_{\mu\nu}\,R = -\kappa\,T_{\mu\nu} \qquad\qquad \text{(Eqn 4.34)}$$

can be represented by the **linearized field equation**

$$\partial_\mu\,\partial_\nu\,h + \Box\,h_{\mu\nu} - \sum_\rho \partial_\nu\,\partial_\rho\,h^\rho{}_\mu - \sum_\rho \partial_\mu\,\partial_\rho\,h^\rho{}_\nu$$

$$- \sum_{\rho,\sigma} \eta_{\mu\nu}(\Box\,h - \partial_\rho\,\partial_\sigma\,h^{\sigma\rho}) = -2\kappa\,T_{\mu\nu}, \qquad\qquad (7.17)$$

where h is defined by

$$h = \sum_\sigma h^\sigma{}_\sigma$$

and the box symbol represents a combination of derivatives that is frequently encountered when dealing with waves that travel with speed c:

$$\Box = \sum_\sigma \partial_\sigma\,\partial^\sigma = \frac{1}{c^2}\frac{\partial^2}{\partial t^2} - \nabla^2 = \frac{1}{c^2}\frac{\partial^2}{\partial t^2} - \frac{\partial^2}{\partial x^2} - \frac{\partial^2}{\partial y^2} - \frac{\partial^2}{\partial z^2}.$$

$\Box$ is known as the d'Alembertian operator and is pronounced 'quad'.

It should be pointed out that the indices in Equation 7.17 are (by definition) raised and lowered using the Minkowski metric tensor $[\eta_{\mu\nu}]$, so Equation 7.17 genuinely is linear in $h_{\mu\nu}$. This linear equation has wave-like solutions, but that is far from obvious, partly due to the effect of *gauge symmetry*.

You may recall that when we discussed the Maxwell equations in Chapter 2, we said that the theory of electromagnetism contained an important symmetry called *gauge symmetry*. A related symmetry arises in general relativity. It is present in Equation 7.17 and prevents us from solving the equation in any simple way. In order to find an explicit solution, it is necessary to impose a condition that removes the effect of this symmetry. This extra condition is said to 'fix' the gauge. There are many ways of fixing the gauge; a common one is to define the quantity

$$\overline{h}_{\mu\nu} = h_{\mu\nu} - \tfrac{1}{2}\eta_{\mu\nu}\,h \qquad\qquad (7.18)$$

and then impose the condition

$$\sum_\mu \partial_\mu\,\overline{h}^{\mu\nu} = 0. \qquad\qquad (7.19)$$

This leads to the greatly simplified linearized field equation

$$\Box\,\overline{h}_{\mu\nu} = -2\kappa\,T_{\mu\nu}. \qquad\qquad (7.20)$$

This kind of differential equation is well known in the study of waves. It is described as an **inhomogeneous wave equation** with a source term $(-2\kappa\,T_{\mu\nu})$. It implies that gravitational waves can be generated by a source that changes in an appropriate way. (The Hulse–Taylor system is such a source, but a body that changes in a spherically symmetric way is not.) In a region where there are no sources, the spacetime disturbances are described by the homogeneous wave equation $\Box\,\overline{h}_{\mu\nu} = 0$, which is satisfied by waves that travel with speed c.

It might appear from what has been said that gauge invariance is simply an unfortunate inconvenience. However, this is far from being true. In both electromagnetism and general relativity, the gauge symmetry is a very deep and fundamental property of the theory.

- Which theorem introduced earlier ensures that a star that collapses in a spherically symmetric way cannot be a source of gravitational waves? Explain the reason for your answer.

○ Birkhoff's theorem. This ensures that the solution exterior to a spherically symmetric body (even one that is collapsing) must be described by the Schwarzschild metric. Since that metric is stationary, it cannot describe a gravitational wave, which will necessarily be described by a non-stationary metric.

7.4.2 Methods of detecting gravitational waves

We have already seen that the indirect observation of gravitational waves has almost certainly been achieved through the study of the Hulse–Taylor binary pulsar. The problem, then, is the direct detection of gravitational waves.

The existence of electromagnetic waves (predicted by Maxwell's equations) was dramatically confirmed by Heinrich Hertz (1857–1894) when he generated such waves in the laboratory using non-steady currents. One could imagine trying to generate gravitational waves in the laboratory by rapidly moving a massive object. Unfortunately, it turns out that if one rotates a bar of steel weighing several tons to the point where it is about to split apart under centrifugal forces, one radiates only about 10^{-30} W. For this reason, current experiments attempt to detect gravitational waves generated by large-scale astronomical events, such as supernovae or mergers of decaying binary systems.

Attempts have been made to detect gravitational waves since the 1960s. All are based on attempting to detect the relative movement of massive bodies caused by the rippling of spacetime as the wave passes through the apparatus. The massive bodies can be either the parts of an elastic body, in which case it is anticipated that the wave would create a resonance akin to the ringing of a bell, or 'free particles', where the relative movement of the individual particles can be detected.

The earliest experiments were of the elastic body type and made use of what is known as a **resonant bar detector** (sometimes called a **Weber bar**) — a large metal bar equipped with sensors to measure tiny movements of the ends (Figure 7.20). The idea was that the effect of a gravitational wave would be amplified by the resonant frequency of the bar and hence produce a measurable change in the distance between the ends. Although modern versions of this device are in operation, they are not sensitive enough to measure anything other than an extremely powerful and therefore very rare gravitational wave.

Figure 7.20 Joseph Weber and his resonant bar detector.

Most modern detectors are of the 'free particle' type since that has a greater potential for detecting the less powerful signals that are almost certainly more common. There are currently several gravitational wave detectors of this type in operation, but the most sensitive is **LIGO**, the **Laser Interferometer Gravitational-Wave Observatory**.

LIGO uses laser interferometers to monitor changes in the separation of suspended mirrors. As shown in Figure 7.21 overleaf, the interferometer consists of two 'light storage arms' at right angles forming an 'L' shape. Each arm has a mirror at either end so that light can repeatedly bounce back and forth. A laser supplies the light, which enters the arms via a beam splitter located at the corner of the L. In simple terms, if the arms are of constant length, the system can be arranged so that interference between the light beams returning to the beam splitter will direct all of the light back towards the laser. However, if either arm changes its length, the interference pattern will change and some light will reach the photodetector, where it can be recorded. When in operation, LIGO seeks changes in the lengths of the arms as revealed by alterations in the signal from the photodetector. The key challenge is to distinguish the very tiny signal from the unavoidable noise.

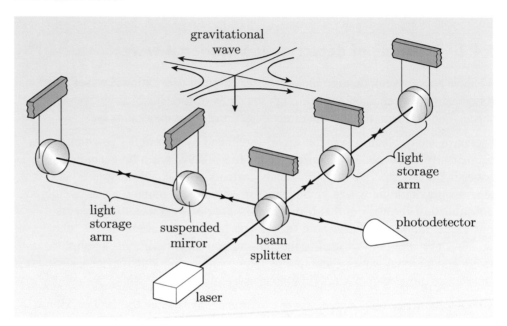

Figure 7.21   A schematic view of LIGO.

The engineering aspects of LIGO are impressive. The laser beams travel in highly evacuated tubes that are $4\,\text{km}$ long, and it is expected that a likely gravitational wave would change the $4\,\text{km}$ mirror spacing by about $10^{-18}\,\text{m}$, which is less than one-thousandth of the 'diameter' of a proton. This is a relative change in distance of approximately one part in 10^{21}.

To detect these tiny changes, LIGO currently uses three interferometers — two at an observatory on the Hanford Nuclear Reservation, in the state of Washington, and one at an observatory in Livingston, Louisiana. Consequently, LIGO has similar detectors separated by a distance of $3002\,\text{km}$. This should enable a gravitational wave to be distinguished from local noise. Since gravitational waves are predicted to travel at the speed of light, the $3002\,\text{km}$ separation corresponds to a difference in arrival times of up to about 10 milliseconds. Triangulation should allow this time difference to be used to determine the direction of the source. Despite its technology, LIGO has still not directly detected any gravitational waves; the sensitivity is still not great enough.

There are plans for an upgrade to LIGO, known as Advanced LIGO, which will increase the sensitivity by a factor of about 100. This is expected to be operational by 2014. Other gravitational wave detectors are also proposed, including **LISA** (the **Laser Interferometer Space Antenna**), which is a joint project between NASA and the European Space Agency to build a laser interferometer consisting of three spacecraft in solar orbit, to form an equilateral triangle with sides of about 5 million kilometres, as shown in Figure 7.22. LISA will be sensitive to gravitational waves at a lower frequency than LIGO, so the two experiments should complement each other. It is currently expected that the spacecraft will be launched in 2019 or 2020 and the project will last about 5 to 8 years.

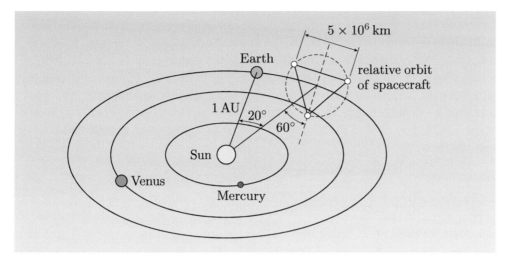

Figure 7.22　The orbit of the LISA spacecraft.

To summarize what has been said so far:

Gravitational waves

Gravitational waves are propagating disturbances in the geometry of spacetime that travel at speed c. Their existence can be predicted on the basis of a linearized version of the Einstein field equations that is appropriate in regions where the gravitational field is weak.

Strong indirect evidence of their existence is provided by the observations of the Hulse–Taylor binary pulsar. Searches for direct evidence using large-scale detectors such as LIGO are proceeding but have not yet succeeded.

7.4.3　Likely sources of gravitational waves

Gravitational waves and supernovae　One of the expected sources of gravitational waves is supernova explosions in neighbouring galaxies. Indeed, the target sensitivities of some existing gravitational detectors have been set with this in mind. Gravitational waves from a supernova explosion in a galaxy in the rich Virgo cluster of galaxies (centred about 60 million light years away) would cause a change of about 1 part in 10^{21} in lengths on Earth, and this is the target sensitivity of LIGO. As mentioned earlier, if the collapse of the star in a supernova is spherically symmetric, then there will be no gravitational radiation. However, it is thought that supernovae, particularly in binary systems, are asymmetric.

Gravitational waves and black holes　A possible source of gravitational waves would involve two black holes in orbit about each other. Such an orbiting pair would steadily emit gravitational radiation, eventually culminating in a huge burst as they fused into a single black hole. While the final black hole would, by virtue of the 'no hair theorem', be indistinguishable from any other black hole of the same mass and angular momentum, the outgoing ripples in spacetime would have encoded in them an account of the process in which they were emitted. This would be a very distinctive signal for the existence of black holes.

Gravitational waves and cosmology Gravitational waves of a wide spectrum of frequencies are expected from the 'quantum fluctuations' in the metric of spacetime that occurred during the Big Bang. The observation of gravitational waves should throw light on a central problem of modern cosmology: the origin of the density fluctuations that eventually led to a lumpy Universe (i.e. one containing galaxies) rather than a perfectly uniform one. The large-scale structure of the Universe is central to the next chapter, which is devoted to relativistic cosmology.

Summary of Chapter 7

1. The four classic tests of general relativity are as follows.

   (a) **The precession of the perihelion of Mercury** The observations, which have an uncertainty of about 1%, are consistent with the predictions of general relativity.

   (b) **Deflection of light by the Sun** The observations, which have an experimental uncertainty of about 10% for optical wavelengths, are in agreement with the predictions of general relativity. The agreement is better than 0.04% for VLBI radio telescope observations.

   (c) **Gravitational redshift** Gravitational redshift has been verified to better than 1% in variants of the Pound–Rebka experiment. Gravity Probe A verified the time dilation due to general relativity to 70 parts per million. The continued functioning of the GPS confirms general relativistic time dilation to about 1% on a daily basis.

   (d) **Time delay of electromagnetic radiation passing the Sun** The Cassini probe confirmed the effect to about 20 parts per million.

2. Satellite-based tests aim to detect two effects:

   - geodesic gyroscope precession (geodetic effect)
   - frame dragging (Lense–Thirring effect).

   Two satellite-based tests are:

   (a) The LAGEOS satellite results, which have been claimed to confirm frame dragging to 10%, but this is disputed.

   (b) Gravity Probe B results, which confirm geodesic gyroscope precession to 1.5%. The expected frame dragging is below the noise level, though there is still some hope that further analysis might improve the situation.

3. There is good evidence for the existence of both stellar mass black holes and supermassive black holes. This includes indirect evidence of black hole

rotation and the presence of an event horizon from analysis of a distorted iron line in the X-ray spectrum. This astronomical evidence gives further support to general relativity but does not provide a precise test.

4. Gravitational energy release through accretion onto black holes provides a plausible mechanism to account for the luminosity of quasars. The extragalactic X-ray sky is dominated by gravitationally powered sources.

5. There are many examples of gravitational lenses. These give additional support to general relativity.

6. Gravitational waves are propagating disturbances in the geometry of spacetime that travel at speed c. Their existence can be predicted on the basis of a linearized version of the Einstein field equations that is appropriate in regions where the gravitational field is weak.

   (a) The orbital decay of the binary pulsar PSR B1913+16 has been observed for over 30 years and is consistent with the expected loss of energy due to the emission of gravitational waves as predicted by general relativity.

   (b) Although no gravitational waves have been directly detected to date (2009), it is expected that they are created by large-scale astronomical events, provided that they are not spherically symmetric.

   (c) Currently, the most sensitive detector is LIGO, the Laser Interferometer Gravitational-Wave Observatory, which has been designed to be able to detect a supernova in the Virgo cluster of galaxies. Advanced LIGO should increase the sensitivity by a factor of 100 and is expected to be operational by 2014.

Chapter 8   Relativistic cosmology

Introduction

Cosmology is the study of the Universe as a whole, including its origin, nature, evolution and eventual fate. It has ancient roots in philosophy and religion, but modern scientific cosmology dates from 1917 when Einstein first used general relativity to formulate a mathematical model of the Universe.

Einstein was not an astronomer, so he sought astronomical advice before attempting to apply general relativity on the cosmic scale. Actually, little was known about the large-scale structure of the Universe at the time, so Einstein was led to formulate a static model, nether expanding nor contracting, that is now known to disagree with observational evidence. As a result, the details of Einstein's original model are mainly of historical interest. Nonetheless, his basic approach, of formulating a mathematical model describing the large-scale features of the Universe, usually called a **cosmological model**, still provides the basis of modern **relativistic cosmology**.

Cosmology is now a booming subject. Much of the subject's recent success has been the result of developments in our understanding of the physics of elementary particles and rapid progress in observational astronomy. It is impossible to do justice to either of these topics in one short chapter. Fortunately, the cosmological aspects of both are covered more fully in this book's companion volume, *Observational cosmology* by Stephen Serjeant. Consequently, the current chapter mainly provides an introduction to those aspects of cosmology that relate directly to general relativity and only includes a minimum of observational information.

The first section concerns the basic principles that underlie modern relativistic cosmology. These are approached from a mainly physical perspective and set the scene for a section devoted to the standard mathematical model of spacetime on the cosmological scale. That model takes the form of a specific metric known as the *Robertson–Walker metric* that, like the Schwarzschild metric, is usually presented as a four-dimensional spacetime line element. Having discussed spacetime on the cosmic scale, we next turn to the contents of that spacetime. As is conventional in cosmology, we treat the contents of spacetime as consisting essentially of matter and radiation, but when we come to write down an energy–momentum tensor for the Universe, we shall also include a contribution from the dark energy or cosmological constant that was mentioned at the end of Chapter 4. Accepting that Einstein's notion of a static Universe was wrong, our main aim in the third section is to use the Einstein field equations to derive the *Friedmann equations* that describe the evolution of the Universe. The Friedmann equations achieve this by relating the large-scale geometric features of spacetime to the large-scale distribution of energy and momentum. The combination of Robertson–Walker spacetime with matter, radiation and dark energy that evolve in accordance with the Friedmann equations results in a class of cosmological models known as the *Friedmann–Robertson–Walker models*. The final section of this chapter considers the observational consequences of supposing that the Universe we inhabit is well described by a Friedmann–Robertson–Walker model, and thereby provides a link to the companion volume on observational cosmology.

8.1 Basic principles and supporting observations

There are many way of approaching relativistic cosmology. Our approach is to recognize three underlying principles that we shall discuss in turn. Those three principles are:

- the applicability of general relativity
- the cosmological principle
- Weyl's postulate.

8.1.1 The applicability of general relativity

The starting point of relativistic cosmology is the supposition that general relativity can be applied to the Universe as a whole. This is a bold assumption, but also a fairly obvious one in view of the nature of general relativity. What it amounts to is the supposition that all of the matter and radiation that exists is 'contained' in a four-dimensional spacetime that can be described by an appropriate metric tensor $[g_{\mu\nu}]$ or by the corresponding spacetime line element $(\mathrm{d}s)^2 = \sum g_{\mu\nu}\,\mathrm{d}x^\mu\,\mathrm{d}x^\nu$. That cosmic spacetime metric can, in principle at least, be determined by solving the field equations of general relativity, and once known will show whether, on the cosmic scale, spacetime is flat or curved, and whether it is finite or infinite. In order to fully determine that cosmic spacetime metric, we need to be able to describe the distribution of energy and momentum on a similarly cosmic scale; that is, we need to be able to write down an energy–momentum tensor $[T_{\mu\nu}]$ for the whole Universe. This sounds like a daunting task and would obviously be quite impossible if we were to attempt a detailed description, planet by planet, star by star, galaxy by galaxy. Being more realistic, what cosmologists try to do is to find a simple prescription for the cosmic energy–momentum tensor that captures the essential large-scale features of the Universe while ignoring the detail that is not relevant to the large-scale concerns of cosmology. You will see examples of this shortly.

As explained in Chapter 4, when dealing with the field equations in the context of cosmology, it is important to be clear about which field equations are being discussed. The field equations that Einstein originally presented in 1915/16 took the form

$$R_{\mu\nu} - \tfrac{1}{2}R\,g_{\mu\nu} = -\kappa\,T_{\mu\nu}, \qquad\qquad \text{(Eqn 4.34)}$$

where $\kappa = 8\pi G/c^4$ is the Einstein constant.

However, when Einstein came to apply general relativity to cosmology in 1917, he recognized the possibility of adding an extra term, sometimes called the *cosmological term*, and therefore introduced the modified field equation

$$R_{\mu\nu} - \tfrac{1}{2}R\,g_{\mu\nu} + \Lambda\,g_{\mu\nu} = -\kappa\,T_{\mu\nu}, \qquad\qquad \text{(Eqn 4.47)}$$

where Λ represents a new universal constant known as the *cosmological constant*.

As Chapter 4 indicated, the modern convention is to retain the original unmodified field equations but to take account of the possibility of a non-zero cosmological constant by accepting that the energy–momentum tensor $[T_{\mu\nu}]$

235

might include a so-called *dark energy* contribution that can be described by its own energy–momentum tensor $[\overline{T}_{\mu\nu}]$ with components

$$\overline{T}_{\mu\nu} = \frac{\Lambda}{\kappa}\, g_{\mu\nu}. \tag{8.1}$$

As noted in Chapter 4, if we suppose that the source of the dark energy contribution can be treated as an ideal fluid with density ρ_Λ and pressure p_Λ, then it would have to be a very strange fluid since we would have

$$\overline{T}_{\mu\nu} = \left(\rho_\Lambda + \frac{p_\Lambda}{c^2}\right) U_\mu U_\nu - p_\Lambda\, g_{\mu\nu} = \frac{\Lambda}{\kappa}\, g_{\mu\nu}, \tag{8.2}$$

so comparing coefficients of $g_{\mu\nu}$ shows that the fluid has a negative pressure

$$p_\Lambda = -\frac{\Lambda}{\kappa}, \tag{8.3}$$

and requiring that the coefficient of $U_\mu U_\nu$ is zero shows that the fluid's density is

$$\rho_\Lambda = -\frac{p_\Lambda}{c^2} = \frac{\Lambda}{c^2\kappa} = \frac{\Lambda c^2}{8\pi G}. \tag{8.4}$$

Note that these are the properties that would ensure that the dark energy contribution precisely replicated the effect of a cosmological constant Λ. Such a contribution would lead to a large-scale repulsion, a kind of 'antigravity', that might be used to balance the gravitational effect of normal matter and radiation in certain circumstances.

Considerations of dark energy are important in modern cosmology. Little is known about its source but it is currently thought to account for about 70% of all the energy in the Universe. Many scientists believe that it is the energy of the vacuum, and therefore a property of empty space, but that interpretation is certainly not firmly established. Indeed, it faces a major problem in that although vacuum energy is expected to exist as a consequence of quantum physics, attempts to estimate its density exceed credible values of the density of dark energy, $\rho_\Lambda c^2$, by about 10^{120}.

To summarize, we have the following.

> **The applicability of general relativity**
>
> It is assumed that Einstein's original (unmodified) field equations of general relativity can be applied to the Universe as a whole, provided that a possible contribution from dark energy is included. We may then speak interchangeably of a Universe characterized by a cosmological constant Λ or one in which there is a dark energy contribution of density ρ_Λ and (negative) pressure $p_\Lambda = -\rho_\Lambda\, c^2 = -\Lambda c^4/8\pi G$.

8.1.2   The cosmological principle

The **cosmological principle** is the name given to a powerful simplifying assumption that makes the formulation of relativistic cosmological models tractable. It amounts to saying that what we learn from large-scale observations of our part of the Universe will be true of the Universe as a whole. The principle can be stated as follows.

The cosmological principle

At any given time, and on a sufficiently large scale, the Universe is **homogeneous** (i.e. the same everywhere) and **isotropic** (i.e. the same in all directions).

At first sight this principle is not at all obvious and it needs to be interpreted with care. It is appropriate that some time is devoted to its justification and explanation.

The first thing to note is that the principle concerns the properties of the Universe on the *large scale*, and in this context that really means a cosmic scale. On the small scale the Universe is certainly not homogeneous, nor is it isotropic. On a scale of hundreds or even thousands of kilometres, the solid Earth is below us, while above there is the air and, beyond that, the near vacuum of outer space. On this scale things are not the same everywhere, nor are they the same in all directions.

Even on much larger scales there is little sign of homogeneity and isotropy. Despite containing several planets and a vast number of minor bodies, the Solar System is dominated by a single star, the Sun, so it is certainly not homogeneous. It is true that the stars that surround the Sun are distributed in a fairly uniform way, with typical separations of a few light-years (where $1\,\mathrm{ly} = 9.46 \times 10^{15}$ m). However, on the $100\,000\,\mathrm{ly}$ scale of our galaxy, the **Milky Way**, it is found that the stars are arranged in a disc, and are gathered more densely at the centre than at the edges. This galactic structure shows that the stars are not, after all, uniformly distributed. On the galactic scale it also becomes apparent that even though stars are responsible for most of a typical galaxy's light emission, they do not account for the majority of its mass. There is good evidence from the rotation of galaxies and elsewhere that galactic mass is mainly attributable to some non-luminous form of matter generally referred to as **dark matter**, which, despite its name, is not thought to bear any relationship to the dark energy mentioned earlier.

On size scales of millions or tens of millions of light-years, galaxies of various shapes and sizes are gathered into groups and clusters. Some are sparsely populated, such as the **Local Group**, the 40 or so members of which include the Milky Way and the nearby Andromeda galaxy, M31. Others, such as the **Virgo Cluster**, are relatively rich, with over 1000 members in a volume not much larger than that of the Local Group.

Another increase in size scale, to about $100\,\mathrm{Mly}$, reveals what are believed to be the largest single structures in the Universe: the clusters of clusters of galaxies known as **superclusters**, and the vast non-luminous regions that separate them, known as giant **voids**. The superclusters and voids form a three-dimensional network that has been compared with a sponge or a cheese with holes, the superclusters occupying about 10% of the total volume and the voids the remaining 90%. It is this three-dimensional network, with a characteristic size scale of about $100\,\mathrm{Mly}$, that constitutes the true *large-scale structure* of the Universe. On any significantly larger scale, several hundred million light-years, say, it is generally believed that any region of the Universe would be much like any other, just as one sponge is just like any other, or one portion of cheese is just like any other. Each typical region would contain several voids and several superclusters, including, of course, the nuclei (mostly hydrogen) that are mainly responsible for the emission of light within the region, and the dark matter that mainly accounts for the region's mass.

Support for this view of a large-scale structure of superclusters and voids has been building over several decades. One important strand of evidence comes from the various large-scale **galaxy surveys** that have been carried out. Among the most recently reported are the two Degree Field Survey (2dF) and the Sloan Digital Sky Survey (SDSS). The 2dF survey provided a detailed view of the distribution of galaxies and clusters in two 'pizza slice' shaped regions, each about 60 degrees across and a few degrees thick, that stretch out to distances of about 2 billion light-years (Figure 8.1). More distant galaxies were recorded, but the sample was limited by the brightness of the observed sources, so it became less representative of the totality of galaxies as the distance increased. It should be noted that Figure 8.1 follows conventional astronomical practice by expressing distances in units of megaparsecs (Mpc), where $1\,\text{Mpc} = 3.26\,\text{Mly} = 3.08 \times 10^{22}\,\text{m}$. We shall have more to say about the precise meaning of these distances in Section 8.4.

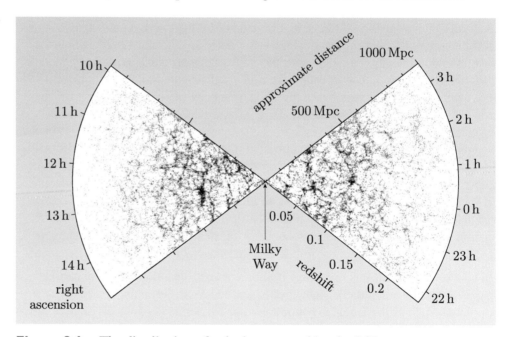

Figure 8.1   The distribution of galaxies reported by the 2dF survey.

Insight into the more remote parts of the Universe was provided by a special part of the 2dF survey devoted to quasars (Figure 8.2). As mentioned earlier, quasars are essentially active galactic nuclei with an exceptional brightness, thought to arise from the release of gravitational potential energy by matter falling into a supermassive black hole. Nearby quasars are too sparsely distributed to show the pattern of superclusters and voids in an obvious way, but on the large scale they can be seen to be distributed isotropically around the Milky Way. Accepting that there is nothing special about our location, the observed isotropic distribution of quasars is evidence that quasars are distributed isotropically about all points, and that is sufficient to ensure that they are also distributed homogeneously at any given time.

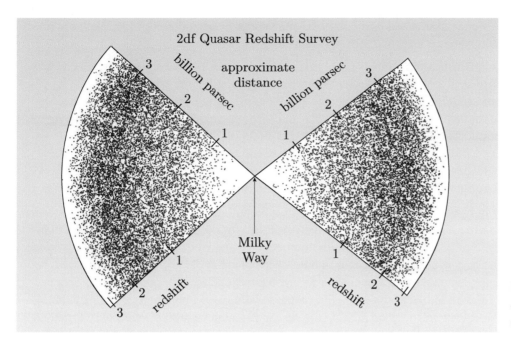

Figure 8.2 The distribution of quasars reported by the 2dF survey.

Looking at Figure 8.2, the distribution of quasars may not look homogeneous but that is because the distances involved are so vast that the more remote quasars are being seen at significantly earlier epochs in the evolution of the Universe, when the average number of quasars per unit volume was quite different from its current value. The observed distribution of quasars therefore provides evidence of cosmic evolution as well as evidence of isotropy and homogeneity. Although the quasars have always been homogeneously distributed since they first appeared on the cosmic scene, their population is believed to have peaked several billion years ago, hence the peak in the observed number density of quasars at a distance of about 3 billion parsecs.

A second, even stronger, strand of evidence concerning isotropy comes from observations of the **cosmic microwave background radiation** (CMBR). This is *thermal* radiation, meaning that it can be characterized by a temperature, in this case about 2.7 K. The CMBR was discovered in the mid-1960s and has been intensively studied ever since, most recently by the Wilkinson Microwave Anisotropy Probe (WMAP), a specialized space observatory that produced its first results in 2003. The CMBR is believed to have originated in the early Universe and is sometimes popularly described as the 'echo of the Big Bang'. It is now known to account for the greater part of all the radiant energy in the Universe, and is a major tool for cosmologists in their efforts to understand the Universe.

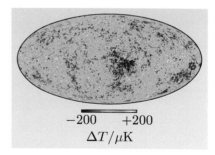

Figure 8.3 An all-sky thermal map of the cosmic microwave background radiation. The intrinsic anisotropies that can be seen in the CMBR amount to less than one part in ten thousand of its mean intensity.

For our present purposes, the most important feature of the CMBR is that, after correcting for the distortions caused by the motion of our observing equipment, it is highly isotropic (see Figure 8.3). The intrinsic mean intensity of the CMBR differs by less than one part in ten thousand in different directions. Since the CMBR is believed to be a universal phenomenon, it can again be argued that the observed isotropy about our location is evidence of isotropy about all locations and is therefore evidence of homogeneity at the present time and, by implication, also evidence of homogeneity at earlier times. It therefore makes good sense to identify the CMBR as a form of 'background radiation' since it should be equally prevalent in all parts of space at any given time, unlike starlight, for example, which is associated with localized sources and would therefore be relatively rare in places such as the voids between superclusters.

It is worth noting at this point that although isotropy about every point is a sufficient condition to ensure homogeneity, the existence of homogeneity is not sufficient to ensure isotropy. It is quite possible for a distribution to be homogeneous but not isotropic. A uniform magnetic field would be a case in point. The field would have a definite direction at every point, so it would not be isotropic, but provided that it had the same direction at every point, it would be homogeneous. So the assertion that on the large scale the Universe is homogeneous *and* isotropic has a real and distinctive meaning.

It is significant that the wording of the cosmological principle includes a reference to time, since this leaves open the possibility of cosmic evolution, provided that the evolution is consistent with homogeneity and isotropy. We have already noted the evolution that is thought to have taken place in the population of quasars, but it is also possible for evolution to involve large-scale motion. Observational evidence that the Universe is in fact *expanding* was published in 1929 by the American astronomer Edwin Hubble (1889–1953). Hubble's data only extended to relatively nearby galaxies and were complicated by the fact that individual galaxies have their own so-called **peculiar motion** relative to the large-scale expansion. However, extensive subsequent studies have confirmed that the average large-scale motion, sometimes called the **Hubble flow**, is isotropic so it can be characterized by a single rate of expansion at any time. Since the mid-1990s it has also become clear that the rate of cosmic expansion is currently increasing with time and has been doing so for at least a billion years. As a result we can say not only that the Universe is expanding but also that its expansion is *accelerating*. The peculiar motions of individual galaxies are generally small and random compared with the overall motion of the Hubble flow. The uniformity of the motion of matter on the large scale provides a third strand of evidence supporting the cosmological principle.

Exercise 8.1   Summarize the three strands of evidence that support the cosmological principle.   ■

8.1.3 Weyl's postulate

Weyl's postulate was advanced in 1923, by the originator of gauge theory, the mathematical physicist Hermann Weyl (1885–1955). It is essentially an assumption about the matter in the Universe, but it came before the nature and distribution of galaxies was well understood, so Weyl treated the material content of the Universe as a fluid and spoke of its constituent particles as forming a *substratum*. Modern statements of Weyl's postulate often replace any mention of the substratum by references to superclusters of galaxies, or even to individual galaxies provided that their peculiar motions are ignored. In this sense, Weyl's postulate is really an assumption about the nature of the Hubble flow that predates the discovery of that flow.

From a modern perspective the significance of Weyl's postulate is that it recognizes the existence of a privileged class of observers who have a particularly simple view of the Universe. These are the observers who move with the Hubble flow. You can think of each such observer as moving with their local supercluster or even with their own local galaxy, as long as its peculiar motion is ignored. It is these observers, sometimes called **fundamental observers**, who will find that the Universe around them (including the CMBR) is isotropic. A non-fundamental observer who moves relative to the local fundamental observer would not find that the Universe was expanding uniformly in all directions, nor would such a non-fundamental observer find the CMBR to be isotropic. In terms of fundamental observers, **Weyl's postulate** can be stated as follows.

Weyl's postulate

In cosmic spacetime there exists a set of privileged fundamental observers whose world-lines form a smooth bundle of time-like geodesics. These geodesics never meet at any event, apart perhaps from an initial singularity in the past and/or a final singularity in the future.

The implications of Weyl's postulate are indicated in Figure 8.4. Essentially, the postulate supposes that the Universe is structured and evolves in a sufficiently orderly way that the proper time measured by each fundamental observer can be correlated with that of every other fundamental observer so that a value of a single, universally meaningful **cosmic time** can be associated with every event.

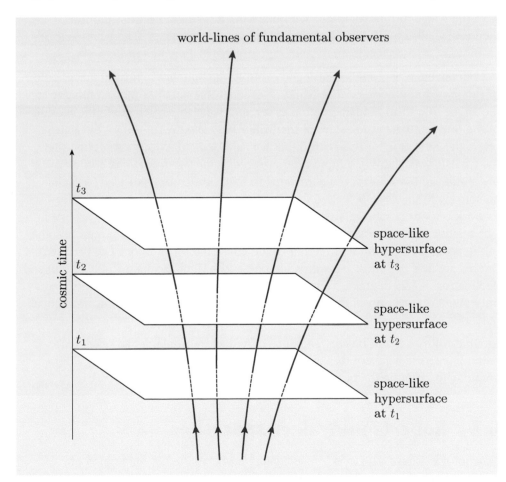

Figure 8.4 The world-lines in cosmic spacetime of the fundamental observers who see the Universe as homogeneous and isotropic. Each world-line can be labelled by fixed co-moving coordinates but intersects successive space-like hypersurfaces at different values of cosmic time.

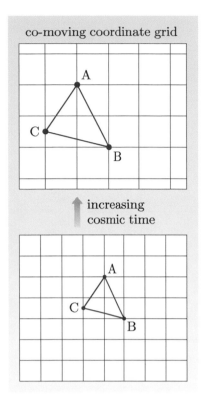

co-moving coordinate grid

↑ increasing
cosmic time

Figure 8.5   Co-moving coordinates expand with the flow that they describe. Points that move with the flow, such as the locations of fundamental observers, will be described by fixed values of the co-moving coordinates.

This might be done, for example, by all fundamental observers agreeing to use the proper time since the Big Bang or, more realistically, the proper time since the CMBR had some particular mean intensity. The ability to define a cosmic time means that we can identify all the events characterized by any particular value of cosmic time. Such a set of events will form a three-dimensional space, technically referred to as a **space-like hypersurface** with geometric properties that are homogeneous and isotropic. Each of the 'surfaces' in Figure 8.4 represents one of these space-like hypersurfaces and can be thought of as the whole of space at a particular moment of cosmic time. The lines threading the surfaces represent the world-lines of the fundamental observers, and may only diverge or converge in such a way that overall homogeneity and isotropy are preserved throughout cosmic time.

Each of the fundamental observer world-lines in Figure 8.4 may be characterized on any particular space-like hypersurface by three spatial coordinates, x^1, x^2 and x^3. Remembering that coordinates have no immediate metrical significance in general relativity, we may, if we wish, choose to define our coordinate system in such a way that the world-line of a fundamental observer is assigned the same values of the three spatial coordinates on every space-like hypersurface. Coordinates of this kind are widely used in cosmology and are called **co-moving coordinates**. In an expanding (or contracting) Universe, the grid of co-moving coordinates must expand or contract with the space-like hypersurfaces. So, in our Universe, a co-moving coordinate grid, like the fundamental observers, must 'go with the flow'. It follows that if we ignore the individual peculiar motions, then every galaxy will have constant co-moving coordinates. The behaviour of co-moving coordinates in an expanding Universe is indicated in Figure 8.5.

We ourselves, living on the Earth and orbiting the Sun, are almost in the situation of fundamental observers. The Milky Way has some peculiar motion relative to the frame of a local fundamental observer, and we also participate in the motion of the Sun relative to the centre of the Milky Way and the motion of the Earth as it orbits the Sun. It is for this reason that we said in the previous subsection that the CMBR was highly isotropic *after correcting for the distortions caused by the motion of our observing equipment*. In fact, observations of a large-scale anisotropy in the CMBR, called the **dipole anisotropy** (see Figure 8.6), allow us to work out our motion relative to the frame of the local fundamental observer. The results show that in such a frame, the Sun is travelling at about a thousandth of the speed of light in the direction of the constellation of Leo. (The precise figures are $368 \pm 2\,\mathrm{km\,s}^{-1}$ towards the point with right ascension 11 h 22 min and declination -7.22 degrees.) The orbital speed of the Earth relative to the Sun is only about one twelfth of the Sun's speed, so it can be ignored for most practical purposes.

In what follows it will be convenient to regard every fundamental observer as being located in a galaxy that exactly follow the isotropic Hubble flow. This amounts to ignoring the peculiar motions that galaxies actually possess.

8.2   Robertson–Walker spacetime

Cosmologists have developed, investigated and classified a wide range of relativistic cosmological models, including some that are neither homogeneous nor isotropic. However, the overwhelming majority of the investigations have concerned models that are homogeneous and isotropic, and therefore conform to the requirements of the cosmological principle. Around 1935, Howard Robertson (1903–1961) of the California Institute of Technology and Arthur Walker (1909–2001) of the University of Liverpool showed, independently, that a single spacetime metric underlies all relativistic models that are homogeneous and isotropic. That metric is now known as the **Robertson–Walker metric**. The Robertson–Walker metric and the spacetime that it describes are the subject of this section.

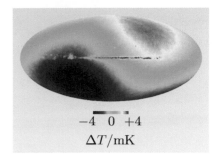

-4   0   +4

$\Delta T/\text{mK}$

Figure 8.6   The large-scale 'dipole' anisotropy in the CMBR. Some 'noise' from sources in the plane of the Milky Way cay be seen crossing the middle of the all-sky map.

8.2.1   The Robertson–Walker metric

Based on the three principles introduced in the previous section, it is natural for a fundamental observer to describe cosmic spacetime using a squared line element of the form

$$(\mathrm{d}s)^2 = c^2(\mathrm{d}t)^2 - \sum_{i,j=1}^{3} g_{ij}\,\mathrm{d}x^i\,\mathrm{d}x^j, \tag{8.5}$$

where t represents cosmic time, x^1, x^2 and x^3 are co-moving coordinates, and the metric coefficients g_{ij} are functions of t, x^1, x^2 and x^3.

Spatial homogeneity and isotropy require that the ratios of distances are the same at all times. So three fundamental observers located at the corners of a triangle at some cosmic time t_1, will also be at the corners of a similar triangle at cosmic time t_2. The triangle may be bigger or smaller, but its angles will be the same, and each side will have increased or decreased its length by the same factor. We can incorporate this requirement into the metric by insisting that the cosmic time enters the metric coefficients g_{ij} only through a common scaling function. For later convenience we shall write this common function as $S^2(t)$, so

$$(\mathrm{d}s)^2 = c^2(\mathrm{d}t)^2 - S^2(t) \sum_{i,j=1}^{3} h_{ij}\,\mathrm{d}x^i\,\mathrm{d}x^j, \tag{8.6}$$

where each of the coefficients $h_{ij} = g_{ij}/S^2(t)$ depends only on x^1, x^2 and x^3.

Now, the curvature tensor of a three-dimensional space generally has $3^4 = 81$ components, of which 6 are independent. However, since the space described by h_{ij} is homogeneous and isotropic, the curvature must be the same everywhere and in all directions. As a result, the curvature must be fixed by a single parameter. If the properties of the space are also independent of time, then that single parameter must be a constant. We shall denote that constant by the upper-case letter K. The metric that describes a three-dimensional space of constant curvature is well known to mathematicians. If we use its most common form to replace the coefficients h_{ij} in Equation 8.6, we obtain the metric

$$(\mathrm{d}s)^2 = c^2(\mathrm{d}t)^2 - S^2(t) \left[\frac{(\mathrm{d}\bar{r})^2}{1 - K\bar{r}^2} + \bar{r}^2(\mathrm{d}\theta)^2 + \bar{r}^2 \sin^2\theta\,(\mathrm{d}\phi)^2 \right], \tag{8.7}$$

where we have replaced the general co-moving coordinates x^1, x^2, x^3 by the co-moving polar coordinates $\bar{r}, \theta, \phi$. You will see why we have called the radial coordinate $\bar{r}$ in just a moment. First, though, note that the expression inside the square brackets represents a space of constant curvature. Its Riemann curvature components are $R_{ijkl} = K(h_{ik}h_{jl} - h_{il}h_{jk})$, the Ricci tensor components are given by $R_{ij} = -2Kh_{ij}$ and the Ricci curvature scalar is $R = -6K$. For the sake of simplicity, such a space is said to have curvature K. The effect of multiplying the expression in square brackets by $S^2(t)$ is to produce a rescaled version of the space that at time t has curvature $K/S^2(t)$. (This is rather like the effect of inflating a spherical balloon, where increasing the balloon's radius by a factor of 2 will make the surface flatter, reducing the (Gaussian) curvature by a factor of 4.)

Equation 8.7 is one form of the Robertson–Walker metric, but not the most common form. It turns out that for many purposes the value of the curvature constant K is less important than whether it is positive or negative. Consequently it is generally convenient to carry out a coordinate transformation that has the effect of replacing the spatial curvature K by a related quantity k, called the **curvature parameter**, that can take only the values $+1$, 0 or -1. This can be achieved by introducing a new rescaled radial coordinate r defined by the relation

$$r = \begin{cases} \overline{r}|K|^{1/2} & \text{if } K \neq 0, \\ \overline{r} & \text{if } K = 0. \end{cases} \qquad (8.8)$$

Using this to eliminate all occurrences of $\overline{r}$ in Equation 8.7, we can rewrite the Robertson–Walker metric in its most common form.

The Robertson–Walker metric

$$(\mathrm{d}s)^2 = c^2(\mathrm{d}t)^2 - R^2(t)\left[\frac{(\mathrm{d}r)^2}{1 - kr^2} + r^2(\mathrm{d}\theta)^2 + r^2\sin^2\theta\,(\mathrm{d}\phi)^2\right]. \quad (8.9)$$

Here r, θ, ϕ are still co-moving coordinates (the rescaling doesn't change that) and the information about distance ratios at different times is now contained in the time-dependent function $R(t)$, which is therefore known as the **scale factor** and is defined by the relation

$$R(t) = \begin{cases} S(t)/|K|^{1/2} & \text{if } K \neq 0, \\ S(t) & \text{if } K = 0. \end{cases} \qquad (8.10)$$

It is important to note that this scale factor $R(t)$ is quite distinct from the Ricci scalar that appears in the field equations and which is also denoted by R. From here on, R will always be the scale factor, never the Ricci scalar.

If the scale factor $R(t)$ increases with time, then the fundamental observers become more widely separated with time, the galaxies containing those fundamental observers get further apart, and the Universe is said to be **expanding**. If $R(t)$ decreases with time, then the fundamental observers and their associated galaxies get closer together, and the Universe may be said to be **contracting**. Remember, though, that throughout this process the co-moving coordinates of any fundamental observer remain fixed at all times. Also remember that the space-like hypersurfaces are homogeneous and isotropic, so although the coordinate system will have some particular origin and some particular orientation, any point may be chosen to be the origin, and the chosen orientation of the axes is equally arbitrary.

As a result of the rescaling, the curvature of the constant-t space-like hypersurface will be $k/R^2(t)$.

Apart from the cosmic time and the co-moving coordinates, the scale factor $R(t)$ and the curvature parameter k are the only quantities that appear in the Robertson–Walker metric. Both are important. The rest of this section will be mainly concerned with the significance of k; the role of $R(t)$ will feature prominently in Section 8.3.

8.2.2 Proper distances and velocities in cosmic spacetime

We already know that in the Robertson–Walker metric, t represents the cosmic time, which can be related to the proper time measured by any fundamental observer. This is the time that might be measured on a clock carried by the fundamental observer. However, we still don't know the precise relationship between the fixed co-moving coordinates of two points and the **proper distance** that would be measured between those points by connecting them with a line of stationary measuring rods at some particular time t.

● Assuming that the measuring rods can be laid along the shortest path between the two points, how would you describe that path?

○ The path of shortest length between two points at a given time would lie in a particular space-like hypersurface, and would be a geodesic of that hypersurface.

For two simultaneous events that occur with infinitesimally separated positions, (r, θ, ϕ) and $(r + \mathrm{d}r, \theta + \mathrm{d}\theta, \phi + \mathrm{d}\phi)$, the proper distance separating them can be read directly from the Robertson–Walker line element. Using the symbol $\mathrm{d}\sigma$ to represent that infinitesimal distance, we have

$$\mathrm{d}\sigma = R(t) \left[\frac{(\mathrm{d}r)^2}{1 - kr^2} + r^2 (\mathrm{d}\theta)^2 + r^2 \sin^2 \theta \, (\mathrm{d}\phi)^2 \right]^{1/2}. \tag{8.11}$$

Note that this proper distance element depends on the proper time at which it is measured. This is to be expected in an expanding or contracting Universe since proper separations will change with time even though (co-moving) coordinates don't change their values.

When dealing with finite separations, the problem of working out proper distances is generally quite challenging. It involves integrating the distance element given in Equation 8.11 along a pathway, and this usually requires the introduction of parameters, just as we did in Chapter 3. However, the problem can be greatly simplified by making use of the homogeneity of the space-like hypersurfaces. Given two points on such a hypersurface, we can always choose one of them to be the origin of coordinates. The other will then be at some specific co-moving radial coordinate value, $r = \chi$ say, in a fixed direction, specified by particular values of θ and ϕ. In such a case, the two points are linked by a purely radial path that will always be a geodesic (we shall not prove this). Along that radial path $\mathrm{d}\theta = 0$ and $\mathrm{d}\phi = 0$, so the element of proper distance is just $\mathrm{d}\sigma = R(t) \, \mathrm{d}r / (1 - kr^2)^{1/2}$. Thus, given two points separated by a fixed radial co-moving coordinate χ, the proper distance between them at time t will be

$$\sigma(t) = \int_0^{\chi} R(t) \frac{\mathrm{d}r}{(1 - kr^2)^{1/2}}. \tag{8.12}$$

Whether k is $+1$, 0 or -1, this is a standard integral with a well-known result.

Proper distance σ related to co-moving coordinate χ

$$\sigma(t) = \begin{cases} R(t)\sin^{-1}\chi & \text{if } k = +1, \\ R(t)\,\chi & \text{if } k = 0, \\ R(t)\sinh^{-1}\chi & \text{if } k = -1. \end{cases} \tag{8.13}$$

These three relationships are illustrated in Figure 8.7.

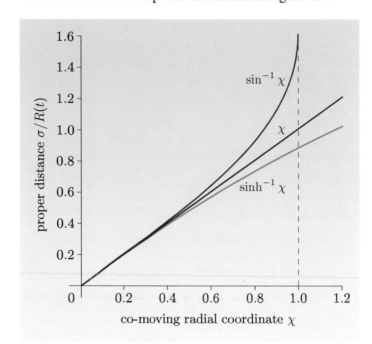

Figure 8.7   The relationship between proper distance and a co-moving radial coordinate χ for the space-like hypersurface corresponding to cosmic time t, in the cases $k = +1, 0, -1$. Note that the proper distance is expressed as a multiple of $R(t)$.

All three of these functions behave in a similar way for small values of χ, but as χ increases, they start to separate until the value $\chi = 1$ is reached, at which point $\sin^{-1}\chi$ reaches its limiting value $\pi/2$. These differences are, of course, a result of the intrinsic curvature of the space-like hypersurfaces. We shall explore this more fully in the next subsection.

An important point to note concerning co-moving coordinates and their relationship to proper distances involves units and dimensions. The proper distance between two points must be a length. However, the co-moving coordinate is not subject to the same restriction. Since all proper lengths are proportional to the scale factor $R(t)$, it is conventional to treat the co-moving coordinate $r = \chi$ as dimensionless and the scale factor $R(t)$ as having the dimensions of length.

Though we now have an expression for proper distance, it will be of interest only for certain theoretical purposes. It's not a distance that can be directly observed astronomically; we can't really set up lines of stationary rulers stretching from one galaxy to another. Nonetheless, it is interesting to ask how quickly the proper distance between fundamental observers would change as a result of any uniform expansion or contraction. (We have to ask about the *proper* distance since the co-moving coordinate χ won't change at all.) Defining the **proper radial velocity** as the rate of change of proper distance with respect to cosmic time, we see from the above that

$$\frac{d\sigma}{dt} = \begin{cases} \dfrac{dR}{dt}\sin^{-1}\chi & \text{if } k = +1, \\[2mm] \dfrac{dR}{dt}\chi & \text{if } k = 0, \\[2mm] \dfrac{dR}{dt}\sinh^{-1}\chi & \text{if } k = -1. \end{cases} \qquad (8.14)$$

In each case we can replace the term involving χ by σ/R. This leads to the same expression for the proper velocity in all three cases:

$$\frac{d\sigma}{dt} = \frac{1}{R}\frac{dR}{dt}\sigma. \qquad (8.15)$$

It is conventional to write this relationship in the more memorable form

$$v_p = H(t)\,d_p, \qquad (8.16)$$

where d_p represents the proper distance between two fundamental observers or their galaxies, v_p represents the proper radial velocity at which they are separating (for positive v_p) or coming together (for negative v_p), and $H(t)$, which is called the **Hubble parameter**, is defined as follows.

The Hubble parameter

$$H(t) = \frac{1}{R}\frac{dR}{dt}. \qquad (8.17)$$

Equation 8.16 tells us that at any cosmic time t, every fundamental observer is moving radially relative to every other fundamental observer at a proper speed that is proportional to the proper distance that separates them. Note that this is an exact consequence of the nature of Robertson–Walker spacetime. Later we shall re-examine this result in connection with Hubble's observations of cosmic expansion. At that stage we shall relate the proper distance to some other distances that really can be measured and also relate the Hubble parameter to an observable quantity known as the *Hubble constant*.

Exercise 8.2   It was claimed above that at any fixed time, a radial line through the origin of a Robertson–Walker spacetime would be a geodesic of the relevant three-dimensional space-like hypersurface. Outline the procedure that you would follow to establish the truth of this claim, starting from the Robertson–Walker metric. ∎

8.2.3 The cosmic geometry of space and spacetime

In general, a homogeneous and isotropic space-like hypersurface has no centre and no boundary. (Do not mistake the point arbitrarily chosen to be the origin of coordinates with a physically significant centre point.) However, such a hypersurface can have a curvature and can be characterized by a curvature parameter (k). In what follows we shall consider the geometrical significance of some particular choices of k and $R(t)$. Remember throughout that k is the curvature parameter, *not* the curvature. As noted earlier, the curvature of any of the fixed-t space-like hypersurfaces is given by $k/R^2(t)$.

Case 1: $k = 0$ **and** $R(t) = \text{constant}$

In this case the constant scale factor can be absorbed into a rescaled radial coordinate with the result that the Robertson–Walker line element of Equation 8.9 reduces to the Minkowski metric of Chapter 3 expressed in spherical coordinates:

$$(\mathrm{d}s)^2 = c^2(\mathrm{d}t)^2 - (\mathrm{d}r)^2 - r^2(\mathrm{d}\theta)^2 - r^2\sin^2\theta\,(\mathrm{d}\phi)^2. \tag{8.18}$$

Each space-like hypersurface (representing space at some particular cosmic time t) will have the geometry of a three-dimensional space with zero curvature (i.e. Euclidean 3-space), and the co-moving coordinate grid will neither expand nor contract. Each fundamental observer would be at rest relative to every other fundamental observer, and each would find that there was no gravity and that special relativity applied everywhere. In this case the Riemann curvature tensor will be zero everywhere and at all times. In short, space would be flat at all times, and the Robertson–Walker spacetime would also be flat.

To be consistent with general relativity, the field equations would demand that this gravity-free, flat spacetime contained no matter, radiation or dark energy, so this really isn't an interesting case from a physical point of view. Nonetheless, it's interesting to see that Minkowski spacetime can emerge as a limiting case of Robertson–Walker spacetime.

Case 2: $k = 0$ **and** $R(t) \neq \text{constant}$

In this case the three-dimensional space-like hypersurfaces will again have the zero-curvature geometry of Euclidean 3-space. The internal angles of a triangle add up to π radians, and the ratio of the circumference of a circle to its radius will be 2π. As we saw in the previous subsection, another indication of the spatial flatness is the proportionality between the co-moving radial coordinate χ and the proper distance σ at any fixed value of t:

$$\sigma(t) = R(t)\,\chi \quad \text{if } k = 0.$$

However, the full four-dimensional Robertson–Walker spacetime will not be flat because the scale factor $R(t)$ will cause the proper distance between co-moving locations to change, and this will generally prevent the Riemann curvature tensor from vanishing.

Exercise 8.3   'The metric used in special relativity is a particular case of the Robertson–Walker metric for which $k = 0$, i.e. for which space is flat.' Comment on the accuracy of this statement.   ∎

Case 3: $k = +1$ **and** $R(t) \neq \text{constant}$

In this case both four-dimensional Robertson–Walker spacetime and its three-dimensional space-like hypersurfaces will have a curved geometry. We have already seen that on any particular hypersurface, the proper distance from the origin is related to the radial co-moving coordinate $r = \chi$ by $\sigma(t) = R(t)\sin^{-1}\chi$, so σ increases more rapidly with increasing χ than in a flat space. Using the proper distance element of Equation 8.11 and the parameterized path method of Chapter 3, an integral around a circle of co-moving coordinate radius χ, centred on the origin and located in the $\theta = \pi/2$ plane for simplicity, shows that the circle has proper circumference $2\pi R(t)\chi$. It follows that the ratio of proper circumference to proper radius for such a circle is

$$\frac{\text{proper circumference of circle}}{\text{proper radius}} = \frac{2\pi R(t)\chi}{R(t)\sin^{-1}\chi} \leq 2\pi.$$

We have also seen that the proper distance reaches a finite limit as χ approaches 1.

All these properties are indications of the positive curvature of the hypersurface. The effects produced are easily remembered by looking at the $k = +1$ case in Figure 8.8.

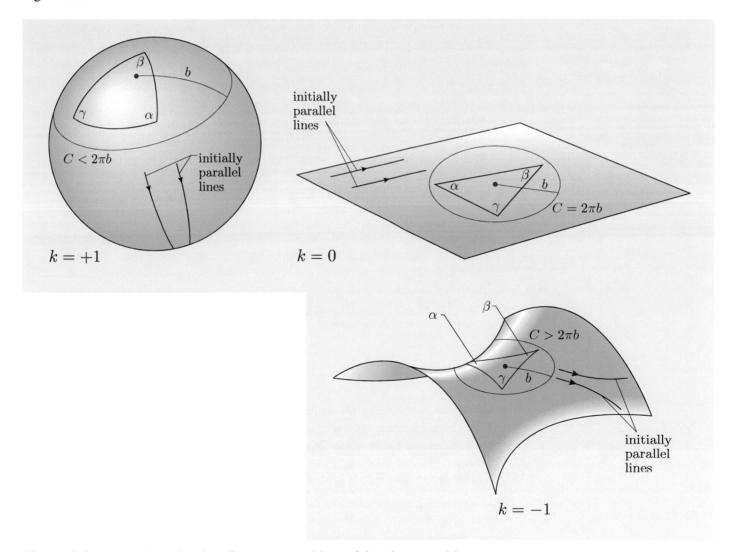

Figure 8.8 Two-dimensional surfaces can provide useful and memorable analogues of the three-dimensional space-like hypersurfaces in the cases $k = +1, 0, -1$. In each case, a circle of proper radius b and proper circumference C is drawn in the surface.

The two-dimensional spherical surface shown there is not supposed to be a picture of the three-dimensional $k = +1$ hypersurface, but it does provide a reminder of some of the non-Euclidean features of the hypersurface. The analogy is quite far reaching. For example, on the surface of the two-dimensional sphere, triangles have interior angles that add up to more than π radians, and geodesics (i.e. 'straight' lines) that are initially parallel will meet at some point; both of these conditions will also hold true on the $k = +1$ space-like hypersurfaces. One other property of the spherical surface is that it has a finite total area. In a similar way, the three-dimensional space-like hypersurface has a finite total proper volume that turns out to be $2\pi^2 R^3(t)$, but like the surface of the sphere, it has no boundary, no edge, and no centre.

Because of its finite volume, the kind of space described by the $k = +1$ hypersurface is often described as **closed**. Sometimes the term **unbounded** is added to emphasize that closure does not imply an edge or any other kind of inhomogeneity. A traveller in such a space would always find it to be homogeneous and isotropic, but following a straight (i.e. geodesic) pathway would eventually bring the traveller back to points that had been visited before.

The surprising effectiveness of the spherical analogy as a source of insight into the $k = +1$ hypersurfaces of Robertson–Walker spacetime is not really an accident. It can be shown that there is a close mathematical relationship between the points on the space-like hypersurface and the points on the three-dimensional surface of a four-dimensional sphere that might be described by the equation $w^2 + x^2 + y^2 + z^2 = a^2$. We shall not pursue this relationship here, but *embedding* a space of three or more dimensions in some space of higher dimensionality is often a source of insight.

Case 4: $k = -1$ **and** $R(t) \neq \text{constant}$

Again, both spacetime and its space-like hypersurfaces will have a curved geometry. In this case, however, the proper distance grows less rapidly with the co-moving coordinate than would be the case in a flat space. In fact, as we saw earlier, $\sigma(t) = R(t)\sinh^{-1}\chi$. A parameterized integral will again show that a circle of co-moving coordinate radius χ has proper circumference $2\pi R(t)\chi$, so in this case

$$\frac{\text{proper circumference of circle}}{\text{proper radius}} = \frac{2\pi R(t)\chi}{R(t)\sinh^{-1}\chi} \geq 2\pi.$$

Again there is an analogous surface shown in Figure 8.8, namely the saddle-shaped surface corresponding to $k = -1$. In this case the angles of a triangle drawn around the saddle point would sum to less than π radians, and there is no restriction on how big χ can be. The $k = -1$ hypersurface does not have a finite proper volume and is said to be **open**.

It is interesting to note that in this case the analogy between the two-dimensional surface and the three-dimensional hypersurface is not as far reaching as it was in the $k = +1$ case. It is simply not possible to embed a three-dimensional surface of constant negative curvature in a four-dimensional (Euclidean) space, so the best that can be achieved is a purely local analogy.

8.3   The Friedmann equations and cosmic evolution

In the previous section we introduced the Robertson–Walker metric and discussed some of its geometric features, giving particular emphasis to the meaning of the coordinates and the significance of the spatial curvature parameter k. We did this on a heuristic basis, guided by general principles such as the cosmological principle. What we did not do was to write down an energy–momentum tensor for the Universe and then look for a solution of the Einstein field equations. That is essentially what we shall do in this section. Already knowing the general form of the Robertson–Walker metric will greatly simplify this task.

In the subsections that follow we first write down an energy–momentum tensor that is designed to represent the large-scale features of the Universe. We then substitute that energy–momentum tensor and the Robertson–Walker metric into the Einstein field equations. The result is a set of differential equations, called the *Friedmann equations*, that relate the Robertson–Walker parameters, k and $R(t)$, to the densities of matter, radiation and dark energy in the Universe and to any associated pressures. Solving those equations leads us to a range of homogeneous cosmological models, each characterized by a particular form of the time-dependent scale factor $R(t)$. In each case the scale factor encapsulates the entire expansion history of the model Universe. These models form the basis of essentially all introductions to relativistic cosmology, and are usually referred to as the *Friedmann–Robertson–Walker models*. It is the task of observational cosmologists to determine which, if any, of these models provides a good description of the Universe that we actually inhabit.

8.3.1   The energy–momentum tensor of the cosmos

In Chapter 4 we saw that in general relativity the sources of gravitation are contained in an energy–momentum tensor $[T^{\mu\nu}]$ that describes the distribution and flow of energy and momentum in a region of spacetime. A reminder of the physical significance of the various parts of the energy–momentum tensor is given in Figure 8.9. Each of the sixteen components of $[T^{\mu\nu}]$ can be measured in units of $\mathrm{J\,m^{-3}}$ though it is often convenient to use other, equivalent, units.

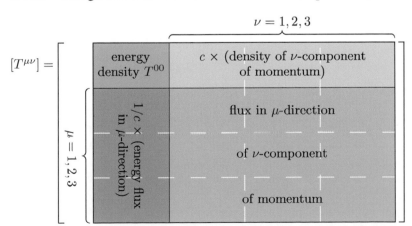

Figure 8.9   A reminder of the significance of the various parts of the energy–momentum tensor $[T^{\mu\nu}]$. 'Flux' implies a measurement per unit time and per unit area at right angles to the specified direction.

Describing in detail the distribution and flow of energy and momentum in the Universe is obviously beyond our capabilities. So, when specifying the cosmic energy–momentum tensor, cosmologists must decide on an acceptable compromise between accuracy and mathematical tractability. Traditionally, the solution is to treat the contents of spacetime as a homogeneous and isotropic ideal fluid that fills the whole of space. Such a fluid can be characterized by a proper density $\rho(t)$ and an associated pressure $p(t)$, each of which may depend only on the cosmic time t. According to a fundamental observer, travelling with the flow of this cosmic fluid, the fluid is locally at rest, so its energy–momentum tensor takes on the simple form that we met in Chapter 4:

$$[T^{\mu\nu}] = \begin{pmatrix} \rho c^2 & 0 & 0 & 0 \\ 0 & p & 0 & 0 \\ 0 & 0 & p & 0 \\ 0 & 0 & 0 & p \end{pmatrix}. \tag{Eqn 4.27}$$

More specifically, the current convention is to treat the contents of spacetime as a multi-component fluid composed of three distinct ideal fluids that respectively represent matter, radiation and the source of dark energy. Thus the homogeneous cosmic density can be written as

$$\rho(t) = \rho_{\rm m}(t) + \rho_{\rm r}(t) + \rho_\Lambda, \tag{8.19}$$

and the corresponding homogeneous and isotropic cosmic pressure is

$$p(t) = p_{\rm m}(t) + p_{\rm r}(t) + p_\Lambda. \tag{8.20}$$

Note that we have already taken account of the fact that the density and pressure due to dark energy are expected to be independent of time by omitting the reference to time in the case of ρ_Λ and p_Λ. It's also worth noting that since the role of dark energy may be nothing more than emulating the effect of a cosmological constant, we shall be quite willing to consider the possibility that ρ_Λ might be negative, even though this would be 'unphysical' in the case of a real fluid.

A few other comments about these various fluid components are in order before we move on. The first point concerns the distinction between matter and radiation. The essential difference is that particles of matter have mass, while particles of radiation (such as photons) do not. Thus, for example, protons are particles of matter but photons are particles of radiation. In the case of matter, the proper density $\rho_{\rm m}$ is just the usual mass density in units of $\rm kg\,m^{-3}$, and the corresponding proper energy density is $\rho_{\rm m}\,c^2$. In the case of radiation, however, there is no mass density; instead, we first determine the energy density of the radiation, $\rho_{\rm r}\,c^2$, and then divide that by c^2 to obtain an 'effective' mass density $\rho_{\rm r}$ for the radiation. It should also be noted that in some situations the mass of a certain kind of particle may be negligible, in which case the particles can be treated as radiation even though they are really particles of matter.

A second point concerns the behaviour of the density of matter and radiation as the Universe expands or contracts. Consider some large cubic region containing particles of matter and radiation. Suppose that a uniform expansion of the Universe causes each side of the cube to increase its proper length by a factor of 2 over some period of cosmic time. As a result the proper volume of the cube will increase by a factor of 8, and the proper number density of particles will decrease by a factor of 8. The expansion won't affect the mass of each particle of matter, so the mass density of matter will also decrease by a factor of 8. In fact, there will be a general relationship between $\rho_{\rm m}$ and $R(t)$ of the form

$$\rho_{\rm m} \propto \frac{1}{R^3}. \tag{8.21}$$

Contrast that with the behaviour of the radiation density $\rho_{\rm r}$, where Planck's law ($E = hf$, where h is Planck's constant) tells us that the energy E of each particle is proportional to its frequency f, and therefore inversely proportional to its wavelength λ. That means that a doubling of $R(t)$ (which will also double the wavelength) halves the energy of each particle and reduces the energy density $\rho_{\rm r} c^2$ and the effective mass density $\rho_{\rm r}$ by a factor of 16. The general relationship for the density of radiation is therefore

$$\rho_{\rm r} \propto \frac{1}{R^4}. \tag{8.22}$$

This difference in behaviour means that in an expanding Universe, the density of radiation will decline more rapidly than the density of matter, but both will decline relative to the constant density of dark energy. Figure 8.10 shows schematically what is believed to have been the history of the various contributions to the cosmic density in our own Universe. As you can see, there may have been past epochs during which radiation and matter were each dominant, but we are now believed to inhabit a Universe that is dominated by dark energy.

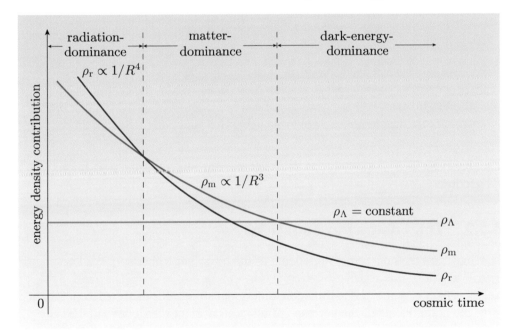

Figure 8.10   The possible evolution of the density of radiation, matter and dark energy over cosmic time in our Universe.

A third point to note concerns the cosmic pressure. We noted earlier that a uniform pressure everywhere acts like an additional source of gravitation. So the homogeneous negative pressure $p_\Lambda = -\rho_\Lambda c^2 = -\Lambda c^4/8\pi G$ associated with dark energy has the same repulsive effect as a cosmological constant Λ. The positive pressure associated with radiation is related by basic physical principles to the density of radiation by $p_{\rm r} = \rho_{\rm r} c^2/3$. The pressure of matter is often ignored (in which case the matter is referred to as dust), but when it is included it is described by a relationship called the **equation of state**, which asserts that

$$p_{\rm m} = w\rho c^2, \tag{8.23}$$

where w takes a constant value that is equal to 0 in the case of dust but would be positive for a real fluid. The concept of an equation of state can be extended to include the radiation fluid (with $w = 1/3$) and the dark energy fluid (with $w = -1$).

Now suppose that there is some particular time t_0 (often taken to be the present time) at which $R(t)$ has a known value $R(t_0) = R_0$. If we use the symbols $\rho_{m,0}$ and $\rho_{r,0}$ to represent the values $\rho_m(t_0)$ and $\rho_r(t_0)$, we can write

$$\rho_m(t) = \rho_{m,0} \left[\frac{R_0}{R(t)} \right]^3 \quad \text{and} \quad \rho_r(t) = \rho_{r,0} \left[\frac{R_0}{R(t)} \right]^4. \tag{8.24}$$

So, in a model Universe where the matter is represented by pressure-free dust, there will be a uniform cosmic density

$$\rho(t) = \rho_{m,0} \left[\frac{R_0}{R(t)} \right]^3 + \rho_{r,0} \left[\frac{R_0}{R(t)} \right]^4 + \rho_\Lambda \tag{8.25}$$

and a corresponding homogeneous and isotropic cosmic pressure

$$p(t) = \frac{\rho_{r,0}\, c^2}{3} \left[\frac{R_0}{R(t)} \right]^4 - \rho_\Lambda\, c^2. \tag{8.26}$$

To summarize, we have the following.

Cosmic composition

At cosmic time $t = t_0$, the sources of cosmic gravitation are specified by just three values: $\rho_{m,0}$, $\rho_{r,0}$ and ρ_Λ. Given these three values, the cosmic density and pressure at any other cosmic time can be determined, provided that the cosmic scale factor $R(t)$ is known as an explicit function of cosmic time.

The determination of the function $R(t)$ is the main subject of the next three subsections.

8.3.2   The Friedmann equations

Starting from the non-zero components of the covariant Robertson–Walker metric tensor, $g_{00} = c^2$, $g_{11} = -R^2(t)/(1 - kr^2)$, $g_{22} = -R^2(t)\, r^2$ and $g_{33} = -R^2(t)\, r^2 \sin^2\theta$, it is time-consuming but straightforward to determine, in turn, the components of the corresponding contravariant metric tensor, the connection coefficients, the Riemann curvature components, the Ricci curvature components and the Ricci scalar (which should not be confused with the scale factor $R(t)$). Once all of this has been done, the Einstein field equations can be written down using the energy–momentum tensor described in the previous subsection. Because of the many terms that vanish and the high degree of symmetry, all this calculation leads to just two independent equations, usually referred to as the **Friedmann equations**.

The Friedmann equations

$$\left[\frac{1}{R} \frac{dR}{dt} \right]^2 = \frac{8\pi G}{3} \rho - \frac{kc^2}{R^2}, \tag{8.27}$$

$$\frac{1}{R} \frac{d^2 R}{dt^2} = -\frac{4\pi G}{3} \left(\rho + \frac{3p}{c^2} \right). \tag{8.28}$$

The first of these equations was derived by Alexander Friedmann (Figure 8.11), a Russian mathematical physicist, in 1922, though he included a cosmological constant Λ that we are representing by dark energy contributions to the density ρ and pressure p. The term in square brackets on the left-hand side of the first equation is the Hubble parameter $H(t)$ that was defined in Equation 8.17.

The Friedmann equations come directly from the formalism of general relativity and can be used as they stand to determine the scale factor $R(t)$ subject to appropriate boundary conditions. However, interestingly, both equations have a very straightforward Newtonian interpretation. The first Friedmann equation is sometimes called the *energy equation*; it looks like a Newtonian energy equation. This impression is strengthened if the equation is rewritten as

$$\frac{1}{2}\left[\frac{dR}{dt}\right]^2 - G\frac{\frac{4}{3}\pi R^3 \rho}{R} = \text{constant}, \qquad (8.29)$$

which, apart from an overall factor representing mass, looks like a statement that the sum of the kinetic and gravitational potential energy of a particle is constant at the surface of a uniform sphere of density ρ and radius R.

Figure 8.11   Alexander Friedmann (1888–1925) published a study of cosmological models with positive curvature in 1922 and negative curvature models in 1924. He died in 1925, aged 37, from typhoid fever.

Similarly, the second Friedmann equation is sometimes called the *acceleration equation* because it involves a second derivative and looks like a Newtonian equation of motion. Again, that impression is greatly strengthened if the equation is rewritten in the form

$$\frac{d^2R}{dt^2} = -G\frac{\frac{4}{3}\pi R^3 \left(\rho + \frac{3p}{c^2}\right)}{R^2}, \qquad (8.30)$$

which looks like a description of the acceleration due to (Newtonian) gravity at the surface of a sphere of radius R and uniform density $\rho + 3p/c^2$.

Returning to general relativity, the Friedmann equations can still be related to energy conservation. Differentiating the energy equation and using the acceleration equation to eliminate the resulting second derivative leads to the following equation, known as the *fluid equation*,

$$\frac{d\rho}{dt} + \left(\rho + \frac{p}{c^2}\right)\frac{3}{R}\frac{dR}{dt} = 0, \qquad (8.31)$$

which can be shown to be an expression of energy conservation, relating changes in the energy of a co-moving volume of fluid to the work done against the external pressure.

The energy, acceleration and fluid equations are not all independent, but different combinations of them may be used to tackle a range of problems in cosmic evolution.

Exercise 8.4   Show that the fluid equation (Equation 8.31) may be derived from the energy equation (Equation 8.27) and the acceleration equation (Equation 8.28). ∎

Of course, when trying to solve the Friedmann equations it is necessary to make explicit the dependence on $R(t)$ that is implicit in $\rho(t)$ and $p(t)$. Accepting the simplifications expressed in Equations 8.25 and 8.26, the equations that we shall use to determine the scale factor $R(t)$ are as follows.

The Friedmann equations — expanded and simplified

$$\left[\frac{1}{R}\frac{dR}{dt}\right]^2 = \frac{8\pi G}{3}\left[\rho_{m,0}\left[\frac{R_0}{R(t)}\right]^3 + \rho_{r,0}\left[\frac{R_0}{R(t)}\right]^4 + \rho_\Lambda\right] - \frac{kc^2}{R^2}, \quad (8.32)$$

$$\frac{1}{R}\frac{d^2R}{dt^2} = -\frac{4\pi G}{3}\left[\rho_{m,0}\left[\frac{R_0}{R(t)}\right]^3 + 2\rho_{r,0}\left[\frac{R_0}{R(t)}\right]^4 - 2\rho_\Lambda\right]. \quad (8.33)$$

Exercise 8.5   Show that the terms in the square brackets on the right of Equation 8.33 arise from the definitions of ρ_m, p_m, ρ_r, p_r, ρ_Λ and p_Λ made earlier.   ∎

8.3.3 Three cosmological models with $k = 0$

As an example of the use of the Friedmann equations, we shall briefly consider three 'unrealistic' single-component cosmological models. These models are chosen primarily because of their mathematical simplicity; none is thought to represent the current state of our Universe, but each still plays an important part in cosmological discussions. All three models have $k = 0$, implying that all (fixed time) space-like hypersurfaces are geometrically flat. (As noted earlier, the flatness of three-dimensional space at fixed times does *not* imply that four-dimensional spacetime is geometrically flat.)

Example 1: the de Sitter model, $k = 0$, $\rho_{m,0} = 0$, $\rho_{r,0} = 0$

In this case, in addition to space being flat, there is no matter and no radiation, only dark energy. Substituting the given values into the first of the Friedmann equations, and taking the positive square root of each side, gives

$$\frac{dR}{dt} = \sqrt{\frac{8\pi G}{3}\rho_\Lambda}\, R. \quad (8.34)$$

This is a first-order differential equation, so its solution requires one initial condition. We adopt the conventional choice that at $t = t_0$ the scale factor $R(t_0)$ has some known value R_0. Subject to this condition, the solution can be written as

$$R(t) = R_0 \exp\left(\sqrt{\frac{8\pi G\rho_\Lambda}{3}}\,(t - t_0)\right). \quad (8.35)$$

In this case the Hubble parameter turns out to be independent of time, since

$$H(t) = \left[\frac{1}{R}\frac{dR}{dt}\right] = \sqrt{\frac{8\pi G\rho_\Lambda}{3}}. \quad (8.36)$$

If we adopt the general convention that $H_0 = H(t_0)$, then in this case we shall have $H_0 = \sqrt{8\pi G\rho_\Lambda/3}$ and we can write the scale factor of this cosmological model as

$$R(t) = R_0 \exp\left(H_0(t - t_0)\right). \quad (8.37)$$

This kind of cosmological model is known as a **de Sitter model**. The model was the second to be formulated and the first to describe an expanding Universe. It was proposed by Willem de Sitter in 1917, though he used a very different approach to its development and presentation. Since the model does not include any matter or radiation, it is not a good model of our current Universe but it has been used to describe a hypothetical epoch in the very early development of our Universe, known as the **inflationary era**, when the Universe is supposed to have undergone a brief period of very rapid expansion. It may also describe the far future of our Universe, when continued cosmic expansion will have reduced the density of matter and radiation to such an extent that those densities will be negligible compared with the (constant) density of dark energy.

Example 2: the flat, pure radiation model, $k = 0$, $\rho_{m,0} = 0$, $\rho_\Lambda = 0$

In this case, space is flat and the Universe contains only radiation. It is thought that our Universe was almost like this during its early evolution, immediately after inflation, when it was strongly dominated by radiation. The first Friedmann equation for such a Universe gives

$$\frac{dR}{dt} = \sqrt{\frac{8\pi G}{3}\rho_{r,0}}\,\frac{R_0^2}{R}. \tag{8.38}$$

Adopting the usual initial condition $R(t_0) = R_0$, the scale factor that satisfies the differential equation can again be written in terms of H_0, the value of the model's Hubble parameter at time t_0. In this case

$$R(t) = R_0(2H_0t)^{1/2}, \tag{8.39}$$

where $H_0 = \sqrt{8\pi G\rho_{r,0}/3}$.

Exercise 8.6   (a) Verify that Equation 8.39 is a solution of Equation 8.38.

(b) Also show that this solution implies that $H(t) = 1/2t$ (so $H_0 = 1/2t_0$), and hence confirm that it satisfies the condition $R(t_0) = R_0$.   ■

Example 3: the Einstein–de Sitter model, $k = 0$, $\rho_{r,0} = 0$, $\rho_\Lambda = 0$

In this case, space is flat and the Universe contains only matter. Einstein and de Sitter agreed to advocate this model in 1932, following Hubble's discovery of cosmic expansion — hence the name **Einstein–de Sitter model**. Having come to disfavour the idea of a cosmological constant, they saw this model as a critical intermediate case, separating open models with $k = -1$ from closed models with $k = +1$. For this reason it is also often referred to as the **critical model**. The critical/Einstein–de Sitter model was regarded by many as providing a good description of our Universe for several decades. Its viability became increasingly suspect as observational data improved in the 1980s, but it wasn't until the late-1990s that it was finally abandoned in favour of models dominated by dark energy.

The first Friedmann equation for an Einstein–de Sitter Universe can be written as

$$\frac{dR}{dt} = \sqrt{\frac{8\pi G}{3}\rho_{m,0}}\,\frac{R_0^{3/2}}{R^{1/2}}. \tag{8.40}$$

With $R(t_0) = R_0$, the solution can be written as

$$R(t) = R_0 \left(\tfrac{3}{2} H_0 t\right)^{2/3}, \tag{8.41}$$

where $H_0 = \sqrt{8\pi G \rho_{\mathrm{m},0}/3}$.

In this case the Hubble parameter is given by $H(t) = 2/(3t)$.

The variation of R with t for all three of the models that we have been discussing is shown in Figure 8.12. Diagrams of this kind provide a useful way of visualizing the expansion history of a cosmological model. You will see more such diagrams in the next section.

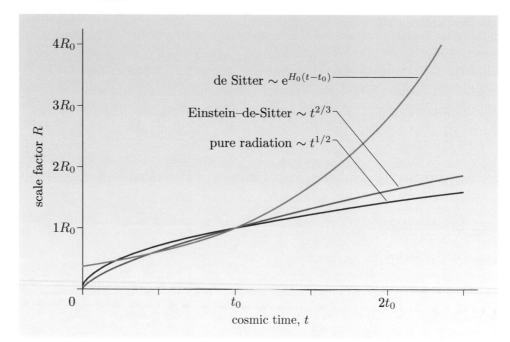

Figure 8.12 Expansion histories of the de Sitter, pure radiation and Einstein–de Sitter cosmological models, all with $k = 0$.

In a Universe where $k = 0$, it follows from the first Friedmann equation and the definition of the Hubble parameter ($H(t) = R^{-1}\,\mathrm{d}R/\mathrm{d}t$) that

$$H^2(t) = \frac{8\pi G}{3} \rho(t). \tag{8.42}$$

So, as a $k = 0$ Universe expands or contracts, the cosmic density must change in proportion to the square of the Hubble parameter. Moreover, for a $k = 0$ Universe, the changing value of the total cosmic density will always have the value implied by Equation 8.42; this value is called the **critical density**. It is denoted by $\rho_{\mathrm{c}}(t)$ and is given by the following.

Critical density

$$\rho_{\mathrm{c}}(t) = \frac{3H^2(t)}{8\pi G}. \tag{8.43}$$

The critical density provides a useful reference density that we shall make use of in the next subsection. The key points of the three flat space models considered in this subsection are summarized in Table 8.1.

Table 8.1 Spatially flat ($k = 0$) single-component models.

Name	de Sitter	Pure radiation	Einstein–de Sitter
Composition	Dark energy only ($w = -1$)	Radiation only ($w = 1/3$)	Matter only ($w = 0$)
Scale factor $R(t)$	$R(t) = R_0 e^{H_0(t-t_0)}$	$R(t) = R_0(2H_0t)^{1/2}$	$R(t) = R_0\left(\frac{3}{2}H_0t\right)^{2/3}$
Hubble parameter $H(t)$	$H(t) = \text{constant}$	$H(t) = \dfrac{1}{2t}$	$H(t) = \dfrac{2}{3t}$
Density at time t_0 ρ_0	$\rho_{\Lambda,0} = \rho_{c,0} = \dfrac{3H_0^2}{8\pi G}$	$\rho_{r,0} = \rho_{c,0} = \dfrac{3H_0^2}{8\pi G}$	$\rho_{m,0} = \rho_{c,0} = \dfrac{3H_0^2}{8\pi G}$
Density at time t $\rho(t) = \rho_c(t)$	$\rho_\Lambda(t) = \rho_{\Lambda,0}$	$\rho_r(t) = \rho_{r,0}\left[\dfrac{t_0}{t}\right]^2$	$\rho_m(t) = \rho_{m,0}\left[\dfrac{t_0}{t}\right]^2$

8.3.4 Friedmann–Robertson–Walker models in general

A relativistic cosmological model based on the Robertson–Walker metric with a scale factor determined by the Friedmann equations is known as a **Friedmann–Robertson–Walker (FRW) model**. The three single-component models with $\rho = \rho_c$ and hence $k = 0$ that we considered in the previous subsection are among the simplest examples of FRW models. When specifying a general FRW model it is conventional to express each of the densities as a fraction of the critical density ρ_c. These fractional densities are called **density parameters** and are defined as follows.

Density parameters

$$\Omega_m(t) = \frac{\rho_m(t)}{\rho_c(t)}, \quad \Omega_r(t) = \frac{\rho_r(t)}{\rho_c(t)}, \quad \Omega_\Lambda(t) = \frac{\rho_\Lambda}{\rho_c(t)}. \tag{8.44}$$

Note that although the density ρ_Λ is independent of time, the density parameter Ω_Λ is not; this is because of the time dependence of ρ_c.

Using the density parameters, the first Friedmann equation can be rewritten as

$$1 = \Omega_m(t) + \Omega_r(t) + \Omega_\Lambda(t) - \frac{c^2 k}{H^2(t)\, R^2(t)}. \tag{8.45}$$

Rearranging this to read

$$\frac{c^2 k}{H^2(t)\, R^2(t)} = \Omega_m(t) + \Omega_r(t) + \Omega_\Lambda(t) - 1, \tag{8.46}$$

it can be seen that at any time the total density parameter determines the cosmic geometry of space, since

$$\text{if } \Omega_m + \Omega_r + \Omega_\Lambda < 1, \quad \text{then } k < 0 \text{ and space will be open,} \qquad (8.47)$$
$$\text{if } \Omega_m + \Omega_r + \Omega_\Lambda = 1, \quad \text{then } k = 0 \text{ and space will be flat,} \qquad (8.48)$$
$$\text{if } \Omega_m + \Omega_r + \Omega_\Lambda > 1, \quad \text{then } k > 0 \text{ and space will be closed.} \qquad (8.49)$$

When it comes to solving the Friedmann equations, a few special cases, such as those considered in the previous subsection, can be treated analytically. However, it is often necessary to resort to numerical methods to find solutions. Some illustrative examples of the kinds of solutions that arise are shown in Figure 8.13.

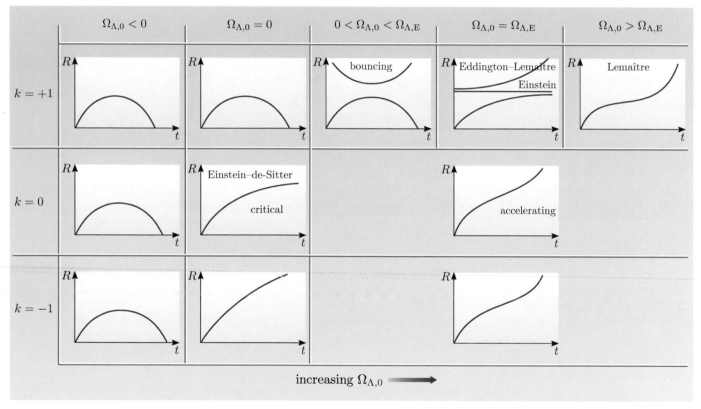

Figure 8.13   A visual catalogue of representative scale factors for a range of FRW models.

The examples are classified according to the value of k (i.e. how $\Omega_{m,0} + \Omega_{r,0} + \Omega_{\Lambda,0}$ compares with 1) and the value of $\Omega_{\Lambda,0}$. In most cases the small graph of R against t that appears in any given cell is intended to be representative of the whole class of specific results that would emerge for different choices of $\Omega_{m,0}$, $\Omega_{r,0}$ and $\Omega_{\Lambda,0}$. Of course, this means that some important cases are not properly illustrated. For instance, the exponentially expanding de Sitter model sits in the cell devoted to $k = 0$ and $\Omega_{\Lambda,0} > 0$, but the graph that appears in that cell is for a model that contains some matter and radiation, which the de Sitter model does not. You can imagine the de Sitter model as a limiting case of the model that is shown.

In fact, the general kind of model shown in the $k = 0$, $\Omega_{\Lambda,0} > 0$ cell is of special interest to cosmologists. It is currently thought to provide a good description of the large-scale features of our Universe. Like many of the models, it starts with $R = 0$ and growing. This is an indication of an early phase in cosmic evolution that would have been dense and hot. It corresponds to the statement that the Universe began with a **Big Bang**. The high density is a simple consequence of the smallness of R at early times; we have already seen that $\rho_{\rm m} \propto 1/R^3$, while for the radiation that dominated the early Universe, $\rho_{\rm r} \propto 1/R^4$. The high temperature, T, follows from the $1/R^4$ dependence of the energy density and the expectation that the radiation was *thermal* radiation, implying (in accordance with *Stefan's law*) that its energy density is proportional to T^4 with the consequence that $T \propto 1/R$. Thus the temperature would also have been higher in the compressed conditions of the early Universe when R was small.

Another interesting feature of this kind of model is that although it indicates continuous expansion (R always gets bigger), it also shows that the rate of expansion initially declines but then begins to increase again. For that reason this is sometimes described as an **accelerating model**. The acceleration in the rate of expansion is a result of the changing densities of matter, radiation and dark energy. The model is characterized by $k = 0$, so the sum of those densities will always be the critical density $\rho_{\rm c}$, but as the critical density itself declines, the proportions contributed by matter, radiation and dark energy will change, with dark energy eventually becoming dominant. (Look again at Figure 8.10.) During the eras when radiation and matter are dominant, the rate of expansion decelerates, but when dark energy becomes dominant, the rate of expansion accelerates. We shall have more to say about this model in the next section.

Looking more generally at the FRW models in Figure 8.13, you can see that if $\Omega_{\Lambda,0} < 0$, as in the column on the left, the model generally starts with a Big Bang but eventually reaches a state of maximum expansion and then recollapses. Its end would involve a state of increasing density as R decreases to zero in a process usually referred to as the **big crunch**. These **recollapsing models** occur with all possible values of k, so their space-like hypersurfaces may be open, flat or closed, depending on which particular variant we choose to study.

The $\Omega_{\Lambda,0} = 0$ models in the middle column include open, ever-expanding models, closed, recollapsing models and, in between, the flat space $k = 0$ models that will include the Einstein–de Sitter model and the flat, pure radiation model.

The set of $\Omega_{\Lambda,0} > 0$ models includes the $k = 0$ accelerating model that we have already discussed, a similar $k = -1$ open model, and several different closed models, including some that do not feature a Big Bang. A particularly interesting case amongst this latter class is the static **Einstein model**, represented by a horizontal R against t graph. This, you will recall, was the first relativistic cosmological model, the one that prompted Einstein to introduce the cosmological constant. Ignoring the effect of radiation (i.e. setting $\Omega_{\rm r,0} = 0$), the Einstein model arises when the effect of dark energy exactly balances the effect of matter to ensure that $dR/dt = d^2R/dt^2 = 0$, so that R has the constant value R_0. For this to be the case, it follows from the second Friedmann equation (Equation 8.33) that $\rho_\Lambda = \rho_{\rm m,0}/2$, or, in terms of density parameters,

$$\Omega_{\Lambda,0} = \frac{\Omega_{\rm m,0}}{2}. \tag{8.50}$$

This is the value of the dark energy density parameter that is indicated by $\Omega_{\Lambda,{\rm E}}$ in Figure 8.13.

One other model that deserves to be mentioned is the **Eddington–Lemaître model** ($k = +1, \Omega_{\Lambda,0} = \Omega_{\Lambda,{\rm E}}$). This was brought to prominence in a 1927 report on expanding-universe models by Georges Lemaître (1894–1966), a Belgian catholic priest and cosmologist. The model was strongly supported by Sir Arthur Eddington – hence the name. It is unusual in that it does not start with a big bang. Rather it can develop from the (static) Einstein model, which is actually unstable against fluctuations in the density. In 1933 Lemaître proposed a primitive variant of Big Bang theory as an explanation of the origin of the Universe, and shifted his allegiance to the model now known as the **Lemaître model** ($k = +1, \Omega_{\Lambda,0} > \Omega_{\Lambda,\mathrm{E}}$).

Exercise 8.7   Using the first Friedmann equation, show that in Einstein's static Universe $R_0 = (c^2/4\pi G\rho_{\mathrm{m},0})^{1/2}$, and evaluate this in light-years and parsecs given that a modern estimate of the current cosmic matter density is $\rho_{\mathrm{m},0} \approx 3 \times 10^{-27}\,\mathrm{kg\,m^{-3}}$.

Exercise 8.8   Using the second Friedmann equation, show that if $\Omega_{\mathrm{r},0}$ is taken to be zero, the condition that distinguishes those FRW Universes that have already started to (positively) accelerate at time t_0 from those that have not is $\Omega_{\Lambda,0} \geq \Omega_{\mathrm{m},0}/2$.

Exercise 8.9   Assuming that $\Omega_{\mathrm{r},0}$ is negligible, the range of FRW models can be represented by points in a plane with coordinates $\Omega_{\mathrm{m},0}$ and $\Omega_{\Lambda,0}$, as indicated in Figure 8.14. Write down the condition that determines the location of the dividing line between models with $k = +1$ and models with $k = -1$, and identify the point or points associated with (i) the de Sitter model, (ii) the Einstein–de Sitter model, and (iii) the Einstein model.   ■

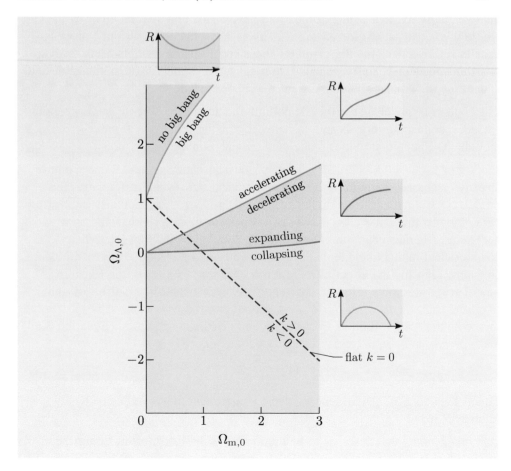

Figure 8.14   Cosmological models in the $\Omega_{\Lambda,0}$–$\Omega_{\mathrm{m},0}$ plane.

8.4 Friedmann–Robertson–Walker models and observations

In this section we consider the relationship between certain observable properties of the Universe in which we live, and the parameters that have played an important part in our discussion of cosmological models, particularly the proper distance (σ or d_p), the Hubble parameter $H(t)$ and the cosmic time t. We said earlier that t_0 is often taken to represent the current cosmic time. From this point on, that will always be the case.

8.4.1 Cosmological redshift and cosmic expansion

Defining redshift

The redshift of spectral lines is a common and useful phenomenon in astronomy. In earlier chapters we have encountered two distinct causes of redshift.

1. The *Doppler effect* of special relativity, which arises when a source of radiation and the observer of that radiation are in relative motion.

2. The *gravitational redshift* of general relativity that is a consequence of the gravitational time dilation that exists between observers who are relatively at rest but located in regions of different spacetime curvature.

You are about to encounter a third cause of redshift, usually referred to as **cosmological redshift**, that arises when the source and the observer are separated by cosmologically large distances in a Universe that is contracting or expanding.

For our present purposes it is useful to introduce a quantitative measure of the redshift of a spectral line. This quantity is widely used in astronomy and is defined as follows.

Quantitative definition of redshift

$$z = \frac{\lambda_{\text{ob}} - \lambda_{\text{em}}}{\lambda_{\text{em}}}. \tag{8.51}$$

Here λ_{em} is the wavelength at which some spectral line is emitted, as measured at the source (or, more realistically, as determined from some laboratory-based experiment involving similar sources), and λ_{ob} is the observed wavelength of the spectral line when it reaches its distant observer. Note that z is a dimensionless ratio, so it's just represented by a number such as 0.1 or 2. A negative value of z is used to indicate a blueshift. In most cases of astronomical interest, all the lines in a spectrum will have the same redshift, so the measured redshift is a property of the body concerned, not just the spectral line.

● Show that when expressed in terms of the emitted and observed frequencies, f_{em} and f_{ob}, the definition of redshift implies that

$$1 + z = \frac{f_{\text{em}}}{f_{\text{ob}}}. \tag{8.52}$$

○ From Equation 8.51 using the general relation $c = f\lambda$,

$$z = \frac{\lambda_{\text{ob}} - \lambda_{\text{em}}}{\lambda_{\text{em}}} = \frac{\lambda_{\text{ob}}}{\lambda_{\text{em}}} - 1 = \frac{f_{\text{em}}}{f_{\text{ob}}} - 1.$$

Adding 1 to each side gives the required result, which we shall use later.

Relating redshift to the scale factor

Suppose that a fundamental observer, at the origin of co-moving coordinates in a Robertson–Walker spacetime, observes a light signal emitted from a distant galaxy at a fixed radial co-moving coordinate $r = \chi$. We can take the coordinates of the emission event to be $(t_{\text{em}}, \chi, 0, 0)$ and the coordinates of the observation event to be $(t_{\text{ob}}, 0, 0, 0)$. The light signal will travel along a null geodesic where $(\mathrm{d}s)^2 = 0$, so it follows from the Robertson–Walker line element that all along that null geodesic,

$$0 = c^2 (\mathrm{d}t)^2 - R^2(t) \frac{(\mathrm{d}r)^2}{1 - kr^2}.$$

Splitting this expression into time-dependent and space-dependent parts, and taking the positive square root, we get

$$\frac{c\,\mathrm{d}t}{R(t)} = \frac{\mathrm{d}r}{\sqrt{1 - kr^2}}.$$

Integrating each part over the whole pathway,

$$\int_{t_{\text{em}}}^{t_{\text{ob}}} \frac{c\,\mathrm{d}t}{R(t)} = \int_0^\chi \frac{\mathrm{d}r}{\sqrt{1 - kr^2}}. \tag{8.53}$$

Now suppose that a second signal is emitted from the same source a short time later, at $t_{\text{em}} + \delta t_{\text{em}}$, and that it is observed a short time after the first signal, at $t_{\text{ob}} + \delta t_{\text{ob}}$. This second signal also travels along a null geodesic, so

$$\int_{t_{\text{em}}+\delta t_{\text{em}}}^{t_{\text{ob}}+\delta t_{\text{ob}}} \frac{c\,\mathrm{d}t}{R(t)} = \int_0^\chi \frac{\mathrm{d}r}{\sqrt{1 - kr^2}}.$$

The spatial integral is the same in both cases since it only involves co-moving coordinates. Consequently we can equate the two time-dependent integrals:

$$\int_{t_{\text{em}}}^{t_{\text{ob}}} \frac{c\,\mathrm{d}t}{R(t)} = \int_{t_{\text{em}}+\delta t_{\text{em}}}^{t_{\text{ob}}+\delta t_{\text{ob}}} \frac{c\,\mathrm{d}t}{R(t)}.$$

Now, each of these integrals can be written as the sum of two parts. For the integral on the left,

$$\int_{t_{\text{em}}}^{t_{\text{ob}}} \frac{c\,\mathrm{d}t}{R(t)} = \int_{t_{\text{em}}}^{t_{\text{em}}+\delta t_{\text{em}}} \frac{c\,\mathrm{d}t}{R(t)} + \int_{t_{\text{em}}+\delta t_{\text{em}}}^{t_{\text{ob}}} \frac{c\,\mathrm{d}t}{R(t)},$$

and for the integral on the right,

$$\int_{t_{\text{em}}+\delta t_{\text{em}}}^{t_{\text{ob}}+\delta t_{\text{ob}}} \frac{c\,\mathrm{d}t}{R(t)} = \int_{t_{\text{em}}+\delta t_{\text{em}}}^{t_{\text{ob}}} \frac{c\,\mathrm{d}t}{R(t)} + \int_{t_{\text{ob}}}^{t_{\text{ob}}+\delta t_{\text{ob}}} \frac{c\,\mathrm{d}t}{R(t)}.$$

Subtracting the corresponding sides of these two equations, we see that

$$0 = \int_{t_{\text{em}}}^{t_{\text{em}}+\delta t_{\text{em}}} \frac{c\,\mathrm{d}t}{R(t)} - \int_{t_{\text{ob}}}^{t_{\text{ob}}+\delta t_{\text{ob}}} \frac{c\,\mathrm{d}t}{R(t)}.$$

Rearranging and cancelling the factor c, we see that

$$\int_{t_{\text{em}}}^{t_{\text{em}}+\delta t_{\text{em}}} \frac{\mathrm{d}t}{R(t)} = \int_{t_{\text{ob}}}^{t_{\text{ob}}+\delta t_{\text{ob}}} \frac{\mathrm{d}t}{R(t)},$$

but each of these integrals covers a very short period of time, so the integrand will be effectively constant for the short duration of the integration, and we can write

$$\frac{\delta t_{\text{em}}}{R(t_{\text{em}})} = \frac{\delta t_{\text{ob}}}{R(t_{\text{ob}})}.$$

It follows that

$$\frac{\delta t_{\text{em}}}{\delta t_{\text{ob}}} = \frac{R(t_{\text{em}})}{R(t_{\text{ob}})}. \tag{8.54}$$

If we now let δt_{em} be the proper period of oscillation of the emitted light, then δt_{ob} will be the period of the observed light and we can use the fact that frequency is inversely proportional to period to replace $\delta t_{\text{em}}/\delta t_{\text{ob}}$ by $f_{\text{ob}}/f_{\text{em}}$, giving

$$\frac{f_{\text{ob}}}{f_{\text{em}}} = \frac{R(t_{\text{em}})}{R(t_{\text{ob}})}. \tag{8.55}$$

Substituting this result into Equation 8.52, we obtain our final result.

Cosmological redshift related to scale factor

$$1 + z = \frac{R(t_{\text{ob}})}{R(t_{\text{em}})}. \tag{8.56}$$

So the redshift of the light is determined by the ratio of the scale factors at the times of observation and emission. In an expanding Universe, $R(t_{\text{ob}})$ will be bigger than $R(t_{\text{em}})$, so Equation 8.56 predicts that the observed light will be positively redshifted. If the Universe expands monotonically, then the more distant the source of the light, the longer the time the light will spend in transit, and, generally speaking, the greater will be the observed redshift.

● A distant quasar has a redshift $z = 6.0$. By what factor has the Universe expanded since the quasar emitted the light that we receive today?

○ Substituting $z = 6.0$ in Equation 8.56 gives $R(t_0)/R(t_{\text{em}}) = 7$.

Note that although galaxies participating in the Hubble flow will have a proper radial velocity away from any fundamental observer, any cosmological redshift that such observers measure is *not* a Doppler effect. The formula for cosmological redshift is quite different from the Doppler formula. However, what might be described as the effect of 'cosmological motion' (i.e. the Hubble flow, not the peculiar motions of individual galaxies or non-fundamental observers) is automatically included in the calculation of cosmological redshift, so there is no need for any kind of additional 'Doppler correction' to account for that motion. A common way of expressing this is to say that cosmological redshift is a result of motion that arises from the expansion *of* space rather than motion *through* space. Figure 8.15 overleaf illustrates this view. It indicates a cosmological redshift that is a consequence of the expansion of space and the corresponding stretching of wavelength while the radiation is in transit between galaxies with fixed co-moving coordinates. The galaxies themselves are supposed to be bound systems, so they are not enlarged by the stretching of space, which can be thought of as a weak 'background' effect that becomes significant only on the cosmic scale.

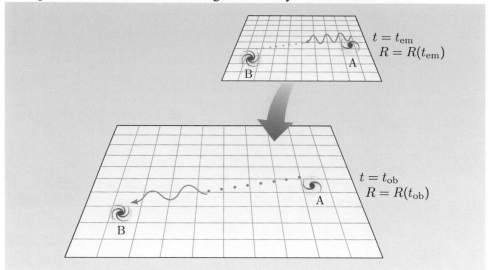

Figure 8.15   A schematic view of the origin of cosmological redshift as a result of the expansion of space.

Exercise 8.10   Can we reasonably expect to measure a change in the value of $R(t)$ by means of local experiments, such as the observation of cosmological redshifts in the spectra of nearby stars?   ∎

Relating redshift to a measurable distance

The relation between redshift and the scale factor is an important step towards linking the cosmological models that we have been developing with observations, but the scale factor itself is not directly measurable. To obtain a relationship that we can test, we still need to relate the redshift to some other quantity that astronomers can actually measure. The most suitable quantity is the **luminosity distance**, d_L. This is defined in terms of the luminosity L of an isotropically radiating source and the energy flux F that reaches the observer, so that

$$F = \frac{L}{4\pi d_L^2}. \tag{8.57}$$

Here $4\pi d_L^2$ represents the area over which the radiation emitted in unit time is spread when it reaches the observer.

In a static Euclidean space d_L would be equal to the coordinate distance of the source. However, in Robertson–Walker spacetime things are not so simple. Consider a fundamental observer making observations from the origin. For a source at radial co-moving coordinate $r = \chi$, the proper area of the sphere over which the radiation is spread when it reaches the observer at time t_{ob} can be shown to be $4\pi R^2(t_{ob})\chi^2$. However, we saw earlier, in Equation 8.54, that in an expanding Universe, radiation emitted over a time period δt_{em} will be observed over a longer time period δt_{ob}, so the observed energy flux will be reduced by a factor

$$\frac{\delta t_{\mathrm{em}}}{\delta t_{\mathrm{ob}}} = \frac{R(t_{\mathrm{em}})}{R(t_{\mathrm{ob}})} = \frac{1}{1+z}. \tag{8.58}$$

We have also seen that in an expanding Universe, the wavelength of each arriving photon will be stretched out, so its energy will be reduced and the observed energy flux will therefore be further reduced by a factor

$$\frac{f_{\mathrm{ob}}}{f_{\mathrm{em}}} = \frac{\lambda_{\mathrm{em}}}{\lambda_{\mathrm{ob}}} = \frac{R(t_{\mathrm{em}})}{R(t_{\mathrm{ob}})} = \frac{1}{1+z}. \tag{8.59}$$

Consequently, in an FRW Universe at time t_{ob},

$$F = \frac{L}{4\pi R^2(t_{\mathrm{ob}}) \, \chi^2 (1+z)^2}. \tag{8.60}$$

Comparing Equations 8.57 and 8.60, it can be seen that

$$d_{\mathrm{L}} = R(t_{\mathrm{ob}}) \, \chi(1+z). \tag{8.61}$$

To obtain a relation between luminosity distance and redshift, we now need to express the quantity $R(t_{\mathrm{ob}}) \, \chi$ in terms of z. This is actually quite tricky, though the method and result are both well known. There is an exact method valid for all values of z and an approximate method valid for $z \ll 1$. Let's deal with the approximate method first; we shall come back to the exact method in the next subsection. The first step is to use Taylor's theorem to expand the scale factor $R(t)$ at some general time t as a power series in the **lookback time**, $(t_0 - t)$, about its current value $R(t_0)$. This series can be written as

$$R(t) \simeq R(t_0) \left[1 - H_0(t_0 - t) - \tfrac{1}{2} q_0 H_0^2 (t_0 - t)^2 + \cdots \right], \tag{8.62}$$

where H_0 is the current value of the Hubble parameter $H(t)$ that was introduced in Equation 8.17,

$$H(t) = \frac{1}{R} \frac{\mathrm{d}R}{\mathrm{d}t}, \tag{Eqn 8.17}$$

and q_0 is the current value of the **deceleration parameter** $q(t)$ defined by

$$q(t) = -\frac{1}{H^2(t)} \frac{1}{R(t)} \frac{\mathrm{d}^2 R}{\mathrm{d}t^2}. \tag{8.63}$$

This series is used in conjunction with Equation 8.61 (which involves the co-moving coordinate χ and the scale parameter $R(t)$) and the relation that we have already found that relates the scale parameter to the redshift, $1 + z = R(t_{\mathrm{ob}})/R(t_{\mathrm{em}})$. The result, after some labour, is that for observations made now, with $t_{\mathrm{ob}} = t_0$,

$$d_{\mathrm{L}} = \frac{c}{H_0} \left[z + \tfrac{1}{2}(1 - q_0) z^2 + \cdots \right]. \tag{8.64}$$

Remembering that this is valid only for small values of z, the relationship tells us that, to a first approximation, and ignoring any peculiar motion, we should expect to find that the redshift of each galaxy is proportional to its luminosity distance.

Predicted relation of redshift to luminosity distance for small z

$$d_{\mathrm{L}} = \frac{c}{H_0} z. \tag{8.65}$$

Here the constant of proportionality H_0 is the current value of the Hubble parameter. In addition, if we make more precise observations, particularly if they involve somewhat larger redshifts (though still significantly less than 1), then we should expect to see deviations from the simple proportional behaviour, and these should, in principle at least, inform us about any acceleration or deceleration of the cosmic expansion via q_0. A graph of the relationship between d_L and z, for a range of values of q_0 and a realistic value of H_0, is shown in Figure 8.16.

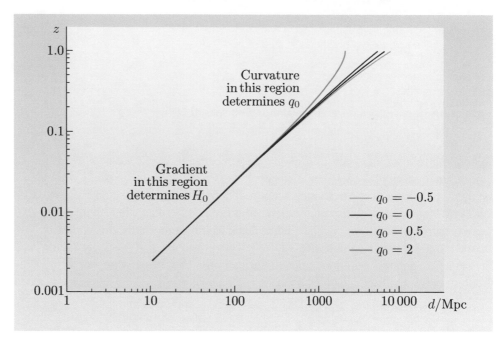

Figure 8.16 The predicted relation between redshift and luminosity distance for various current values of the deceleration parameter q_0.

Relating observations to the FRW models

In 1929 Edwin Hubble announced his discovery, based on a small sample of relatively nearby galaxies (all with $z < 0.004$), that redshift increased roughly in proportion to distance. Actually, he sowed the seeds of much future confusion by using the approximate Doppler formula, $v = cz$, to convert the redshifts into recession velocities and then expressing his finding in terms of an increase of recession velocity with distance, but redshift is what was actually measured. This publication is usually hailed as marking the discovery of the expansion of the Universe.

Hubble himself was always very cautious about the interpretation of his findings, but he was aware of de Sitter's 1917 paper about an expanding Universe, and he knew that de Sitter had suggested that systematic increases in observed redshifts would be a consequence. In fact, towards the end of his 1929 paper, Hubble said:

> The outstanding feature, however, is the possibility that the velocity–distance relation may represent the de Sitter effect, and hence that numerical data may be introduced into discussions of the general curvature of space.
>
> Hubble E., (1929) *A relation between distance and radial velocity among extra-galactic nebulae*, Proc. of the National Academy of Sciences of the United States of America, Vol. 15, Issue 3, pp. 168–173

Ironically, de Sitter was also cautious about the significance of the redshifts that he predicted in his empty Universe, describing the associated positive radial velocities as 'spurious'. As a result, there continues to be a mild academic debate about who should really be credited as the 'discoverer' of cosmic expansion.

Among Hubble's original sample of galaxies, the highest radial velocity that he found was not much more than $1000 \, \text{km s}^{-1}$. As a result, his original findings were badly affected by peculiar velocities that are typically of the order of hundreds of km s^{-1}. Nonetheless, he had recognized the basic nature of cosmic expansion, and within a few years had extended his studies to more distant galaxies with sufficiently high recessional velocities that their peculiar velocities were relatively unimportant compared with the effect of the large-scale (Hubble) flow. Subsequent studies, by Hubble and many others, have confirmed these general findings and led to a consensus that for moderately nearby galaxies, the observed relationship between redshift and luminosity distance can be described as follows.

Observed redshift–distance relation

$$d_{\text{L}} = \frac{c}{H_0} z, \qquad (8.66)$$

where, according to recent estimates, $H_0 = (70.4 \pm 1.5) \, \text{km s}^{-1} \, \text{Mpc}^{-1}$. It is conventional to refer to the currently observed proportionality constant H_0 as the **Hubble constant**, but note that we have deliberately tailored our notation so that the (observational) Hubble constant can be seen as the current value of the (theoretical) Hubble parameter $H(t)$.

An acceptable SI unit of H_0 is the inverse second (s^{-1}), but it is traditional to quote the Hubble constant in units of $\text{km s}^{-1} \, \text{Mpc}^{-1}$, harking back to Hubble's decision to present his results as a velocity–distance relation. Indeed, it's still the case that when astronomers invoke **Hubble's law**, they usually write it in the form $v = H_0 d$, despite the potential ambiguity of v and d.

As data have accumulated, it has become increasingly clear that there are indeed deviations from the simple linear relation between redshift and luminosity distance. However, much of the evidence relates to observations of distant supernovae and involves sources with redshifts between 0.5 and 1. As a result, the approximate treatment that led to the deceleration parameter is not particularly useful. For that reason the use of the deceleration parameter has fallen into disfavour and has been replaced by other methods that we shall take up in the next subsection.

8.4.2 Density parameters and the age of the Universe

We saw in Subsection 8.3.3 that we could specify the Friedmann equations relevant to a particular FRW model by giving the current values of three density parameters, $\Omega_{\text{m},0}$, $\Omega_{\text{r},0}$ and $\Omega_{\Lambda,0}$, and we were able to specify a particular solution of those equations by imposing an appropriate boundary condition such as the value of $R(t)$ at time t_0. In practice the condition most often used is the current value of the Hubble parameter H_0. The value $(+1, 0 \text{ or } -1)$ of the curvature parameter k does not need to be specified because it is determined by the sign of $\Omega_0 - 1$, where $\Omega_0 = \Omega_{m,0} + \Omega_{r,0} + \Omega_{\Lambda,0}$. So the set of parameters $(\Omega_{m,0}, \Omega_{r,0}, \Omega_{\Lambda,0}, H_0)$ specifies a particular FRW model with a specific expansion history and, in the case that it starts with a Big Bang, a definite age at time t_0.

In such a Universe the Friedmann equations can be used to supply a direct but complicated link between the co-moving coordinate of any source and the redshift of radiation from that source when it arrives at the origin at time t_0. The model will also relate the co-moving coordinate of the source to its luminosity distance at time t_0. Thus, provided that Hubble's constant is known, it is possible to acquire information about the current values of the cosmic density parameters from measurements of redshift and luminosity distance.

In fact, there are several other ways of obtaining information about these parameters, particularly through detailed measurements of the anisotropies in the CMBR. We shall not pursue those here since they are discussed in detail in the companion volume on observational cosmology. We shall, however, note that as a result of a wide range of cosmological studies, primarily but not exclusively based on observations of the CMBR, there is now widespread agreement that the following set of parameter values provides a reasonable description of the large-scale features of our Universe.

> **Key cosmological parameters**
>
> $$\Omega_{m,0} \approx 0.27, \quad \Omega_{r,0} \approx 0.00, \quad \Omega_{\Lambda,0} \approx 0.73,$$
> $$H_0 = (70.4 \pm 1.5)\,\mathrm{km\,s^{-1}\,Mpc^{-1}}.$$

The implication is that the total density parameter is close to 1, so the Universe has a nearly flat spatial geometry with $k = 0$ and a total density that is close to the current critical density $\rho_{c,0} = 3H_0^2/(8\pi G)$, roughly $1 \times 10^{-26}\,\mathrm{kg\,m^{-3}}$.

This is an *accelerating Universe* of the kind that we discussed earlier. It started with a Big Bang, and light reaching us now (at time t_0) with redshift z can be shown to have been emitted at time

$$t(z) = \frac{1}{H_0} \int_0^{1/(1+z)} \frac{\mathrm{d}x}{x\sqrt{\Omega_{\Lambda,0} + (\Omega_0 - 1)x^{-2} + \Omega_{m,0}\,x^{-3} + \Omega_{r,0}\,x^{-4}}}, \quad (8.67)$$

so, the current age of the Universe, t_0 (corresponding to $z = 0$), is given by

$$t_0 = \frac{1}{H_0} \int_0^1 \frac{\mathrm{d}x}{x\sqrt{\Omega_{\Lambda,0} + (\Omega_0 - 1)x^{-2} + \Omega_{m,0}\,x^{-3} + \Omega_{r,0}\,x^{-4}}}. \quad (8.68)$$

With the currently favoured key values for the various parameters, this indicates a value for t_0 of about $(13.73 \pm 0.15) \times 10^9$ years. This is the age of the Universe.

As observational data improve, it will be interesting to see if these values continue to be upheld and if the use of a FRW cosmological model continues to be regarded as appropriate.

8.4.3   Horizons and limits

We end with a short discussion of two diagrams that provide a general view of some general observational features of the kind of expanding, accelerating FRW model that is currently thought to describe our Universe. The diagrams are complicated and will repay detailed study. They are shown as Figures 8.17 and 8.18, and are based on diagrams produced by Mark Whittle of the University of Virginia, though they are also strongly related to diagrams published by C. H. Lineweaver and T. M. Davis in *Publications of the Astronomical Society of Australia*, vol. 21, pages 97–109 (2004).

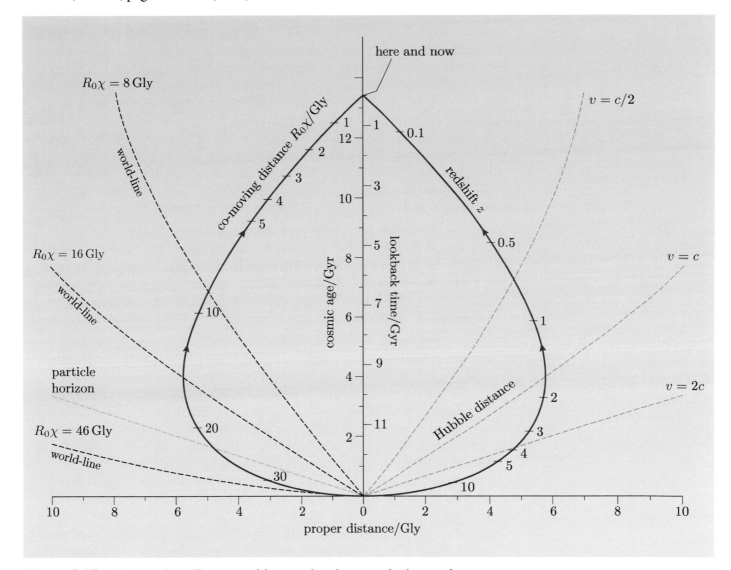

Figure 8.17 A spacetime diagram, with axes showing cosmic time and proper distance, for a Friedmann–Robertson–Walker Universe with $\Omega_{\Lambda,0} = 0.7$, $\Omega_{m,0} = 0.3$ and $H_0 = 70\,\mathrm{km\,s^{-1}\,Mpc^{-1}}$ (which implies $t_0 = 13.5 \times 10^9$ years).

Looking at Figure 8.17, the first thing to note is that this is a spacetime diagram with cosmic time, in billions of years since the Big Bang, on the vertical axis, and proper distance, in billions of light-years, on the horizontal axis. The red teardrop is the past lightcone of observers on the Earth now. (Peculiar velocities are ignored and Earth-based observers are treated as though they are fundamental observers.) Everything that we observe *at the present time* is located on this past lightcone. The right half of that lightcone is marked with redshifts, the left half with **co-moving distances** that are simply co-moving coordinates multiplied by the current value of the scale factor. (We shall have more to say about these when we consider Figure 8.18.)

The curved black lines originating at $(0, 0)$ that cut across the left-hand side of the past lightcone are the world-lines of 'galaxies' (or more accurately, fundamental observers) that travel along geodesics of the Robertson–Walker spacetime as they fall freely under the gravitational influence of the matter and dark energy that shape that spacetime. Each of these world-lines is marked with the co-moving distance of the corresponding 'galaxy'. Also shown cutting across the left half of the past lightcone is a green line called the **particle horizon**. This represents the location in spacetime of a signal that travels with speed c from the $(0, 0)$ event. At any cosmic time t, that line marks the location of the most distant object that can be observed. In this sense the particle horizon is the edge of the **observable Universe**. Currently the particle horizon is at a proper distance of about 46 billion light-years, though that is too far out to be shown on the diagram. Also shown crossing the left side of the diagram immediately below the particle horizon, is the world-line of a galaxy that is currently on the particle horizon. Up until now that galaxy has been outside the observable Universe. It is only now entering the observable Universe as the particle horizon moves outwards.

There is a second horizon, called the **cosmological event horizon** that is not shown on the diagram. This represents the past lightcone for observers at our position infinitely far in the future. It separates events that we might observe at some finite time from those that we will never be able to see, no matter how long we wait. That ultimate limit of observability is now at about 60 billion light-years. No event occurring beyond that event horizon will ever be seen from our location.

Another set of curves cuts across the right-hand half of the past lightcone. These lines connect points at which the Hubble flow has a specific proper radial velocity relative to fundamental observers on the vertical axis (i.e. us). Note in particular the middle (orange) line marked **Hubble distance**. This shows the proper distance at which an object participating in the Hubble flow would have a proper radial velocity of c. Note in particular that for the galaxies that we see now (i.e. those at the events that make up the past lightcone), all those with a redshift greater than about 1.6 are receding at a proper radial speed that is greater than c. All those with redshift less than 1.6 are receding at a sub-light speed. These 'faster-than-light' proper speeds are not in any way in conflict with the special relativistic prohibition on faster-than-light signals, because they are not carrying information between observers at faster-than-light speeds; rather, they concern the speed at which observers are being separated by the expansion of the Universe. Although it cannot be easily seen from the diagram, in order for an object to be receding from us at the speed of light, it would currently have to be at a proper distance of about 15 billion light-years.

Figure 8.18 shows essentially the same information but presents it using differently scaled axes. The horizontal axis now shows co-moving distance $R_0 \chi = R(t_0)\chi$, while the vertical axis uses a variable called **conformal time** that, when combined with the use of co-moving distance, has the effect of making the past lightcone take on a form that is familiar in the flat spacetime of special relativity. The world-lines of galaxies are now simple vertical lines, reflecting their fixed co-moving coordinates. The definition of co-moving distance ensures that it is equal to the present value of the proper distance.

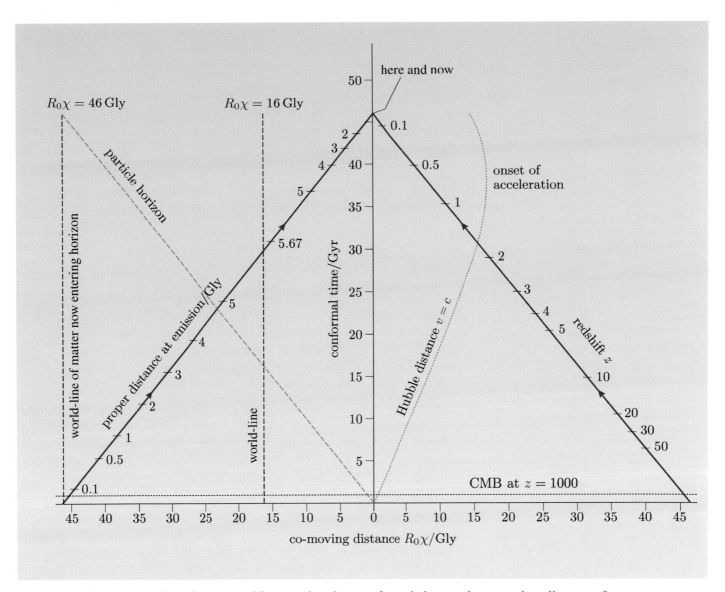

Figure 8.18 A spacetime diagram, with axes showing conformal time and co-moving distance, for a Friedmann–Robertson–Walker Universe with $\Omega_{\Lambda,0} = 0.7$, $\Omega_{m,0} = 0.3$ and $H_0 = 70\,\mathrm{km\,s^{-1}\,Mpc^{-1}}$. The past lightcone is shown in red, the particle horizon in green, the Hubble distance in orange and world-lines of fundamental observers (or their galaxies) in black.

As before, the past lightcone links all the events that we see now from the Earth. Marked along the left half of the past lightcone are the proper distances of those events when the light that we see now left them. Note that those figures rise and fall. The greatest proper distance from which any signal is currently reaching us is about 5.7 billion light-years. The objects responsible for those signals are currently at a co-moving distance of about 16 billion light-years. This diagram shows quite clearly that a galaxy at a co-moving distance of 46 billion light-years is only now entering the particle horizon and becoming part of the observable Universe. The CMBR anisotropy map shown in Figure 8.3 is based on radiation emitted about 400 000 years after the start of cosmic expansion and comes to us from events with a redshift of about 1000. It represents the actual current limit of cosmic visibility and is thought to pre-date the formation of any galaxy. It was emitted at a very small proper distance, less than 0.1 billion light-years but would currently be at a co-moving distance of about 45 billion light-years, close to the particle horizon.

Exercise 8.11   Figure 8.1 and more particularly Figure 8.2 showed information about the large-scale distribution of galaxies and quasars that extended to distances of order 10 billion light-years, yet Figure 8.17 indicates that we do not receive any signals from events at proper distances greater than about 5 billion light-years. Comment on this apparent inconsistency.

Exercise 8.12   To complete your work in this book, summarize the historical development of the Friedmann–Robertson–Walker models for the Universe.   ■

Summary of Chapter 8

1. A starting assumption of modern relativistic cosmology is that Einstein's original (unmodified) field equations of general relativity can be applied to the Universe as a whole, provided that a possible contribution from dark energy is included. We may then speak interchangeably of a Universe characterized by a cosmological constant Λ or one in which there is a dark energy contribution of density ρ_Λ and (negative) pressure $p_\Lambda = -\rho_\Lambda c^2 = -\Lambda c^4/8\pi G$.

2. According to the cosmological principle, at any given time, and on a sufficiently large scale, the Universe is homogeneous (i.e. the same everywhere) and isotropic (i.e. the same in all directions). This is supported by a range of evidence, including the low level of intrinsic anisotropies in the cosmic microwave background radiation.

3. According to the Weyl postulate, in cosmic spacetime there exists a set of privileged fundamental observers whose world-lines form a smooth bundle of time-like geodesics. These geodesics never meet at any event, apart perhaps from an initial singularity in the past and/or a final singularity in the future. The motion of the Earth relative to the frame of a local fundamental observer can be deduced from the dipole anisotropy in the CMBR.

4. The Robertson–Walker metric that describes a homogeneous and isotropic spacetime is

$$(\mathrm{d}s)^2 = c^2(\mathrm{d}t)^2 - R^2(t)\left[\frac{(\mathrm{d}r)^2}{1 - kr^2} + r^2(\mathrm{d}\theta)^2 + r^2\sin^2\theta\,(\mathrm{d}\phi)^2\right],$$
(Eqn 8.9)

where t is the cosmic time, r, θ and ϕ are co-moving spherical coordinates, $R(t)$ is the cosmic scale factor, and k is the spatial curvature parameter.

5. In Robertson–Walker spacetime, proper distance $\sigma(t)$ (as measured by a line of stationary rulers at some fixed cosmic time) is related to co-moving coordinate position χ by

$$\sigma(t) = \int_0^\chi R(t)\,\frac{\mathrm{d}r}{(1 - kr^2)^{1/2}},$$
(Eqn 8.12)

leading to the relations

$$\sigma(t) = \begin{cases} R(t)\sin^{-1}\chi & \text{if } k = +1, \\ R(t)\chi & \text{if } k = 0, \\ R(t)\sinh^{-1}\chi & \text{if } k = -1. \end{cases} \qquad \text{(Eqn 8.13)}$$

6. A further consequence at any time t is the exact relationship

$$v_{\mathrm{p}} = H(t)\,d_{\mathrm{p}}, \qquad \text{(Eqn 8.16)}$$

where d_{p} represents the proper distance between two fundamental observers (or their galaxies), v_{p} represents the proper radial velocity at which they are separating, and $H(t)$ is the Hubble parameter, defined by

$$H(t) = \frac{1}{R}\frac{\mathrm{d}R}{\mathrm{d}t}. \qquad \text{(Eqn 8.17)}$$

7. The space-like hypersurfaces of a Robertson–Walker spacetime may be described as open, flat or closed (and unbounded) according to the value of the curvature parameter k and the corresponding total volume of space, which may be infinite or finite.

8. In homogeneous and isotropic cosmological models, where the contents of spacetime are represented by ideal fluids corresponding to matter, radiation and the source of dark energy, the uniform cosmic density $\rho(t)$ and pressure $p(t)$ are specified at time $t = t_0$ by the quantities $\rho_{\mathrm{m},0}$, $\rho_{\mathrm{r},0}$ and ρ_Λ (and the appropriate equations of state linking them to pressure). Given these three values, the cosmic density and pressure at any other cosmic time can be determined, provided that the cosmic scale factor $R(t)$ is known as an explicit function of cosmic time.

9. The evolution of the cosmic scale factor is determined by the Friedmann equations

$$\left[\frac{1}{R}\frac{\mathrm{d}R}{\mathrm{d}t}\right]^2 = \frac{8\pi G}{3}\rho - \frac{kc^2}{R^2}, \qquad \text{(Eqn 8.27)}$$

$$\frac{1}{R}\frac{\mathrm{d}^2R}{\mathrm{d}t^2} = -\frac{4\pi G}{3}\left(\rho + \frac{3p}{c^2}\right). \qquad \text{(Eqn 8.28)}$$

10. In practical applications, the Friedmann equations take the form

$$\left[\frac{1}{R}\frac{\mathrm{d}R}{\mathrm{d}t}\right]^2 = \frac{8\pi G}{3}\left[\rho_{\mathrm{m},0}\left[\frac{R_0}{R(t)}\right]^3 + \rho_{\mathrm{r},0}\left[\frac{R_0}{R(t)}\right]^4 + \rho_\Lambda\right] - \frac{kc^2}{R^2},$$
$$\text{(Eqn 8.32)}$$

$$\frac{1}{R}\frac{\mathrm{d}^2R}{\mathrm{d}t^2} = -\frac{4\pi G}{3}\left[\rho_{\mathrm{m},0}\left[\frac{R_0}{R(t)}\right]^3 + 2\rho_{\mathrm{r},0}\left[\frac{R_0}{R(t)}\right]^4 - 2\rho_\Lambda\right].$$
$$\text{(Eqn 8.33)}$$

11. In flat space ($k = 0$), single-component models dominated respectively by dark energy, radiation and matter, the cosmic scale factor evolves as follows:

de Sitter model

$$R(t) = R_0\exp\left(H_0(t - t_0)\right); \qquad \text{(Eqn 8.37)}$$

pure radiation model

$$R(t) = R_0 \left(2H_0 t\right)^{1/2};$$ (Eqn 8.39)

Einstein–de Sitter model

$$R(t) = R_0 \left(\tfrac{3}{2}H_0 t\right)^{2/3}.$$ (Eqn 8.41)

12. A relativistic cosmological model based on the Robertson–Walker metric with a scale factor determined by the Friedmann equations is known as a Friedmann–Robertson–Walker (FRW) model. When specifying a general FRW model it is conventional to express each of the densities as a fraction of the critical density $\rho_c = 3H^2(t)/8\pi G$. These fractional densities are called density parameters and are defined as follows:

$$\Omega_m(t) = \frac{\rho_m(t)}{\rho_c(t)}, \quad \Omega_r(t) = \frac{\rho_r(t)}{\rho_c(t)}, \quad \Omega_\Lambda(t) = \frac{\rho_\Lambda}{\rho_c(t)}.$$ (Eqn 8.44)

13. The Friedmann equations imply that

if $\Omega_m + \Omega_r + \Omega_\Lambda < 1$, then $k < 0$ and space is open, (Eqn 8.47)
if $\Omega_m + \Omega_r + \Omega_\Lambda = 1$, then $k = 0$ and space is flat, (Eqn 8.48)
if $\Omega_m + \Omega_r + \Omega_\Lambda > 1$, then $k > 0$ and space is closed. (Eqn 8.49)

14. A quantitative measure of redshift is

$$z = \frac{\lambda_{ob} - \lambda_{em}}{\lambda_{em}}.$$ (Eqn 8.51)

In a Friedmann–Robertson–Walker model, observed redshift is related to the scale factor by

$$1 + z = \frac{R(t_{ob})}{R(t_{em})}.$$ (Eqn 8.56)

15. The luminosity distance of an isotropically radiating source is defined by

$$F = \frac{L}{4\pi d_L^2}$$ (Eqn 8.57)

and is related to redshift at small z by the approximate relation

$$d_L = \frac{c}{H_0} \left[z + \tfrac{1}{2}(1 - q_0)z^2 + \cdots\right],$$ (Eqn 8.64)

where H_0 and q_0 represent the current values of the Hubble and deceleration parameters. To a first approximation this is consistent with Hubble's (observational) law ($v = H_0 d$) and allows the observed Hubble constant to be identified with $H(t_0)$.

16. Currently observed values of the key cosmological parameters include

$$\Omega_{m,0} \approx 0.27, \quad \Omega_{r,0} \approx 0.00, \quad \Omega_{\Lambda,0} \approx 0.73,$$
$$H_0 = (70.4 \pm 1.5)\,\text{km}\,\text{s}^{-1}\,\text{Mpc}^{-1}.$$

The implication is that the total density parameter is close to 1, so the Universe has a nearly flat spatial geometry with $k = 0$ and a total density that is close to $1 \times 10^{-26}\,\text{kg}\,\text{m}^{-3}$. Such a Universe originated with a Big Bang, is accelerating its expansion and has an expansion age of about (13.73 ± 0.15) billion years.

Appendix

Table A.1 Common SI unit conversions and derived units.

Quantity	Unit	Conversion
speed	$\mathrm{m\,s^{-1}}$	
acceleration	$\mathrm{m\,s^{-2}}$	
angular speed	$\mathrm{rad\,s^{-1}}$	
angular acceleration	$\mathrm{rad\,s^{-2}}$	
linear momentum	$\mathrm{kg\,m\,s^{-1}}$	
angular momentum	$\mathrm{kg\,m^2\,s^{-1}}$	
force	newton (N)	$1\,\mathrm{N} = 1\,\mathrm{kg\,m\,s^{-2}}$
energy	joule (J)	$1\,\mathrm{J} = 1\,\mathrm{N\,m} = 1\,\mathrm{kg\,m^2\,s^{-2}}$
power	watt (W)	$1\,\mathrm{W} = 1\,\mathrm{J\,s^{-1}} = 1\,\mathrm{kg\,m^2\,s^{-3}}$
pressure	pascal (Pa)	$1\,\mathrm{Pa} = 1\,\mathrm{N\,m^{-2}} = 1\,\mathrm{kg\,m^{-1}\,s^{-2}}$
frequency	hertz (Hz)	$1\,\mathrm{Hz} = 1\,\mathrm{s^{-1}}$
charge	coulomb (C)	$1\,\mathrm{C} = 1\,\mathrm{A\,s}$
potential difference	volt (V)	$1\,\mathrm{V} = 1\,\mathrm{J\,C^{-1}} = 1\,\mathrm{kg\,m^2\,s^{-3}\,A^{-1}}$
electric field	$\mathrm{N\,C^{-1}}$	$1\,\mathrm{N\,C^{-1}} = 1\,\mathrm{V\,m^{-1}} = 1\,\mathrm{kg\,m\,s^{-3}\,A^{-1}}$
magnetic field	tesla (T)	$1\,\mathrm{T} = 1\,\mathrm{N\,s\,m^{-1}\,C^{-1}} = 1\,\mathrm{kg\,s^{-2}\,A^{-1}}$

Table A.2 Other unit conversions.

wavelength
1 nanometre (nm) $= 10\,\text{Å} = 10^{-9}\,\mathrm{m}$
1 ångstrom $= 0.1\,\mathrm{nm} = 10^{-10}\,\mathrm{m}$

mass–energy equivalence
$1\,\mathrm{kg} = 8.99 \times 10^{16}\,\mathrm{J}/c^2$ (c in $\mathrm{m\,s^{-1}}$)
$1\,\mathrm{kg} = 5.61 \times 10^{35}\,\mathrm{eV}/c^2$ (c in $\mathrm{m\,s^{-1}}$)

angular measure
$1° = 60\,\mathrm{arcmin} = 3600\,\mathrm{arcsec}$
$1° = 0.017\,45\,\mathrm{radian}$
$1\,\mathrm{radian} = 57.30°$

distance
1 astronomical unit (AU) $= 1.496 \times 10^{11}\,\mathrm{m}$
1 light-year (ly) $= 9.461 \times 10^{15}\,\mathrm{m} = 0.307\,\mathrm{pc}$
1 parsec (pc) $= 3.086 \times 10^{16}\,\mathrm{m} = 3.26\,\mathrm{ly}$

temperature
absolute zero: $0\,\mathrm{K} = -273.15\,°\mathrm{C}$
$0\,°\mathrm{C} = 273.15\,\mathrm{K}$

energy
$1\,\mathrm{eV} = 1.602 \times 10^{-19}\,\mathrm{J}$
$1\,\mathrm{J} = 6.242 \times 10^{18}\,\mathrm{eV}$

spectral flux density
1 jansky (Jy) $= 10^{-26}\,\mathrm{W\,m^{-2}\,Hz^{-1}}$
$1\,\mathrm{W\,m^{-2}\,Hz^{-1}} = 10^{26}\,\mathrm{Jy}$

cross-sectional area
1 barn $= 10^{-28}\,\mathrm{m^2}$
$1\,\mathrm{m^2} = 10^{28}\,\mathrm{barn}$

cgs units
$1\,\mathrm{erg} = 10^{-7}\,\mathrm{J}$
$1\,\mathrm{dyne} = 10^{-5}\,\mathrm{N}$
$1\,\mathrm{gauss} = 10^{-4}\,\mathrm{T}$
$1\,\mathrm{emu} = 10\,\mathrm{C}$

pressure
$1\,\mathrm{bar} = 10^5\,\mathrm{Pa}$
$1\,\mathrm{Pa} = 10^{-5}\,\mathrm{bar}$
1 atmosphere $= 1.013\,25\,\mathrm{bar}$
1 atmosphere $= 1.013\,25 \times 10^5\,\mathrm{Pa}$

Table A.3 Constants.

Name of constant	Symbol	SI value
Fundamental constants		
gravitational constant	G	$6.673 \times 10^{-11}\,\mathrm{N\,m^2\,kg^{-2}}$
Boltzmann's constant	k	$1.381 \times 10^{-23}\,\mathrm{J\,K^{-1}}$
speed of light in vacuum	c	$2.998 \times 10^{8}\,\mathrm{m\,s^{-1}}$
Planck's constant	h	$6.626 \times 10^{-34}\,\mathrm{J\,s}$
	$\hbar = h/2\pi$	$1.055 \times 10^{-34}\,\mathrm{J\,s}$
fine structure constant	$\alpha = e^2/4\pi\varepsilon_0\hbar c$	$1/137.0$
Stefan–Boltzmann constant	σ	$5.671 \times 10^{-8}\,\mathrm{J\,m^{-2}\,K^{-4}\,s^{-1}}$
Thomson cross-section	σ_{T}	$6.652 \times 10^{-29}\,\mathrm{m^2}$
permittivity of free space	ε_0	$8.854 \times 10^{-12}\,\mathrm{C^2\,N^{-1}\,m^{-2}}$
permeability of free space	μ_0	$4\pi \times 10^{-7}\,\mathrm{T\,m\,A^{-1}}$
Particle constants		
charge of proton	e	$1.602 \times 10^{-19}\,\mathrm{C}$
charge of electron	$-e$	$-1.602 \times 10^{-19}\,\mathrm{C}$
electron rest mass	m_{e}	$9.109 \times 10^{-31}\,\mathrm{kg}$
		$= 0.511\,\mathrm{MeV}/c^2$
proton rest mass	m_{p}	$1.673 \times 10^{-27}\,\mathrm{kg}$
		$= 938.3\,\mathrm{MeV}/c^2$
neutron rest mass	m_{n}	$1.675 \times 10^{-27}\,\mathrm{kg}$
		$= 939.6\,\mathrm{MeV}/c^2$
atomic mass unit	u	$1.661 \times 10^{-27}\,\mathrm{kg}$
Astronomical constants		
mass of the Sun	$\mathrm{M}_\odot$	$1.99 \times 10^{30}\,\mathrm{kg}$
radius of the Sun	$\mathrm{R}_\odot$	$6.96 \times 10^{8}\,\mathrm{m}$
luminosity of the sun	$\mathrm{L}_\odot$	$3.83 \times 10^{26}\,\mathrm{W}$
mass of the Earth	$\mathrm{M}_\oplus$	$5.97 \times 10^{24}\,\mathrm{kg}$
radius of the Earth	$\mathrm{R}_\oplus$	$6.37 \times 10^{6}\,\mathrm{m}$
mass of Jupiter	M_{J}	$1.90 \times 10^{27}\,\mathrm{kg}$
radius of Jupiter	R_{J}	$7.15 \times 10^{7}\,\mathrm{m}$
astronomical unit	AU	$1.496 \times 10^{11}\,\mathrm{m}$
light-year	ly	$9.461 \times 10^{15}\,\mathrm{m}$
parsec	pc	$3.086 \times 10^{16}\,\mathrm{m}$
Hubble parameter	H_0	$(70.4 \pm 1.5)\,\mathrm{km\,s^{-1}\,Mpc^{-1}}$
		$(2.28 \pm 0.05) \times 10^{-18}\,\mathrm{s^{-1}}$
age of Universe	t_0	$(13.73 \pm 0.15) \times 10^{9}\,\mathrm{years}$
current critical density	$\rho_{\mathrm{c},0}$	$(9.30 \pm 0.40) \times 10^{-27}\,\mathrm{kg\,m^{-3}}$
current dark energy density	$\Omega_{\Lambda,0}$	$(73.2 \pm 1.8)\%$
current matter density	$\Omega_{\mathrm{m},0}$	$(26.8 \pm 1.8)\%$
current baryonic matter density	$\Omega_{\mathrm{b},0}$	$(4.4 \pm 0.2)\%$
current non-baryonic matter density	$\Omega_{\mathrm{c},0}$	$(22.3 \pm 0.9)\%$
current curvature density	$\Omega_{\mathrm{k},0}$	$(-1.4 \pm 1.7)\%$
current deceleration	q_0	-0.595 ± 0.025

Solutions to exercises

Exercise 1.1 A 'stationary' particle in any laboratory on the Earth is actually subject to gravitational forces due to the Earth and the Sun. These help to ensure that the particle moves with the laboratory. If steps were taken to counterbalance these forces so that the particle was really not subject to any net force, then the rotation of the Earth and the Earth's orbital motion around the Sun would carry the laboratory away from the particle, causing the force-free particle to follow a curving path through the laboratory. This would clearly show that the particle did not have constant velocity in the laboratory (i.e. constant speed in a fixed direction) and hence that a frame fixed in the laboratory is not an inertial frame. More realistically, an experiment performed using the kind of long, freely suspended pendulum known as a *Foucault pendulum* could reveal the fact that a frame fixed on the Earth is rotating and therefore cannot be an inertial frame of reference. An even more practical demonstration is provided by the winds, which do not flow directly from areas of high pressure to areas of low pressure because of the Earth's rotation.

Exercise 1.2 The Lorentz factor is $\gamma(V) = 1/\sqrt{1 - V^2/c^2}$.

(a) If $V = 0.1c$, then

$$\gamma = \frac{1}{\sqrt{1 - (0.1c)^2/c^2}} = 1.01 \quad \text{(to 3 s.f.)}.$$

(b) If $V = 0.9c$, then

$$\gamma = \frac{1}{\sqrt{1 - (0.9c)^2/c^2}} = 2.29 \quad \text{(to 3 s.f.)}.$$

Note that it is often convenient to write speeds in terms of c instead of writing the values in $\mathrm{m\,s^{-1}}$, because of the cancellation between factors of c.

Exercise 1.3 The inverse of a 2×2 matrix $M = \begin{pmatrix} A & B \\ C & D \end{pmatrix}$ is

$$M^{-1} = \frac{1}{AD - BC} \begin{pmatrix} D & -B \\ -C & A \end{pmatrix}.$$

Taking $A = \gamma(V)$, $B = -\gamma(V)V/c$, $C = -\gamma(V)V/c$ and $D = \gamma(V)$, and noting that $AD - BC = [\gamma(V)]^2(1 - V^2/c^2) = 1$, we have

$$[\Lambda]^{-1} = \begin{pmatrix} \gamma(V) & +\gamma(V)V/c \\ +\gamma(V)V/c & \gamma(V) \end{pmatrix}.$$

This is the correct form of the inverse Lorentz transformation matrix.

Exercise 1.4 First compute the Lorentz factor:

$$\gamma(V) = 1/\sqrt{1 - V^2/c^2}$$
$$= 1/\sqrt{1 - 9/25} = 1/\sqrt{16/25} = 5/4.$$

Thus the measured lifetime is $\Delta T = 5 \times 2.2/4\,\mu\mathrm{s} = 2.8\,\mu\mathrm{s}$. Note that not all muons live for the same time; rather, they have a range of lifetimes. But a large group of muons travelling with a common speed does have a well-defined *mean lifetime*, and it is the dilation of this quantity that is easily demonstrated experimentally.

Exercise 1.5 The alternative definition of length can't be used in the rest frame of the rod as the rod does not move in its own rest frame. The proper length is therefore defined as before and related to the positions of the two events as observed in the rest frame. (This works, because event 1 and event 2 still occur at the end-points of the rod and the rod never moves in the rest frame S'.)

As before, it is helpful to write down all the intervals that are known in a table.

Event	S (laboratory)	S' (rest frame)
2	$(t_2, 0)$	(t'_2, x'_2)
1	$(t_1, 0)$	(t'_1, x'_1)
Intervals	$(t_2 - t_1, 0)$	$(t'_2 - t'_1, x'_2 - x'_1)$
	$\equiv (\Delta t, \Delta x)$	$\equiv (\Delta t', \Delta x')$
Relation to intervals	$(L/V, 0)$	$(?, L_P)$

By examining the intervals, it can be seen that Δx, Δt and $\Delta x'$ are known. From the interval transformation rules, only Equation 1.33 relates the three known intervals. Substituting the known intervals into that equation gives $L_P = \gamma(V)(0 - V(L/V))$. In this way, length contraction is predicted as before:

$$L = L_P/\gamma(V).$$

Exercise 1.6 The received wavelength is less than the emitted wavelength. This means that the jet is approaching. We can therefore use Equation 1.42 provided that we change the sign of V. Combining it with the formula $f\lambda = c$ shows that $\lambda' = \lambda\sqrt{(c - V)/(c + V)}$. Squaring both sides and rearranging gives

$$(\lambda'/\lambda)^2 = (c - V)/(c + V).$$

From this it follows that

$$(\lambda'/\lambda)^2(c + V) = (c - V),$$

so

$$V(1 + (\lambda'/\lambda)^2) = c(1 - (\lambda'/\lambda)^2),$$

thus

$$V = c(1 - (\lambda'/\lambda)^2)/(1 + (\lambda'/\lambda)^2).$$

Substituting $\lambda' = 4483 \times 10^{-10}$ m and $\lambda = 5850 \times 10^{-10}$ m, the speed is found to be $v = 0.26c$ (to 2 s.f.).

Exercise 1.7 Let the spacestation be the origin of frame S, and the nearer of the spacecraft the origin of frame S', which therefore moves with speed $V = c/2$ as measured in S. Let these two frames be in standard configuration. The velocity of the further of the two spacecraft, as observed in S, is then $v = (3c/4, 0, 0)$. It follows from the velocity transformation that the velocity of the further spacecraft as observed from the nearer will be $v' = (v'_x, 0, 0)$, where

$$v'_x = \frac{v_x - V}{1 - v_x V/c^2} = \frac{3c/4 - c/2}{1 - (3c/4)(c/2)/c^2} = 2c/5.$$

Exercise 1.8 $\Delta x = (5 - 7)\,\text{m} = -2\,\text{m}$ and $c\,\Delta t = (5 - 3)\,\text{m} = 2\,\text{m}$. Since the spacetime separation is $(\Delta s)^2 = (c\,\Delta t)^2 - (\Delta x)^2$ in this case, it follows that

$(\Delta s)^2 = (2\,\mathrm{m})^2 - (2\,\mathrm{m})^2 = 0$. The value $(\Delta s)^2 = 0$ is permitted; it describes situations in which the two events could be linked by a light signal. In fact, any such separation is said to be *light-like*.

Exercise 1.9 Start with $(\Delta s')^2 = (c\,\Delta t')^2 - (\Delta x')^2$. The aim is to show that $(\Delta s')^2 = (\Delta s)^2$.

Substitute $\Delta x' = \gamma(\Delta x - V\,\Delta t)$ and $c\,\Delta t' = \gamma(c\,\Delta t - V\,\Delta x/c)$ so that

$$(\Delta s')^2 = \gamma^2\left(c^2(\Delta t)^2 - 2V\,\Delta x \Delta t + V^2(\Delta x)^2/c^2\right)$$
$$- \gamma^2\left((\Delta x)^2 - 2V\,\Delta x \Delta t + V^2(\Delta t)^2\right).$$

Cross terms involving $\Delta x\,\Delta t$ cancel. Collecting common terms in $c^2(\Delta t)^2$ and $(\Delta x)^2$ gives

$$(\Delta s')^2 = \gamma^2 c^2(\Delta t)^2(1 - V^2/c^2) - \gamma^2(\Delta x)^2(1 - V^2/c^2).$$

Finally, noting that $\gamma^2 = [1 - V^2/c^2]^{-1}$, there is a cancellation of terms, giving

$$(\Delta s')^2 = c^2(\Delta t)^2 - (\Delta x)^2 = (\Delta s)^2,$$

thus showing that $(\Delta s')^2 = (\Delta s)^2$.

Exercise 1.10 Since $(\Delta s)^2 = (c\,\Delta t)^2 - (\Delta l)^2$, and $(\Delta s)^2$ is invariant, it follows that all inertial observers will find $(c\,\Delta t)^2 = (\Delta s)^2 + (\Delta l)^2$, where $(\Delta l)^2$ cannot be negative. Since $(\Delta l)^2 = 0$ in the frame in which the proper time is measured, it follows that no other inertial observer can find a smaller value for the time between the events.

Exercise 1.11 In Terra's frame, Stella's ship has velocity $(v_x, v_y, v_z) = (-V, 0, 0)$. It follows from the velocity transformation that in Astra's frame, the velocity of Stella's ship will be $(v_x', 0, 0)$, where $v_x' = (v_x - V)/(1 - v_x V/c^2)$. Taking $v_x = -V$ gives

$$v_x' = \frac{(-V - V)}{(1 - (-V)V/c^2)} = \frac{-2V}{1 + V^2/c^2}.$$

Taking the magnitude of this single non-zero velocity component gives the speed of approach, $2V/(1 + V^2/c^2)$, as required.

Exercise 1.12 In Terra's frame, the signals would have an emitted frequency $f_{\mathrm{em}} = 1\,\mathrm{Hz}$. In Astra's frame, the Doppler effect tells us that the signals would be received with a different frequency f_{rec}. On the outward leg of the journey, the signals would be redshifted and the received frequency would be

$$f_{\mathrm{rec}} = f_{\mathrm{em}}\sqrt{(c - V)/(c + V)}.$$

On the return leg of the journey, the signals would be blueshifted and the received frequency would be

$$f_{\mathrm{rec}} = f_{\mathrm{em}}\sqrt{(c + V)/(c - V)}.$$

Exercise 2.1 The Lorentz factor is

$$\gamma = 1/\sqrt{1 - v^2/c^2} = 1/\sqrt{1 - 16c^2/25c^2} = 1/\sqrt{9/25} = 5/3.$$

The electron has mass $m = 9.11 \times 10^{-31}$ kg. Thus the magnitude of the electron's momentum is

$$p = 5/3 \times 4c/5 \times m = (5/3) \times (4 \times 3.00 \times 10^8 \text{ m s}^{-1}/5) \times 9.11 \times 10^{-31} \text{ kg} = 3.6 \times 10^{-22} \text{ kg m s}^{-1}.$$

Exercise 2.2 The kinetic energy is $E_K = (\gamma - 1)mc^2$. Taking the speed to be $9c/10$, the Lorentz factor is

$$\gamma = 1/\sqrt{1 - v^2/c^2} = 1/\sqrt{1 - (9/10)^2} = 2.29.$$

Noting that $m = 1.88 \times 10^{-28}$ kg, the kinetic energy is

$$E_K = (2.29 - 1) \times 1.88 \times 10^{-28} \text{ kg} \times (3.00 \times 10^8 \text{ m s}^{-1})^2 = 2.2 \times 10^{-11} \text{ J}.$$

Exercise 2.3 $v = 3c/5$ corresponds to a Lorentz factor

$$\gamma(v) = 1/\sqrt{1 - v^2/c^2} = 1/\sqrt{1 - 9/25} = 5/4.$$

The proton has mass $m_p = 1.67 \times 10^{-27}$ kg, therefore the total energy is

$$E = \gamma(v)mc^2 = (5/4) \times 1.67 \times 10^{-27} \text{ kg} \times (3.00 \times 10^8 \text{ m s}^{-1})^2 = 1.88 \times 10^{-10} \text{ J}.$$

Exercise 2.4 Since the total energy is $E = \gamma mc^2$, it is clear that the total energy is twice the mass energy when $\gamma = 2$. This means that $2 = 1/\sqrt{1 - v^2/c^2}$. Squaring and inverting both sides, $1/4 = 1 - v^2/c^2$, so $v^2/c^2 = 3/4$. Taking the positive square root, $v/c = \sqrt{3}/2$.

Exercise 2.5 (a) The energy difference is $\Delta E = \Delta m\, c^2$, where $\Delta m = 3.08 \times 10^{-28}$ kg. Thus

$$\Delta E = 3.08 \times 10^{-28} \text{ kg} \times (3.00 \times 10^8 \text{ m s}^{-1})^2 = 2.77 \times 10^{-11} \text{ J}.$$

Converting to electronvolts, this is

$$2.77 \times 10^{-11} \text{ J}/1.60 \times 10^{-19} \text{ J eV}^{-1} = 1.73 \times 10^8 \text{ eV} = 173 \text{ MeV}.$$

(b) From $\Delta E = \Delta m\, c^2$, the mass difference is $\Delta m = \Delta E/c^2$. Now, $\Delta E = 13.6$ eV or, converting to joules,

$$\Delta E = 13.6 \text{ eV} \times 1.60 \times 10^{-19} \text{ J eV}^{-1} = 2.18 \times 10^{-18} \text{ J}.$$

Therefore

$$\Delta m = 2.18 \times 10^{-18} \text{ J}/(3.00 \times 10^8 \text{ m s}^{-1})^2 = 2.42 \times 10^{-35} \text{ kg}.$$

Note that the masses of the electron and proton are 9.11×10^{-31} kg and 1.67×10^{-27} kg, respectively, so the mass difference from chemical binding is small enough to be negligible in most cases. However, mass–energy equivalence is not unique to nuclear reactions.

Exercise 2.6 The transformations are $E' = \gamma(V)(E - Vp_x)$ and $p'_x = \gamma(V)(p_x - VE/c^2)$. In this case, $E = 3m_ec^2$ and $p_x = \sqrt{8}m_ec^2$. For relative speed $V = 4c/5$ between the two frames, the Lorentz factor is $\gamma = 1/\sqrt{1 - (4/5)^2} = 5/3$. Substituting the values,

$$E' = 5/3(3m_ec^2 - 4c/5 \times \sqrt{8}m_ec) = 1.23m_ec^2$$

and

$$p' = 5/3(\sqrt{8}m_e c - 4c/5 \times 3m_e c^2/c^2) = 0.714 m_e c.$$

Exercise 2.7 (a) For a photon $m = 0$, so

$$p = E/c = hf/c = \frac{6.63 \times 10^{-34}\,\text{J s} \times 5.00 \times 10^{14}\,\text{s}^{-1}}{3.00 \times 10^8\,\text{m s}^{-1}} = 1.11 \times 10^{-27}\,\text{kg m s}^{-1}.$$

(b) Using the Newtonian relation that the force is equal to the rate of change of momentum (we shall have more to say about this later), the magnitude of the force on the sail will be $F = np$, where n is the rate at which photons are absorbed by the sail (number of photons per second). Thus

$$n = F/p = 10\,\text{N}/1.11 \times 10^{-27}\,\text{kg m s}^{-1} = 9.0 \times 10^{27}\,\text{s}^{-1}.$$

Exercise 2.8 To be a valid energy/momentum combination, the energy–momentum relation must be satisfied, i.e. $E_f^2 - p_f^2 c^2 = m_f^2 c^4$. For the given values of energy and momentum,

$$E_f^2 - p_f^2 c^2 = 9m_f^2 c^4 - 49m_f^2 c^4 = -40m_f^2 c^4 \neq m_f^2 c^4.$$

So they are not valid values.

Exercise 2.9 It follows directly from the transformation rules for the last three components of the four-force F^μ that

$$\gamma(v')f_x' = \gamma(V)\left[\gamma(v)f_x - V\gamma(v)\boldsymbol{f}\cdot\boldsymbol{v}/c^2\right],$$
$$\gamma(v')f_y' = \gamma(v)f_y,$$
$$\gamma(v')f_z' = \gamma(v)f_z.$$

Note that the transformation of f_x involves both the speed of the particle v as measured in frame S and the speed V of frame S' as measured in frame S. Both $\gamma(v)$ and $\gamma(V)$ appear in the transformation.

Exercise 2.10 Since the four-vector is contravariant, it transforms just like the four-displacement. Thus

$$c\rho' = \gamma(V)(c\rho - VJ_x/c),$$
$$J_x' = \gamma(V)(J_x - V(c\rho)/c),$$
$$J_y' = J_y,$$
$$J_z' = J_z,$$

where V is the speed of frame S' as measured in frame S.

The covariant counterpart to $(c\rho, J_x, J_y, J_z)$ is $(c\rho, -J_x, -J_y, -J_z)$.

Exercise 2.11 The components of a contravariant four-vector transform differently from those of a covariant four-vector. The former transform like the components of a displacement, according to the matrix $[\Lambda^\mu{}_\nu]$ that implements the Lorentz transformation. The latter transform like derivatives, according to the inverse of the Lorentz transformation matrix, $[(\Lambda^{-1})_\mu{}^\nu]$. Since one matrix 'undoes' the effect of the other in the sense that their product is the unit matrix, it is to be expected that combinations such as $\sum_{\mu=0}^{3} J_\mu J^\mu$ will transform as invariants, while other combinations, such as $\sum_{\mu=0}^{3} J_\mu J_\mu$ and $\sum_{\mu=0}^{3} J^\mu J^\mu$, will not.

Exercise 2.12 The indices must balance. They do this in both cases, but in the former case the lowering of indices can be achieved by the legitimate process of multiplying by the Minkowski metric and summing over a common index. In the latter case an additional step is required, the replacement of $F_{\mu\nu}$ by $F_{\nu\mu}$. This would be allowable if $[F_{\nu\mu}]$ was symmetric — that is, if $F_{\mu\nu} = F_{\nu\mu}$ for all values of μ and ν — but it is not. Making such an additional change will alter some of the signs in an unacceptable way. The general lesson is clear: indices may be raised and lowered in a balanced way, but the order of indices is important and should be preserved. This is why elements of the mixed version of the field tensor may be written as $F^\mu{}_\nu$ or $F_\mu{}^\nu$ but should not be written as F^μ_ν.

Exercise 2.13 The field component of interest is given by cF'^{10}, so we need to evaluate

$$F'^{10} = \sum_{\alpha,\beta} \Lambda^1{}_\alpha \Lambda^0{}_\beta F^{\alpha\beta}.$$

$\Lambda^1{}_\alpha$ is non-zero only when $\alpha = 0$ and $\alpha = 1$. Similarly, $\Lambda^0{}_\beta$ is non-zero only when $\beta = 0$ and $\beta = 1$. This makes the sum much shorter, so it can be written out explicitly:

$$F'^{10} = \Lambda^1{}_0\Lambda^0{}_0F^{00} + \Lambda^1{}_0\Lambda^0{}_1F^{01} + \Lambda^1{}_1\Lambda^0{}_0F^{10} + \Lambda^1{}_1\Lambda^0{}_1F^{11}.$$

Since $F^{00} = 0$ and $F^{11} = 0$, the sum reduces to

$$F'^{10} = \Lambda^1{}_0\Lambda^0{}_1F^{01} + \Lambda^1{}_1\Lambda^0{}_0F^{10}.$$

It is now a matter of substituting known values: $F^{10} = -F^{01} = \mathcal{E}_x/c$, $\Lambda^0{}_0 = \Lambda^1{}_1 = \gamma(V)$ and $\Lambda^0{}_1 = \Lambda^1{}_0 = -V\gamma(V)/c$, which leads to

$$\mathcal{E}'_x/c = \gamma^2(1 - V^2/c^2)\mathcal{E}_x/c.$$

Since $1 - V^2/c^2 = \gamma^{-2}$, we have

$$\mathcal{E}'_x = \mathcal{E}_x,$$

as required.

With patience, all the other field transformation rules can be determined in the same way.

Exercise 2.14 $H'_{\alpha\beta\gamma\delta} = \sum_{\mu,\nu,\rho,\eta=0}^{3} (\Lambda^{-1})_\alpha{}^\mu (\Lambda^{-1})_\beta{}^\nu (\Lambda^{-1})_\gamma{}^\rho (\Lambda^{-1})_\delta{}^\eta H_{\mu\nu\rho\eta}.$

Exercise 3.1 (a) You could note that $y/x = 4/3$ for all values of u, and also $u = 0$ gives $y = x = 0$, so this is the part of the straight line with positive u values and gradient $4/3$ through the origin. Or you could work out x and y for a few values of u, as shown in the table below.

u	0	1	2	3
x	0	3	12	27
y	0	4	16	36

Either way, your sketch should look like Figure S3.1.

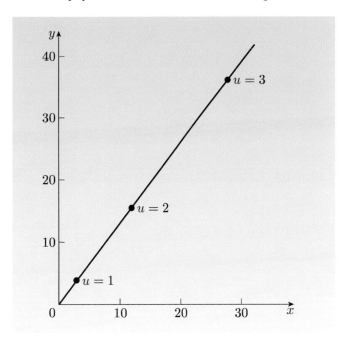

Figure S3.1 Sketch of the line $x = 3u^2$, $y = 4u^2$.

(b) We have

$$\frac{\mathrm{d}x}{\mathrm{d}u} = 6u \quad \text{and} \quad \frac{\mathrm{d}y}{\mathrm{d}u} = 8u,$$

so

$$L = \int_0^3 \left((6u)^2 + (8u)^2\right)^{1/2} \mathrm{d}u = \int_0^3 10u \, \mathrm{d}u = \left[5u^2\right]_0^3 = 45.$$

Exercise 3.2 Since $r = R$ and $\phi = u$, we have $\mathrm{d}r = 0$ and $\mathrm{d}\phi = \mathrm{d}u$, so

$$C = \int_0^{2\pi} \mathrm{d}l = \int_0^{2\pi} (\mathrm{d}r^2 + r^2 \, \mathrm{d}\phi^2)^{1/2} = \int_0^{2\pi} (0^2 + R^2 \, \mathrm{d}u^2)^{1/2}$$

$$= \int_0^{2\pi} R \, \mathrm{d}u = [Ru]_0^{2\pi} = 2\pi R.$$

Exercise 3.3 (a) Like the cylinder, the cone can be formed by rolling up a region of the plane. Once again this won't change the geometry; the circles and triangles will have the same properties as they have on the plane. So the cone has flat geometry.

(b) In this case, distances for the bugs are shorter towards the edge of the disc, so the shortest distance from P to Q, as measured by the bugs, will appear to us to curve outwards. The angles of the triangle PQR add up to more than $180°$, as shown in Figure 3.12, so for this inverse hotplate the results are qualitatively similar to the geometry of the sphere, and the hotplate again has intrinsically curved geometry despite the lack of any extrinsic curvature.

Exercise 3.4 From Equation 3.10, we have

$$\mathrm{d}l^2 = R^2 \, \mathrm{d}\theta^2 + R^2 \sin^2 \theta \, \mathrm{d}\phi^2.$$

Again there are only squared coordinate differentials, so $g_{ij} = 0$ for $i \neq j$. We can also see that $g_{11} = R^2$ and $g_{22} = R^2 \sin^2 x^1$, so

$$[g_{ij}] = \begin{pmatrix} R^2 & 0 \\ 0 & R^2 \sin^2 x^1 \end{pmatrix}.$$

Exercise 3.5   In this case we only have squared coordinate differentials, so $g_{ij} = 0$ for $i \neq j$. Also, $g_{11} = 1$, $g_{22} = (x^1)^2$, $g_{33} = (x^1)^2 \sin^2 x^2$, and therefore

$$[g_{ij}] = \begin{pmatrix} 1 & 0 & 0 \\ 0 & (x^1)^2 & 0 \\ 0 & 0 & (x^1)^2 \sin^2 x^2 \end{pmatrix}.$$

Note that the final entry involves the coordinate x^2, *not* x squared.

Exercise 3.6   Defining $x^1 = r$ and $x^2 = \phi$, we have

$$[g_{ij}] = \begin{pmatrix} 1 & 0 \\ 0 & (x^1)^2 \end{pmatrix}.$$

Exercise 3.7   (a) Since the line element is $dl^2 = (dx^1)^2 + (dx^2)^2$, we have

$$[g_{ij}] = \begin{pmatrix} 1 & 0 \\ 0 & 1 \end{pmatrix}.$$

From Equation 3.23, the connection coefficients are defined by

$$\Gamma^i{}_{jk} = \frac{1}{2} \sum_l g^{il} \left(\frac{\partial g_{lk}}{\partial x^j} + \frac{\partial g_{jl}}{\partial x^k} - \frac{\partial g_{jk}}{\partial x^l} \right),$$

and since $\partial g_{ij}/\partial x^k = 0$ for all values of i, j, k, it follows that $\Gamma^i{}_{jk} = 0$ for all i, j, k.

Comment: This argument generalizes to any n-dimensional Euclidean space; consequently, when Cartesian coordinates are used, such spaces have vanishing connection coefficients.

(b) From Exercise 3.4, the metric is

$$[g_{ij}] = \begin{pmatrix} R^2 & 0 \\ 0 & R^2 \sin^2 x^1 \end{pmatrix},$$

and the dual metric is the inverse matrix

$$[g^{ij}] = \begin{pmatrix} 1/R^2 & 0 \\ 0 & 1/R^2 \sin^2 x^1 \end{pmatrix}.$$

But in this case $R = 1$, so

$$[g^{ij}] = \begin{pmatrix} 1 & 0 \\ 0 & 1/\sin^2 x^1 \end{pmatrix}.$$

Since

$$\Gamma^i{}_{jk} = \frac{1}{2} \sum_l g^{il} \left(\frac{\partial g_{lk}}{\partial x^j} + \frac{\partial g_{jl}}{\partial x^k} - \frac{\partial g_{jk}}{\partial x^l} \right),$$

there are six independent connection coefficients:

$$\Gamma^1{}_{11}, \quad \Gamma^1{}_{12}(= \Gamma^1{}_{21}), \quad \Gamma^1{}_{22},$$
$$\Gamma^2{}_{11}, \quad \Gamma^2{}_{12}(= \Gamma^2{}_{21}), \quad \Gamma^2{}_{22}.$$

However,

$$\frac{\partial g_{22}}{\partial x^1} = 2 \sin x^1 \cos x^1, \qquad \text{while} \qquad \frac{\partial g_{ij}}{\partial x^k} = 0$$

for all other values of i, j, k. Also, $g^{il} = 0$ for $i \neq l$, from which we can see that

$$\Gamma^1{}_{11} = \frac{1}{2} g^{11} \left(\frac{\partial g_{11}}{\partial x^1} + \frac{\partial g_{11}}{\partial x^1} - \frac{\partial g_{11}}{\partial x^1} \right) = 0,$$

$$\Gamma^1{}_{12} = \frac{1}{2} g^{11} \left(\frac{\partial g_{12}}{\partial x^1} + \frac{\partial g_{11}}{\partial x^2} - \frac{\partial g_{12}}{\partial x^1} \right) = 0,$$

$$\Gamma^1{}_{22} = \frac{1}{2} g^{11} \left(\frac{\partial g_{12}}{\partial x^2} + \frac{\partial g_{21}}{\partial x^2} - \frac{\partial g_{22}}{\partial x^1} \right) = -\frac{1}{2} g^{11} \frac{\partial g_{22}}{\partial x^1},$$

$$\Gamma^2{}_{11} = \frac{1}{2} g^{22} \left(\frac{\partial g_{21}}{\partial x^1} + \frac{\partial g_{12}}{\partial x^1} - \frac{\partial g_{11}}{\partial x^2} \right) = 0,$$

$$\Gamma^2{}_{12} = \frac{1}{2} g^{22} \left(\frac{\partial g_{22}}{\partial x^1} + \frac{\partial g_{12}}{\partial x^2} - \frac{\partial g_{12}}{\partial x^2} \right) = \frac{1}{2} g^{22} \frac{\partial g_{22}}{\partial x^1},$$

$$\Gamma^2{}_{22} = \frac{1}{2} g^{22} \left(\frac{\partial g_{22}}{\partial x^2} + \frac{\partial g_{22}}{\partial x^2} - \frac{\partial g_{22}}{\partial x^2} \right) = 0.$$

Consequently, the only non-zero values of the six independent connection coefficients listed above are

$$\Gamma^1{}_{22} = -\frac{1}{2} g^{11} \frac{\partial g_{22}}{\partial x^1} = -\sin x^1 \cos x^1 \qquad \text{and} \qquad \Gamma^2{}_{12} = \frac{1}{2} g^{22} \frac{\partial g_{22}}{\partial x^1} = \frac{\cos x^1}{\sin x^1} = \cot x^1.$$

(The only other non-zero connection coefficient is $\Gamma^2{}_{21} = \Gamma^2{}_{12}$.)

Exercise 3.8 From Exercise 3.7(a), $\Gamma^i{}_{jk} = 0$ for all i, j, k in this metric, so Equation 3.27 reduces to

$$\frac{\mathrm{d}^2 x^i}{\mathrm{d}\lambda^2} = 0,$$

giving the solutions $x^i = a_i \lambda + b_i$ for constants a_i, b_i. Writing this as $x(\lambda) = a\lambda + b$ and $y(\lambda) = c\lambda + d$, we see that these equations parameterize the straight line through (b, d) with gradient c/a.

Exercise 3.9 Using our usual coordinates for the surface of a sphere, $x^1 = \theta$, $x^2 = \phi$, and the results of Exercise 3.7(b) for the connection coefficients, Equation 3.27 becomes

$$\frac{\mathrm{d}^2 \theta}{\mathrm{d}\lambda^2} - \sin \theta \cos \theta \left(\frac{\mathrm{d}\phi}{\mathrm{d}\lambda} \right)^2 = 0 \tag{S3.1}$$

and

$$\frac{\mathrm{d}^2 \phi}{\mathrm{d}\lambda^2} + 2 \frac{\cos \theta}{\sin \theta} \frac{\mathrm{d}\theta}{\mathrm{d}\lambda} \frac{\mathrm{d}\phi}{\mathrm{d}\lambda} = 0. \tag{S3.2}$$

(a) The portion of a meridian A can be parameterized by

$$\theta(\lambda) = \lambda, \quad 0 \leq \lambda \leq \frac{\pi}{2},$$
$$\phi(\lambda) = 0,$$

so we have

$$\frac{d\theta}{d\lambda} = 1, \quad \frac{d^2\theta}{d\lambda^2} = \frac{d^2\phi}{d\lambda^2} = \frac{d\phi}{d\lambda} = 0,$$
$$\sin\theta = \sin(\lambda), \quad \cos\theta = \cos(\lambda).$$

Equation S3.1 becomes

$$0 - \sin(\lambda)\cos(\lambda) \times 0 = 0,$$

and Equation S3.2 becomes

$$0 + 2\cot(\lambda) \times 1 \times 0 = 0.$$

So A satisfies the geodesic equations and is a geodesic.

Comment: This is what we would expect, because A is part of a great circle.

(b) B can be parameterized by

$$\theta(\lambda) = \frac{\pi}{2},$$
$$\phi(\lambda) = \lambda, \quad 0 \le \lambda < 2\pi.$$

So we have

$$\frac{d\phi}{d\lambda} = 1, \quad \frac{d^2\phi}{d\lambda^2} = \frac{d^2\theta}{d\lambda^2} = \frac{d\theta}{d\lambda} = 0,$$
$$\sin\theta = 1, \quad \cos\theta = 0.$$

Equation S3.1 becomes $0 - 1 \times 0 \times 1 = 0$, and Equation S3.2 becomes $0 + 2 \times 0 \times 1 \times 0 = 0$. So B satisfies the geodesic equations and is a geodesic.

(c) C can be parameterized by

$$\theta(\lambda) = \frac{\pi}{4},$$
$$\phi(\lambda) = \lambda, \quad 0 \le \lambda < 2\pi.$$

So we have

$$\frac{d\phi}{d\lambda} = 1, \quad \frac{d^2\phi}{d\lambda^2} = \frac{d^2\theta}{d\lambda^2} = \frac{d\theta}{d\lambda} = 0,$$
$$\sin\theta = \cos\theta = 1/\sqrt{2}.$$

Equation S3.1 becomes $0 - (1/\sqrt{2}) \times (1/\sqrt{2}) \times 1 = -1/2 \ne 0$, and Equation S3.2 becomes $0 + 2 \times 1 \times 0 \times 1 = 0$. So C is not a geodesic because it doesn't satisfy both geodesic equations.

Exercise 3.10 (a) Since k is constant at every point on the curve and $k = 1/R$, we have

$$R = \frac{1}{k} = \frac{1}{0.2\,\text{cm}^{-1}} = 5\,\text{cm}.$$

So the best approximating circle at every point on the curve is a circle of radius 5 cm, and the curve itself is a circle of radius 5 cm.

(b) Here again k will be constant, as the straight line has constant 'curvature'. However big we draw the circle, a larger circle will approximate the straight line better, so the curvature of a straight line must be smaller than $1/R$ for all possible R. Hence k must be zero. In other words,

$$k = \lim_{R \to \infty} \frac{1}{R} = 0.$$

Exercise 3.11 The parabola can be parameterized by $x(\lambda) = \lambda$ and $y(\lambda) = \lambda^2$. Consequently,

$$\dot{x} = 1, \quad \ddot{x} = 0, \quad \dot{y} = 2\lambda, \quad \ddot{y} = 2,$$

and for $\lambda = 0$ we have

$$\dot{x} = 1, \quad \ddot{x} = 0, \quad \dot{y} = 0, \quad \ddot{y} = 2.$$

So the curvature at $\lambda = 0$ is

$$k = \frac{|\dot{x}\ddot{y} - \dot{y}\ddot{x}|}{(\dot{x}^2 + \dot{y}^2)^{3/2}} = \frac{|1 \times 2 - 0 \times 0|}{(1^2 + 0^2)^{3/2}} = 2,$$

and the approximating circle has the radius

$$R = \frac{1}{k} = \frac{1}{2}.$$

The centre of the circle is at $x = 0$, $y = 0.5$.

Exercise 3.12 The derivatives of x and y are given by

$$\dot{x} = -a\sin\lambda, \quad \ddot{x} = -a\cos\lambda, \quad \dot{y} = b\cos\lambda, \quad \ddot{y} = -b\sin\lambda,$$

so the curvature is given by

$$k = \frac{|\dot{x}\ddot{y} - \dot{y}\ddot{x}|}{(\dot{x}^2 + \dot{y}^2)^{3/2}} = \frac{ab\sin^2\lambda + ab\cos^2\lambda}{(a^2\sin^2\lambda + b^2\cos^2\lambda)^{3/2}} = \frac{ab}{(a^2\sin^2\lambda + b^2\cos^2\lambda)^{3/2}}.$$

For the circle of radius R we have $a = R$ and $b = R$, so

$$k = \frac{ab}{(a^2\sin^2\lambda + b^2\cos^2\lambda)^{3/2}} = \frac{R^2}{(R^2\sin^2\lambda + R^2\cos^2\lambda)^{3/2}} = \frac{1}{R},$$

which is as expected.

Exercise 3.13 Interchanging the j, k indices in Equation 3.35, we get

$$R^l{}_{ikj} = \frac{\partial\Gamma^l{}_{ij}}{\partial x^k} - \frac{\partial\Gamma^l{}_{ik}}{\partial x^j} + \sum_m \Gamma^m{}_{ij}\Gamma^l{}_{mk} - \sum_m \Gamma^m{}_{ik}\Gamma^l{}_{mj}.$$

Swapping the first and second terms, and the third and fourth terms, leads to

$$R^l{}_{ikj} = -\frac{\partial\Gamma^l{}_{ik}}{\partial x^j} + \frac{\partial\Gamma^l{}_{ij}}{\partial x^k} - \sum_m \Gamma^m{}_{ik}\Gamma^l{}_{mj} + \sum_m \Gamma^m{}_{ij}\Gamma^l{}_{mk}.$$

Comparison with Equation 3.35 shows that the expression on the right-hand side of this equation is $-R^l{}_{ijk}$, hence proving that $R^l{}_{ijk} = -R^l{}_{ikj}$.

Exercise 3.14 From Exercise 3.7(a), all connection coefficients for this space are zero, and hence from Equation 3.35, we have

$$R^l{}_{ijk} = 0.$$

Since the connection coefficients also vanish for an n-dimensional Euclidean space, it follows that the Riemann tensor is zero for such spaces.

Exercise 3.15 From Equation 3.35 and Exercise 3.7(b), we have

$$R^1{}_{212} = \frac{\partial\Gamma^1{}_{22}}{\partial x^1} - \frac{\partial\Gamma^1{}_{21}}{\partial x^2} + \sum_\lambda \Gamma^\lambda{}_{22}\Gamma^1{}_{\lambda 1} - \sum_\lambda \Gamma^\lambda{}_{21}\Gamma^1{}_{\lambda 2}$$

$$= \frac{\partial\Gamma^1{}_{22}}{\partial x^1} - \frac{\partial\Gamma^1{}_{21}}{\partial x^2} + \Gamma^1{}_{22}\Gamma^1{}_{11} + \Gamma^2{}_{22}\Gamma^1{}_{21} - \Gamma^1{}_{21}\Gamma^1{}_{12} - \Gamma^2{}_{21}\Gamma^1{}_{22}.$$

But from Exercise 3.7(b),

$$\Gamma^1_{11} = \Gamma^1_{12} = \Gamma^1_{21} = \Gamma^2_{11} = \Gamma^2_{22} = 0,$$

so

$$
\begin{aligned}
R^1_{212} &= \frac{\partial \Gamma^1_{22}}{\partial x^1} - \Gamma^2_{21}\,\Gamma^1_{22} \\
&= \frac{\partial}{\partial x^1}(-\sin x^1 \cos x^1) - \frac{\cos x^1}{\sin x^1}(-\sin x^1 \cos x^1) \\
&= -\cos^2(x^1) + \sin^2(x^1) + \cos^2(x^1) \\
&= \sin^2 x^1.
\end{aligned}
$$

Exercise 3.16 From the earlier in-text question, we know that $K = a^{-2}$, and from Exercise 3.15,

$$R^1_{212} = \sin^2 x^1.$$

However, from Exercise 3.7(b),

$$[g_{ij}] = \begin{pmatrix} a^2 & 0 \\ 0 & a^2 \sin^2 x^1 \end{pmatrix},$$

so

$$g = \det[g_{ij}] = a^4 \sin^2 x^1.$$

Also, from Chapter 2 we know that lowering the first index on R^1_{212} gives

$$R_{1212} = \sum_{i=1}^{2} g_{1i} R^i_{212} = g_{11} R^1_{212} + g_{12} R^2_{212}.$$

However, $g_{12} = 0$, hence

$$\frac{R_{1212}}{g} = \frac{a^2 \times \sin^2 x^1}{a^4 \sin^2 x^1} = \frac{1}{a^2},$$

which is the same as K.

Exercise 3.17 (a) Just as in Exercise 3.7(a), the connection coefficients are zero since the metric is constant.

(b) Since the connection coefficients for a Minkowski spacetime are zero, as shown in part (a), and each term in the Riemann tensor defined by Equation 3.35 involves at least one connection coefficient, it follows that all components of the Riemann tensor are zero.

Exercise 3.18 (a) The metric is

$$[g_{ij}] = \begin{pmatrix} c^2 & 0 \\ 0 & -f^2(t) \end{pmatrix}$$

and the dual metric is

$$[g^{ij}] = \begin{pmatrix} 1/c^2 & 0 \\ 0 & -1/f^2(t) \end{pmatrix}.$$

As in Exercise 3.7(b), there are only six independent connection coefficients:

$$\Gamma^0_{00}, \quad \Gamma^0_{01}(= \Gamma^0_{10}), \quad \Gamma^0_{11},$$
$$\Gamma^1_{00}, \quad \Gamma^1_{01}(= \Gamma^1_{10}), \quad \Gamma^1_{11}.$$

Moreover,

$$\frac{\partial g_{11}}{\partial x^0} = -2f\dot{f}, \quad \text{where } \dot{f} \equiv \frac{\mathrm{d}f(t)}{\mathrm{d}t},$$

and

$$\frac{\partial g_{ij}}{\partial x^k} = 0$$

for all other values of i, j, k. Also, $g^{il} = 0$ for $i \neq l$, from which we can see that

$$\Gamma^0{}_{00} = \frac{1}{2}g^{00}\left(\frac{\partial g_{00}}{\partial x^0} + \frac{\partial g_{00}}{\partial x^0} - \frac{\partial g_{00}}{\partial x^0}\right) = 0,$$

$$\Gamma^0{}_{01} = \frac{1}{2}g^{00}\left(\frac{\partial g_{01}}{\partial x^0} + \frac{\partial g_{00}}{\partial x^1} - \frac{\partial g_{01}}{\partial x^0}\right) = 0,$$

$$\Gamma^0{}_{11} = \frac{1}{2}g^{00}\left(\frac{\partial g_{01}}{\partial x^1} + \frac{\partial g_{10}}{\partial x^1} - \frac{\partial g_{11}}{\partial x^0}\right) = -\frac{1}{2}g^{00}\frac{\partial g_{11}}{\partial x^0},$$

$$\Gamma^1{}_{00} = \frac{1}{2}g^{11}\left(\frac{\partial g_{10}}{\partial x^0} + \frac{\partial g_{01}}{\partial x^0} - \frac{\partial g_{00}}{\partial x^1}\right) = 0,$$

$$\Gamma^1{}_{01} = \frac{1}{2}g^{11}\left(\frac{\partial g_{11}}{\partial x^0} + \frac{\partial g_{01}}{\partial x^1} - \frac{\partial g_{01}}{\partial x^1}\right) = \frac{1}{2}g^{11}\frac{\partial g_{11}}{\partial x^0},$$

$$\Gamma^1{}_{11} = \frac{1}{2}g^{11}\left(\frac{\partial g_{11}}{\partial x^1} + \frac{\partial g_{11}}{\partial x^1} - \frac{\partial g_{11}}{\partial x^1}\right) = 0.$$

Consequently, the only non-zero values of the six independent connection coefficients listed above are

$$\Gamma^0{}_{11} = -\frac{1}{2}g^{00}\frac{\partial g_{11}}{\partial x^0} = -\frac{1}{2} \times \frac{1}{c^2} \times (-2f\dot{f}) = \frac{f\dot{f}}{c^2}$$

and

$$\Gamma^1{}_{01} = \frac{1}{2}g^{11}\frac{\partial g_{11}}{\partial x^0} = \frac{1}{2} \times \frac{-1}{f^2} \times (-2f\dot{f}) = \frac{\dot{f}}{f}.$$

The only other non-zero connection coefficient is $\Gamma^1{}_{10} = \Gamma^1{}_{01}$.

(b) As in Exercise 3.15,

$$R^0{}_{101} = \frac{\partial \Gamma^0{}_{11}}{\partial x^0} - \frac{\partial \Gamma^0{}_{10}}{\partial x^1} + \sum_\lambda \Gamma^\lambda{}_{11}\Gamma^0{}_{\lambda 0} - \sum_\lambda \Gamma^\lambda{}_{10}\Gamma^0{}_{\lambda 1}$$

$$= \frac{\partial \Gamma^0{}_{11}}{\partial x^0} - \frac{\partial \Gamma^0{}_{10}}{\partial x^1} + \Gamma^0{}_{11}\Gamma^0{}_{00} + \Gamma^1{}_{11}\Gamma^0{}_{10} - \Gamma^0{}_{10}\Gamma^0{}_{01} - \Gamma^1{}_{10}\Gamma^0{}_{11}.$$

Since $\Gamma^0{}_{00} = \Gamma^0{}_{01} = \Gamma^0{}_{10} = \Gamma^1{}_{00} = \Gamma^1{}_{11} = 0$, we have

$$R^0{}_{101} = \frac{\partial \Gamma^0{}_{11}}{\partial x^0} - \Gamma^1{}_{10}\Gamma^0{}_{11} = \frac{\partial}{\partial x^0}\left[\frac{f\dot{f}}{c^2}\right] - \frac{\dot{f}}{f} \times \frac{f\dot{f}}{c^2}$$

$$= \frac{1}{c^2}\frac{\partial}{\partial t}\left[f\dot{f}\right] - \frac{\dot{f}^2}{c^2} = \frac{1}{c^2}\left[\dot{f}\dot{f} + f\ddot{f}\right] - \frac{\dot{f}^2}{c^2}$$

$$= \frac{f\ddot{f}}{c^2}.$$

Exercise 4.1   (a) Suppose that the separation is l and the distance from the centre of the Earth is R, as shown in Figure S4.1.

Then the magnitude of the horizontal acceleration of each object is $g \sin \theta \approx g\theta$, so the total (relative) acceleration is $g2\theta$. However, $2\theta = l/R$, so the magnitude of the total acceleration, a, is given by

$$a = \frac{gl}{R} = \frac{9.81 \times 2.00}{6.38 \times 10^6} \, \text{m s}^{-2} = 3.08 \times 10^{-6} \, \text{m s}^{-2}.$$

(b) Suppose that one object is a distance l vertically above the other object. Since Newtonian gravity is an inverse square law, the magnitudes of acceleration at R and $R + l$ are related by

$$\frac{g(R)}{g(R + l)} = \frac{(R + l)^2}{R^2} = \left(1 + \frac{l}{R}\right)^2 \approx 1 + \frac{2l}{R}.$$

Hence Δg, the difference between the magnitudes of acceleration at R and $R + l$, is given by

$$\Delta g = \frac{2gl}{R} = \frac{2 \times 9.81 \times 2.00}{6.38 \times 10^6} \, \text{m s}^{-2} = 6.15 \times 10^{-6} \, \text{m s}^{-2}.$$

Exercise 4.2

(a) As indicated by Figure S4.2, the coordinates are related by $x = r \cos \theta$, $y = r \sin \theta$.

Setting $(x'^1, x'^2) = (x, y)$ and $(x^1, x^2) = (r, \theta)$, we have

$$\frac{\partial x'^1}{\partial x^1} = \frac{\partial x}{\partial r} = \cos \theta, \qquad \frac{\partial x'^1}{\partial x^2} = \frac{\partial x}{\partial \theta} = -r \sin \theta$$

and

$$\frac{\partial x'^2}{\partial x^1} = \frac{\partial y}{\partial r} = \sin \theta, \qquad \frac{\partial x'^2}{\partial x^2} = \frac{\partial y}{\partial \theta} = r \cos \theta.$$

In this case, the general tensor transformation law reduces to

$$A'^1 = \sum_\nu \frac{\partial x'^1}{\partial x^\nu} A^\nu, \quad \text{and} \quad A'^2 = \sum_\nu \frac{\partial x'^2}{\partial x^\nu} A^\nu.$$

This means that A'^μ and A^μ must be related by

$$A'^1 = \cos \theta \, A^1 - r \sin \theta \, A^2, \quad \text{and} \quad A'^2 = \sin \theta \, A^1 + r \cos \theta \, A^2.$$

(b) In the case of the infinitesimal displacement, this general transformation rule implies that

$$dx = \cos \theta \, dr - r \sin \theta \, d\theta, \quad \text{and} \quad dy = \sin \theta \, dr + r \cos \theta \, d\theta.$$

But this is exactly the relationship between these different sets of coordinates given by the chain rule, so the infinitesimal displacement does transform as a contravariant rank 1 tensor.

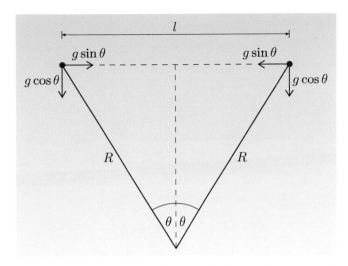

Figure S4.1 Accelerations of horizontally separated masses in a freely falling lift.

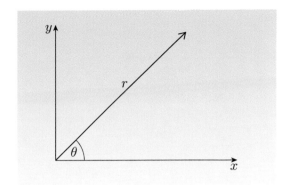

Figure S4.2 Polar coordinates.

Exercise 4.3   We know that

$$A_\mu = \sum_{\alpha=0}^{3} g_{\mu\alpha} A^\alpha.$$

Multiplying by $g^{\nu\mu}$ and summing over μ, we have

$$\sum_{\mu=0}^{3} g^{\nu\mu} A_\mu = \sum_{\mu=0}^{3} \sum_{\alpha=0}^{3} g^{\nu\mu} g_{\mu\alpha} A^\alpha.$$

Reversing the order in which we do the summation on the right-hand side of this equation enables us to write it as

$$\sum_{\mu=0}^{3} g^{\nu\mu} A_\mu = \sum_{\alpha=0}^{3} A^\alpha \sum_{\mu=0}^{3} g^{\nu\mu} g_{\mu\alpha}.$$

However,

$$\sum_{\mu=0}^{3} g^{\nu\mu} g_{\mu\alpha} = \delta^\nu{}_\alpha.$$

Since $\delta^\nu{}_\alpha = 1$ when $\nu = \alpha$ and $\delta^\nu{}_\alpha = 0$ when $\nu \neq \alpha$, we have

$$\sum_{\mu=0}^{3} g^{\nu\mu} A_\mu = A^\nu.$$

Exercise 4.4   (a) There are two reasons. The μ index is up on A^μ but down on B_μ. The K term has no μ index.

(b) The ν index cannot be up on both $Y^{\mu\nu}$ and Z^ν; it must be up on one term and down on the other.

(c) There cannot be three instances of the ν index on the right-hand side of this equation.

Exercise 4.5   Being a scalar, this quantity has no contravariant or covariant indices. So in this particular case, covariant differentiation simply gives

$$\nabla_\lambda S = \frac{\partial S}{\partial x^\lambda}.$$

Exercise 4.6   We know that

$$[\eta_{\mu\nu}] = [\eta^{\mu\nu}] = \begin{pmatrix} 1 & 0 & 0 & 0 \\ 0 & -1 & 0 & 0 \\ 0 & 0 & -1 & 0 \\ 0 & 0 & 0 & -1 \end{pmatrix}$$

and

$$[U^\mu] = \gamma(v)(c, \boldsymbol{v}) = \gamma(v)\left(c, \frac{dx^1}{dt}, \frac{dx^2}{dt}, \frac{dx^3}{dt}\right).$$

Since $U^0 = c$ in the instantaneous rest frame, we have $T^{00} = \rho c^2$. Also, $T^{0i} = 0$ since $\eta^{0i} = 0$ and $U^i = 0$ in this frame. Likewise,

$$T^{ii} = \left(\rho + \frac{p}{c^2}\right)U^i U^i + p = p.$$

Finally, for $i \neq j$,

$$T^{ij} = \left(\rho + \frac{p}{c^2}\right)U^i U^j - p\eta^{ij} = 0$$

since $\eta^{ij} = 0$ for $i \neq j$ and $U^i = 0$ in the instantaneous rest frame.

Exercise 4.7   Multiplying Equation 4.34 by $g^{\mu\nu}$ and summing over both indices, we obtain

$$\sum_{\mu,\nu} g^{\mu\nu} R_{\mu\nu} - \sum_{\mu,\nu} \tfrac{1}{2} R\, g_{\mu\nu}\, g^{\mu\nu} = \sum_{\mu,\nu} -\kappa\, g^{\mu\nu}\, T_{\mu\nu}.$$

Now using the fact that

$$\sum_{\mu,\nu} g^{\mu\nu}\, g_{\mu\nu} = \sum_\nu \delta^\nu{}_\nu = 4,$$

this becomes

$$R - 2R = -\kappa T.$$

Hence $R = \kappa T$, which we can substitute in Equation 4.34 to obtain Equation 4.35:

$$R_{\mu\nu} - \tfrac{1}{2}\kappa T\, g_{\mu\nu} = -\kappa\, T_{\mu\nu},$$

so

$$R_{\mu\nu} = -\kappa\left(T_{\mu\nu} - \tfrac{1}{2}g_{\mu\nu}\, T\right).$$

Exercise 5.1   From the definition of the Einstein tensor,

$$G_{00} = R_{00} - \tfrac{1}{2}g_{00} R$$

and we have

$$R_{00} = -e^{2(A-B)}\left(A'' + (A')^2 - A'B' + \frac{2A'}{r}\right),$$

$$g_{00} = e^{2A}$$

and

$$R = -2e^{-2B}\left(A'' + (A')^2 - A'B' + \frac{2}{r}(A' - B') + \frac{1}{r^2}\right) + \frac{2}{r^2}.$$

So

$$
\begin{aligned}
G_{00} &= R_{00} - \tfrac{1}{2}g_{00}R \\
&= -e^{2(A-B)}\left(A'' + (A')^2 - A'B' + \frac{2A'}{r}\right) \\
&\quad + e^{2(A-B)}\left(A'' + (A')^2 - A'B' + \frac{2}{r}(A' - B') + \frac{1}{r^2}\right) - \frac{e^{2A}}{r^2} \\
&= -e^{2(A-B)}\left(\frac{2B'}{r} - \frac{1}{r^2}\right) - \frac{e^{2A}}{r^2},
\end{aligned}
$$

as required.

Exercise 5.2 (a) The only place where the coordinate ϕ appears in the Schwarzschild line element is in the term $r^2 \sin^2\theta\,(\mathrm{d}\phi)^2$. But since $\phi' = \phi + \phi_0$, the difference in the ϕ-coordinates of any two events will be equal to the difference in the ϕ'-coordinates of those events, and in the limit, for infinitesimally separated events, $\mathrm{d}\phi' = \mathrm{d}(\phi + \phi_0) = \mathrm{d}\phi$. So the Schwarzschild line element is unaffected by the change of coordinates apart from the replacement of ϕ by ϕ'. This establishes the form-invariance of the metric under the change of coordinates.

(b) In a system of spherical coordinates, a given value of the coordinate ϕ corresponds to a meridian of the kind shown in Figure S5.1.

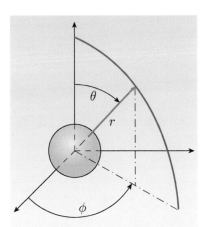

Figure S5.1 Radial coordinates with a (meridian) line of constant ϕ.

The replacement of ϕ by ϕ' effectively shifts every such meridian by the same angle ϕ_0. Since the body that determines the Schwarzschild metric is spherically symmetric, the displacement of the meridians will have no physical significance. Moreover, since each meridian is replaced by another, all that really happens in this case is that each meridian is relabelled, and this will not even change the form of the metric.

Exercise 5.3 We require

$$\frac{\mathrm{d}\tau}{\mathrm{d}t} \geq 1 - 10^{-8}.$$

With $dr = d\theta = d\phi = 0$ the metric reduces to

$$\frac{d\tau}{dt} = \left(1 - \frac{2GM}{c^2 r}\right)^{1/2} \approx 1 - \frac{GM}{c^2 r}, \qquad \text{so} \qquad \frac{GM}{c^2 r} \leq 10^{-8}.$$

Rearranging gives

$$r \geq \frac{GM}{c^2 \times 10^{-8}} = 1.5 \times 10^{11} \text{ metres}.$$

We have not yet found the relationship between the Schwarzschild coordinate r and physical (proper) distance — that is the subject of the next section. Nonetheless it is interesting to note that a proper distance of 1.5×10^{11} metres is about the distance from the Earth to the Sun.

Exercise 5.4   The proper distance $d\sigma$ between two neighbouring events that happen at the same time ($dt = 0$) is given by the metric via the relationship $(ds)^2 = -(d\sigma)^2$. Thus

$$(d\sigma)^2 = \frac{(dr)^2}{1 - \frac{2GM}{c^2 r}} + r^2 (d\theta)^2 + r^2 \sin^2\theta\, (d\phi)^2.$$

For the circumference at a given r-coordinate in the $\theta = \pi/2$ plane, $dr = d\theta = 0$, hence

$$(d\sigma)^2 = r^2 (d\phi)^2.$$

So

$$d\sigma = r\, d\phi \qquad \text{and therefore} \qquad C = \int_0^{2\pi} r\, d\phi = 2\pi r,$$

as required.

Exercise 5.5   It follows from the general equation for an affinely parameterized geodesic that

$$\frac{d^2 x^0}{d\lambda^2} + \sum_{\nu,\rho} \Gamma^0{}_{\nu\rho} \frac{dx^\nu}{d\lambda} \frac{dx^\rho}{d\lambda} = 0.$$

Since the only non-zero connection coefficients with a raised index 0 are $\Gamma^0{}_{01} = \Gamma^0{}_{10}$, the sum may be expanded to give

$$\frac{d^2 x^0}{d\lambda^2} + 2\Gamma^0{}_{01} \frac{dx^0}{d\lambda} \frac{dx^1}{d\lambda} = 0.$$

Identifying $x^0 = ct$, $x^1 = r$ and $\Gamma^0{}_{01} = \dfrac{GM}{r^2 c^2 \left(1 - \frac{2GM}{c^2 r}\right)}$, we see that

$$\frac{d^2 t}{d\lambda^2} + \frac{2GM}{c^2 r^2 \left(1 - \frac{2GM}{c^2 r}\right)} \frac{dr}{d\lambda} \frac{dt}{d\lambda} = 0,$$

as required.

Exercise 5.6   For circular motion at a given r-coordinate in the equatorial plane, u is constant, so

$$\frac{du}{d\phi} = \frac{d^2 u}{d\phi^2} = 0 \quad \text{and also} \quad \frac{dr}{d\tau} = 0.$$

(a) It follows from the orbital shape equation (Equation 5.36) that for a circular orbit with $J^2/m^2 = 12G^2M^2/c^2$,

$$\frac{3GMu^2}{c^2} - u + GM\left(\frac{12G^2M^2}{c^2}\right)^{-1} = 0,$$

that is

$$\frac{3GMu^2}{c^2} - u + \frac{c^2}{12GM} = 0.$$

Solving this quadratic equation in u gives $u = c^2/6GM$, so $r = 6GM/c^2$ is the minimum radius of a stable circular orbit.

(b) The corresponding value of E may be determined from the radial motion equation (Equation 5.32), remembering that $dr/d\tau = 0$:

$$\left(\frac{dr}{d\tau}\right)^2 + \frac{J^2}{m^2r^2}\left(1 - \frac{2GM}{c^2r}\right) - \frac{2GM}{r} = c^2\left[\left(\frac{E}{mc^2}\right)^2 - 1\right].$$

So

$$0 + \frac{12G^2M^2}{c^2}\left(\frac{c^2}{6GM}\right)^2\left(1 - \frac{2GM}{c^2}\frac{c^2}{6GM}\right) - 2GM\frac{c^2}{6GM}$$

$$= c^2\left[\left(\frac{E}{mc^2}\right)^2 - 1\right].$$

Simplifying this, we have

$$\frac{c^2}{3}\left(1 - \frac{2}{6}\right) - \frac{c^2}{3} = c^2\left[\left(\frac{E}{mc^2}\right)^2 - 1\right]$$

that is

$$-\frac{c^2}{9} = c^2\left[\left(\frac{E}{mc^2}\right)^2 - 1\right],$$

which can be rearranged to give $E = \sqrt{8}mc^2/3$.

Exercise 6.1 (a) For the Sun, $R_S = 3\,\text{km}$. So for a black hole with three times the Sun's mass, the Schwarzschild radius is $9\,\text{km}$. Substituting this value into Equation 6.10, we find that the proper time required for the fall is just

$$\tau_{\text{fall}} = 6 \times 10^3/(3 \times 10^8)\,\text{s} = 2 \times 10^{-5}\,\text{s}.$$

(b) For a $10^9\,M_\odot$ galactic-centre black hole, the Schwarzschild radius and the in-fall time are both greater by a factor of $10^9/3$. A calculation similar to that in part (a) therefore gives a free fall time of $6700\,\text{s}$, or about 112 minutes. (Note that these results apply to a body that starts its fall from far away, not from the horizon.)

Exercise 6.2 According to Equation 6.12, for events on the world-line of an outward radially travelling photon,

$$\frac{dr}{dt} = c(1 - R_S/r).$$

For a stationary local observer, i.e. an observer at rest at r, we saw in Chapter 5 that intervals of proper time are related to intervals of coordinate time by $d\tau = dt\,(1 - R_S/r)^{1/2}$, while intervals of proper distance are related to intervals of coordinate distance by $d\sigma = dr\,(1 - R_S/r)^{-1/2}$. It follows that the speed of light as measured by a local observer, irrespective of their location, will always be

$$\frac{d\sigma}{d\tau} = \frac{dr}{dt}\frac{1}{1 - R_S/r}.$$

So, in the case that the intervals being referred to are those between events on the world-line of a radially travelling photon, we see that the locally observed speed of the photon is

$$\frac{d\sigma}{d\tau} = c(1 - R_S/r)\frac{1}{1 - R_S/r} = c.$$

Exercise 6.3   According to the reciprocal of Equation 6.17, for events on the world-line of a freely falling body,

$$\frac{dr}{dt} = -cR_S^{1/2}\frac{1 - R_S/r}{(1 - R_S/r_0)^{1/2}}\left(\frac{r_0 - r}{rr_0}\right)^{1/2}.$$

We already know from the previous exercise that for a stationary local observer,

$$\frac{d\sigma}{d\tau} = \frac{dr}{dt}\frac{1}{1 - R_S/r}.$$

So, in the case of a freely falling body, the measured inward radial velocity will be

$$\frac{d\sigma}{d\tau} = -cR_S^{1/2}\frac{1 - R_S/r}{(1 - R_S/r_0)^{1/2}}\left(\frac{r_0 - r}{rr_0}\right)^{1/2}\frac{1}{1 - R_S/r} = -cR_S^{1/2}\frac{1}{(1 - R_S/r_0)^{1/2}}\left(\frac{r_0 - r}{rr_0}\right)^{1/2}$$

$$= -c\left(\frac{R_S}{(r_0 - R_S)}\times\frac{r_0 - r}{r}\right)^{1/2}.$$

In the limit as $r \to R_S$, the locally observed speed is given by $|d\sigma/d\tau| \to c$.

Exercise 6.4   Initially, the fall would look fairly normal with the astronaut apparently getting smaller and picking up speed as the distance from the observer increased. At first the frequency of the astronaut's waves would also look normal, though detailed measurements would reveal a small decrease due to the Doppler effect. As the distance increased, the astronaut's speed of fall would continue to increase and the frequency of waving would decrease. This would be accompanied by a similar change in the frequency of light received from the falling astronaut, so the astronaut would appear to become redder as well as more distant. The reddening would be increased due to gravitational redshift, though the astronaut's motion would continue to contribute. As the astronaut approached the event horizon, the effect of spacetime distortion would become dominant. The astronaut's rate of fall would be seen to decrease, but the image would become very red and would rapidly dim, causing the departing astronaut to fade away.

Though something along these lines is the expected answer, there is another effect to take into account, that depends on the mass of the black hole. This is a consequence of tidal forces and will be discussed in the next section.

Exercise 6.5   The increasing narrowness and gradual tipping of the lightcones as they approach the event horizon indicates the difficulty of outward escape for photons and, by implication, for any particles that travel slower than light. This effect reaches a critical stage at the event horizon, where the outgoing edge of the lightcone becomes vertical, indicating that even photons emitted in the outward direction are unable to make progress in that direction. A diagrammatic study of lightcones alone is unable to prove the impossibility of escape from within the event horizon, but the progressive narrowing and tipping of lightcones in that region is at least suggestive of the impossibility of escape, and it is indeed a fact that all affinely parameterized geodesics that enter the event horizon of a non-rotating black hole reach the central singularity at some finite value of the affine parameter.

Exercise 6.6   The time-like geodesic for the Schwarzschild case has already been given in Figure 6.11. The nature of the lightcones is also represented in that figure, so the expected answer is shown in Figure S6.1a. In the case of Eddington–Finkelstein coordinates, Figure 6.13 plays a similar role, suggesting (rather than showing) the form of the time-like geodesic and indicating the form of the lightcones. The expected answer is shown in Figure S6.1b.

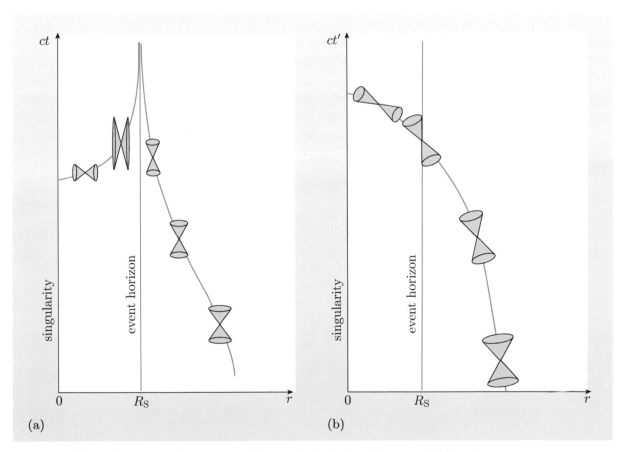

Figure S6.1   Lightcones along a time-like geodesic in (a) Schwarzschild and (b) advanced Eddington–Finkelstein coordinates.

Exercise 6.7   (a) When $J = GM^2/c$, we have $a = J/Mc = GM/c^2 = R_S/2$. Inserting this into Equations 6.32 and 6.33, the second term vanishes and we find

$r_\pm = R_S/2$.

(b) When $J = 0$, we have $a = 0$ and we obtain $r_+ = R_S$, $r_- = 0$.

In both cases (a) and (b), there is only one event horizon as the inner horizon vanishes.

Exercise 6.8 (a) The path indicated by the dashed line in Figure 6.20 shows no change in angle as it approaches the static limit. Space outside the static limit is also dragged around, even though rotation is no longer compulsory. However, a particle in free fall must be affected by this dragging, and so a particle in free fall could not fall in on the dashed line. The path of free fall would have to curve in the direction of rotation of the black hole.

(b) It is possible to follow the dashed path, but the spacecraft would have to exert thrust to counteract the effects of the spacetime curvature of the rotating black hole that make the paths of free fall have a decreasing angular coordinate.

(c) The dotted path represents an impossible trip for the spacecraft. Inside the ergosphere, no amount of thrust in the anticlockwise direction can make the spacecraft maintain a constant angular coordinate while decreasing the radial coordinate.

Exercise 6.9 The discovery of a mini black hole would imply (contrary to most expectations) that conditions during the Big Bang were such as to lead to the production of mini black holes. This would be an important development for cosmology.

Such a discovery would also open up the possibility of confirming the existence of Hawking radiation, thus giving some experimental support to attempts to weld together quantum theory and general relativity, such as string theory.

Exercise 7.1 We first need to decide how many days make up a century. This is not entirely straightforward because leap years don't simply occur every 4 years in the Gregorian calendar. However, it is the Julian year that is used in astronomy and this is defined so that one year is precisely 365.25 days. Consequently we have $36\,525$ days per century, which we denote by d. If we use T to denote the period of the orbit in (Julian) days, then the number of orbits per century is d/T. Equation 7.1 gives the angle in radians, but it is more usual to express the observations in seconds of arc so we need to use the fact that π radians equals 180×3600 seconds of arc. Putting all this together, we find that the general relativistic contribution to the mean rate of precession of the perihelion in seconds of arc per century is given by

$$\frac{\mathrm{d}\phi}{\mathrm{d}t} = \frac{d}{T} \times \frac{6\pi GM_\odot}{a(1-e^2)c^2} \times \frac{648\,000}{\pi} \text{ seconds of arc} = \frac{dGM_\odot}{Ta(1-e^2)c^2} \times 3\,888\,000 \text{ seconds of arc}$$

$$= \frac{36\,525 \times 6.673 \times 10^{-11} \times 1.989 \times 10^{30} \times 3\,888\,000}{87.969 \times 5.791 \times 10^{10} \times (1-(0.2067)^2) \times (2.998 \times 10^8)^2} \text{ seconds of arc per century}$$

$$= 42''.99 \text{ per century}.$$

Exercise 7.2 For rays just grazing the Sun, b is the radius of the Sun, which is $R_\odot = 6.96 \times 10^8$ m, and M is $M_\odot = 1.989 \times 10^{30}$ kg. Hence the deflection in seconds of arc is given by

$$\Delta\theta = \frac{4GM_\odot}{c^2 b} \times \frac{648\,000}{\pi} \text{ seconds of arc} = \frac{6.674 \times 10^{-11} \times 1.989 \times 10^{30}}{(2.998 \times 10^8)^2 \times 6.96 \times 10^8} \times \frac{2\,592\,000}{\pi} \text{ seconds of arc}$$

$$= 1''.75.$$

Exercise 7.3   (a) Let $R_\oplus = 6371.0$ km be the mean radius of the Earth, $M_\oplus = 5.9736 \times 10^{24}$ kg be the mass of the Earth, and $h = 20\,200$ km be the height of the satellite above the Earth. From Equation 5.14, the coordinate time interval at $R_\oplus$ and the coordinate time interval at $R_\oplus + h$ are related by

$$\frac{\Delta t_{R_\oplus + h}}{\Delta t_{R_\oplus}} = \left(\frac{1 - \frac{2M_\oplus G}{c^2(R_\oplus + h)}}{1 - \frac{2M_\oplus G}{c^2 R_\oplus}} \right)^{-1/2}.$$

Since the time dilation is small, we can use the first few terms of a Taylor expansion to evaluate this. Putting $2M_\oplus G/c^2(R_\oplus + h) = x$ and $2M_\oplus G/c^2 R_\oplus = y$, the right-hand side above becomes $(1 - x)^{-1/2} \times (1 - y)^{1/2}$. By a Taylor expansion, this is approximately $(1 + \frac{x}{2})(1 - \frac{y}{2}) \approx 1 + \frac{x}{2} - \frac{y}{2}$. So we have

$$\Delta t_{R_\oplus + h} \approx \left(1 + \frac{M_\oplus G}{c^2(R_\oplus + h)} - \frac{M_\oplus G}{c^2 R_\oplus} \right) \Delta t_{R_\oplus} = \Delta t_{R_\oplus} - \frac{M_\oplus G h}{c^2 R_\oplus (R_\oplus + h)} \Delta t_{R_\oplus}.$$

The discrepancy over 24 hours is given by

$$\Delta t_{R_\oplus + h} - \Delta t_{R_\oplus} = -\frac{5.9736 \times 10^{24} \times 6.673 \times 10^{-11} \times 2.02 \times 10^7}{(2.998 \times 10^8)^2 \times 6.371 \times 10^6 \times (6.371 + 20.2) \times 10^6} \times 24 \times 3600\,\text{s}$$

$$= -45.7\,\mu\text{s}.$$

The negative sign indicates that the effect of general relativity is that the satellite clock runs more rapidly than a ground-based one.

(b) Special relativity relates a time interval Δt for a clock moving at speed v with the time interval Δt_0 for one at rest by

$$\Delta t = \left(1 - \frac{v^2}{c^2} \right)^{-1/2} \Delta t_0.$$

For a satellite of mass m orbiting the Earth at a distance h from the Earth's surface, its speed v is given by

$$\frac{GM_\oplus m}{(R_\oplus + h)^2} = \frac{mv^2}{R_\oplus + h} \qquad \text{therefore} \qquad v^2 = \frac{GM_\oplus}{R_\oplus + h}$$

and hence

$$\Delta t = \left(1 - \frac{GM_\oplus}{c^2(R_\oplus + h)} \right)^{-1/2} \Delta t_0 \approx \left(1 + \frac{GM_\oplus}{2c^2(R_\oplus + h)} \right) \Delta t_0.$$

Hence the discrepancy over 24 hours between satellite- and ground-based clocks is

$$\Delta t - \Delta t_0 \approx \frac{GM_\oplus}{2c^2(R_\oplus + h)} \Delta t_0 = \frac{6.673 \times 10^{-11} \times 5.9736 \times 10^{24}}{2 \times (2.998 \times 10^8)^2 \times (6.371 + 20.2) \times 10^6} \times 24 \times 3600\,\text{s}$$

$$= 7.2\,\mu\text{s}.$$

The positive result indicates that the effect of special relativity is that the satellite clock runs slower than a ground-based one.

(c) The total effect of the results obtained in parts (a) and (b) is a discrepancy between ground-based and satellite-based clocks of $(-45.7 + 7.2) = -38.5\,\mu s$ per day. Since the basis of the GPS is the accurate timing of radio pulses, over 24 hours this could lead to an error in distance of up to

$$c(\Delta t - \Delta t_0) = 2.998 \times 10^8 \times 38.5 \times 10^{-6}\,\mathrm{m} = 11.5\,\mathrm{km}.$$

Exercise 7.4 We can approximate the radius of the satellite's orbit by the Earth's radius. Hence the period of the orbit, T, is given by

$$T = 2\pi\sqrt{\frac{R_\oplus^3}{GM_\oplus}}.$$

Since

$$\frac{GM_\oplus}{c^2 R_\oplus} \approx 10^{-9} \ll 1,$$

Equation 7.13 can be approximated by

$$\alpha \approx 2\pi\left[1 - \left(1 - \frac{3GM_\oplus}{2c^2 R_\oplus}\right)\right] \approx 3\pi\frac{GM_\oplus}{c^2 R_\oplus}.$$

After a time Y, the number of orbits is Y/T and the total precession is given by

$$\alpha_{\text{total}} = \frac{Y}{T} \times 3\pi\frac{GM_\oplus}{c^2 R_\oplus} = \frac{Y}{2\pi}\left(\frac{GM_\oplus}{R_\oplus^3}\right)^{1/2} \times 3\pi\frac{GM_\oplus}{c^2 R_\oplus} = \frac{3Y}{2c^2}\sqrt{\frac{G^3 M_\oplus^3}{R_\oplus^5}}.$$

Converting from radians to seconds of arc, we find that the total precession angle for one year is

$$\alpha_{\text{total}} = \frac{3 \times 365.25 \times 24 \times 3600}{2 \times (2.998 \times 10^8)^2} \times \sqrt{\frac{(6.673 \times 10^{-11})^3 \times (5.974 \times 10^{24})^3}{(6.371 \times 10^6)^5}} \times \frac{180 \times 3600}{\pi} = 8''.44.$$

Exercise 7.5 We have previously carried out a similar calculation for low Earth orbit, the only difference here being that the radius of the orbit is now $R = (6.371 \times 10^6\,\mathrm{m}) + (642 \times 10^3\,\mathrm{m})$ instead of $6.371 \times 10^6\,\mathrm{m}$. Consequently, the expected precession is

$$8''.44 \times \left(\frac{6.371}{7.013}\right)^{5/2} = 6''.64.$$

Exercise 7.6 When considering light rays travelling from a distant source to a detector, it is not just one ray that travels from the source to the detector, but a cone of rays. Gravitational lensing effectively increases the size of the cone of rays that reach the detector. The light is *not* concentrated in the same way as in Figure 7.15, but it is concentrated.

Exercise 8.1 (i) On size scales significantly greater than 100 Mly, the large-scale structure of voids and superclusters (i.e. clusters of clusters of galaxies) does indeed appear to be homogeneous and isotropic.

(ii) After removing distortions due to local motions, the mean intensity of the cosmic microwave background radiation differs by less than one part in ten thousand in different directions. This too is evidence of isotropy and homogeneity.

(iii) The uniformity of the motion of galaxies on large scales, known as the Hubble flow, is a third piece of evidence in favour of a homogeneous and isotropic Universe.

Exercise 8.2   Geodesics are found using the geodesic equation. The first step is to identify the covariant metric coefficients of the relevant space-like hypersurface (only g_{11}, g_{22} and g_{33} will be non-zero). The contravariant form of the metric coefficients will follow immediately from the requirement that $[g_{ij}]$ is the matrix inverse of $[g^{ij}]$. The covariant and contravariant components can then be used to determine the connection coefficients $\Gamma^i{}_{jk}$. Once the connection coefficients for the hypersurface have been determined, the spatial geodesics may be found by solving the geodesic equation for the hypersurface. At that stage it would be sufficient to demonstrate that a parameterized path of the form $r = r(\lambda)$, $\theta = $ constant, $\phi = $ constant does indeed satisfy the geodesic equation for the hypersurface.

Exercise 8.3   The Minkowski metric differs in that it does not feature the scale factor $R(t)$. It is true that $k = 0$ for both cases, and this means that space is flat. But the presence of the scale factor in the Robertson–Walker metric allows spacetime to be non-flat.

Exercise 8.4   We start with the energy equation

$$\frac{1}{R^2}\left(\frac{\mathrm{d}R}{\mathrm{d}t}\right)^2 = \frac{8\pi G}{3}\rho - \frac{kc^2}{R^2}, \tag{Eqn 8.27}$$

and differentiate it with respect to time t. We use the product rule on the left-hand side and obtain

$$\left(\frac{\mathrm{d}R}{\mathrm{d}t}\right)^2 \frac{\mathrm{d}}{\mathrm{d}t}\left(\frac{1}{R^2}\right) + \frac{1}{R^2}\frac{\mathrm{d}}{\mathrm{d}t}\left(\frac{\mathrm{d}R}{\mathrm{d}t}\right)^2 = \frac{8\pi G}{3}\left(\frac{\mathrm{d}\rho}{\mathrm{d}t}\right) - kc^2\frac{\mathrm{d}}{\mathrm{d}t}\left(\frac{1}{R^2}\right).$$

We then use the chain rule to replace $\frac{\mathrm{d}}{\mathrm{d}t}$ with $\left(\frac{\mathrm{d}R}{\mathrm{d}t}\right)\frac{\mathrm{d}}{\mathrm{d}R}$, which gives

$$\left(\frac{\mathrm{d}R}{\mathrm{d}t}\right)^2\left(\frac{\mathrm{d}R}{\mathrm{d}t}\right)\frac{\mathrm{d}}{\mathrm{d}R}\left(\frac{1}{R^2}\right) + \frac{2}{R^2}\left(\frac{\mathrm{d}R}{\mathrm{d}t}\right)\frac{\mathrm{d}}{\mathrm{d}t}\left(\frac{\mathrm{d}R}{\mathrm{d}t}\right) = \frac{8\pi G}{3}\left(\frac{\mathrm{d}\rho}{\mathrm{d}t}\right) - kc^2\left(\frac{\mathrm{d}R}{\mathrm{d}t}\right)\frac{\mathrm{d}}{\mathrm{d}R}\left(\frac{1}{R^2}\right).$$

Then carrying out the various differentiations with respect to R and t, we get

$$-\frac{2}{R^3}\left(\frac{\mathrm{d}R}{\mathrm{d}t}\right)^2\left(\frac{\mathrm{d}R}{\mathrm{d}t}\right) + \frac{2}{R^2}\left(\frac{\mathrm{d}R}{\mathrm{d}t}\right)\left(\frac{\mathrm{d}^2R}{\mathrm{d}t^2}\right) = \frac{8\pi G}{3}\left(\frac{\mathrm{d}\rho}{\mathrm{d}t}\right) + \frac{2kc^2}{R^3}\left(\frac{\mathrm{d}R}{\mathrm{d}t}\right).$$

We then substitute back in for $\frac{1}{R^2}\left(\frac{\mathrm{d}R}{\mathrm{d}t}\right)^2$ in the first term on the left-hand side, using the energy equation again, to get

$$-\frac{2}{R}\left(\frac{\mathrm{d}R}{\mathrm{d}t}\right)\left(\frac{8\pi G\rho}{3} - \frac{kc^2}{R^2}\right) + \frac{2}{R^2}\left(\frac{\mathrm{d}R}{\mathrm{d}t}\right)\left(\frac{\mathrm{d}^2R}{\mathrm{d}t^2}\right) = \frac{8\pi G}{3}\left(\frac{\mathrm{d}\rho}{\mathrm{d}t}\right) + \frac{2kc^2}{R^3}\left(\frac{\mathrm{d}R}{\mathrm{d}t}\right).$$

We now substitute for $\frac{1}{R}\left(\frac{\mathrm{d}^2R}{\mathrm{d}t^2}\right)$ in the second term on the left-hand side, using the acceleration equation (Equation 8.28), to get

$$-\frac{2}{R}\left(\frac{\mathrm{d}R}{\mathrm{d}t}\right)\left(\frac{8\pi G\rho}{3} - \frac{kc^2}{R^2}\right) + \frac{2}{R}\left(\frac{\mathrm{d}R}{\mathrm{d}t}\right)\left[-\frac{4\pi G}{3}\left(\rho + \frac{3p}{c^2}\right)\right] = \frac{8\pi G}{3}\left(\frac{\mathrm{d}\rho}{\mathrm{d}t}\right) + \frac{2kc^2}{R^3}\left(\frac{\mathrm{d}R}{\mathrm{d}t}\right).$$

Now we collect all terms with $\frac{1}{R}\left(\frac{dR}{dt}\right)$ as a common factor, to get

$$\frac{8\pi G}{3}\left(\frac{d\rho}{dt}\right) + \frac{1}{R}\left(\frac{dR}{dt}\right)\left[\frac{2kc^2}{R^2} + \frac{16\pi G\rho}{3} - \frac{2kc^2}{R^2} + \frac{8\pi G\rho}{3} + \frac{8\pi Gp}{c^2}\right] = 0.$$

The terms in $2kc^2/R^2$ cancel out, and dividing through by $\frac{8\pi G}{3}$ gives

$$\left(\frac{d\rho}{dt}\right) + \frac{1}{R}\left(\frac{dR}{dt}\right)\left[2\rho + \rho + \frac{3p}{c^2}\right] = 0,$$

which clearly yields the fluid equation as required:

$$\frac{d\rho}{dt} + \left(\rho + \frac{p}{c^2}\right)\frac{3}{R}\frac{dR}{dt} = 0. \qquad \text{(Eqn 8.31)}$$

Exercise 8.5   The density and pressure term in the original version of the second of the Friedmann equations (Equation 8.28) may be written as

$$\rho + \frac{3p}{c^2} = \rho_m + \rho_r + \rho_\Lambda + \frac{3}{c^2}\left(p_m + p_r + p_\Lambda\right).$$

The dark energy density term is constant (ρ_Λ), and the other density terms may be written as

$$\rho_m = \rho_{m,0}\left[\frac{R_0}{R(t)}\right]^3, \quad \rho_r = \rho_{r,0}\left[\frac{R_0}{R(t)}\right]^4.$$

The pressure due to matter is assumed to be zero (i.e. dust), the pressure due to radiation is $p_r = \rho_r c^2/3$, and the pressure due to dark energy is $p_\Lambda = -\rho_\Lambda c^2$. Putting all this together, we have

$$\rho + \frac{3p}{c^2} = \rho_{m,0}\left[\frac{R_0}{R(t)}\right]^3 + \rho_{r,0}\left[\frac{R_0}{R(t)}\right]^4 + \rho_\Lambda + \frac{3}{c^2}\left(0 + \frac{\rho_r c^2}{3} - \rho_\Lambda c^2\right)$$

$$= \rho_{m,0}\left[\frac{R_0}{R(t)}\right]^3 + \rho_{r,0}\left[\frac{R_0}{R(t)}\right]^4 + \rho_\Lambda + \frac{3}{c^2}\left(\frac{\rho_{r,0} c^2}{3}\left[\frac{R_0}{R(t)}\right]^4 - \rho_\Lambda c^2\right)$$

$$= \rho_{m,0}\left[\frac{R_0}{R(t)}\right]^3 + 2\rho_{r,0}\left[\frac{R_0}{R(t)}\right]^4 - 2\rho_\Lambda, \quad \text{as required.}$$

Exercise 8.6   (a) Substituting the proposed solution into the differential equation, we have

$$\frac{d}{dt}\left(R_0(2H_0 t)^{1/2}\right) = \sqrt{\frac{8\pi G}{3}\rho_{r,0}}\,\frac{R_0^2}{R_0(2H_0 t)^{1/2}}.$$

Evaluating the derivative, we get

$$R_0(2H_0)^{1/2}\frac{1}{2t^{1/2}} = \sqrt{\frac{8\pi G}{3}\rho_{r,0}}\,\frac{R_0}{(2H_0)^{1/2}\,t^{1/2}}.$$

Cancelling the factor $R_0/t^{1/2}$ on both sides and collecting terms in H_0, this yields

$$H_0 = \sqrt{\frac{8\pi G}{3}\rho_{r,0}}, \quad \text{as required.}$$

(b) Using the definition of the Hubble parameter,

$$H(t) = \frac{1}{R}\frac{dR}{dt},$$

we substitute in for $R(t)$ from the proposed solution to get

$$H(t) = \left(\frac{1}{R_0(2H_0 t)^{1/2}} \right) \frac{\mathrm{d}}{\mathrm{d}t} \left(R_0(2H_0 t)^{1/2} \right) = \left(\frac{1}{R_0(2H_0 t)^{1/2}} \right) \frac{R_0(2H_0)^{1/2}}{2t^{1/2}} = \frac{1}{2t},$$

as required. Hence $H_0 = 1/2t_0$, and substituting this into the proposed solution gives

$$R(t_0) = R_0(2H_0 t_0)^{1/2} = R_0 \left(\frac{2t_0}{2t_0} \right)^{1/2} = R_0,$$

again as required.

Exercise 8.7 Setting $\mathrm{d}R/\mathrm{d}t = 0$ and $\rho_{r,0} = 0$ in the first Friedmann equation implies that

$$0 = \frac{8\pi G}{3} \left[\rho_{m,0} \left[\frac{R_0}{R(t)} \right]^3 + \rho_\Lambda \right] - \frac{kc^2}{R^2}.$$

But we already know from Equation 8.50 that ρ_Λ and $\rho_{m,0}$ must have the same sign in this case. Consequently, k must be positive and hence equal to $+1$. Using Equation 8.50, and the first Friedmann equation at $t = t_0$, we can therefore write

$$\frac{8\pi G}{3} \left[\frac{3\rho_{m,0}}{2} \right] = \frac{c^2}{R_0^2},$$

leading immediately to the required result

$$R_0 = \left(\frac{c^2}{4\pi G \rho_{m,0}} \right)^{1/2}.$$

Inserting values for G and c, along with the quoted approximate value for the current mean cosmic density of matter, gives $R_0 = 1.8 \times 10^{26}$ m. Since $1\,\mathrm{ly} = 9.46 \times 10^{15}$ m, it follows that, in round figures, $R_0 = 20\,000$ Mly in this static model. Recalling that a parsec is 3.26 light-years, we can also say, roughly speaking, that in the Einstein model, for the given matter density, R_0 is about 6000 Mpc.

Exercise 8.8 The condition for an expanding FRW model to be accelerating at time t_0 is that $\frac{1}{R} \frac{\mathrm{d}^2 R}{\mathrm{d}t^2}$ should be positive at that time. We already know from Equation 8.50 that the condition for it to vanish is that

$$\Omega_{\Lambda,0} = \frac{\Omega_{m,0}}{2}.$$

Examining the equation, it is clear that the condition that we now seek is

$$\Omega_{\Lambda,0} \geq \frac{\Omega_{m,0}}{2}.$$

Exercise 8.9 In the $\Omega_{\Lambda,0}$–$\Omega_{m,0}$ plane, the dividing line between the $k = +1$ and $k = -1$ models corresponds to the condition for $k = 0$. This is the condition that the density should have the critical value $\rho_c(t) = 3H^2(t)/8\pi G$, and may be expressed in terms of $\Omega_{\Lambda,0}$ and $\Omega_{m,0}$ as

$$\Omega_{m,0} + \Omega_{\Lambda,0} = 1.$$

(i) The de Sitter model is at the point $\Omega_{m,0} = 0$, $\Omega_{\Lambda,0} = 1$.

(ii) The Einstein–de Sitter model is at the point $\Omega_{m,0} = 1$, $\Omega_{\Lambda,0} = 0$.

(iii) The Einstein model has a location that depends on the value of $\Omega_{m,0}$, so in the $\Omega_{\Lambda,0}$–$\Omega_{m,0}$ plane it is represented by the line $\Omega_{\Lambda,0} = \Omega_{m,0}/2$, which coincides with the dividing line between accelerating and decelerating models. The Einstein model also requires $k > 0$, so this corresponds to any location along the green line, to the right of the red dashed line.

Exercise 8.10   The scale change $R(t_{ob})/R(t_{em})$ shows up in extragalactic redshift measurements because the light has been 'in transit' for a long time as space has expanded. To measure changes in $R(t)$ locally requires our measuring equipment to be in free fall, far from any non-gravitational forces that would mask the effects of general relativity. However, the large aggregates of matter within our galaxy distort spacetime locally and create a gravitational redshift that would almost certainly mask the effects of cosmic expansion on the wavelength of light. Nearby stars simply will not participate in the cosmic expansion due to these local effects. Thus a local measurement would not be expected to reveal the changing scale factor — any more than a survey of the irregularities on your kitchen floor would reveal the curvature of the Earth.

Exercise 8.11   The figure of 5 billion light-years relates to the proper distances of sources at the time of emission. For sources at redshifts of 2 or 3, as in the case of Figure 8.2, the current proper distances of the sources are between about 16 and 25 billion light-years. The distances quoted in Figure 8.2 indicate that, in a field such as relativistic cosmology where there are many different kinds of distance, there is a problem of converting measured quantities such as redshifts into 'deduced' quantities such as distances. When such deduced quantities are used, it is always necessary to provide clear information about their precise meaning if they are to be properly interpreted.

Exercise 8.12   Historically, the discovery of the Friedmann–Robertson–Walker models was a rather tortuous process. Einstein initiated relativistic cosmology with his 1917 proposal of a static cosmological model. Einstein's model featured a positively curved space ($k = +1$) and used the repulsive effect of a positive cosmological constant Λ to balance the gravitational effect of a homogeneous distribution of matter of density ρ_m. Later in the same year, Willem de Sitter introduced the first model of an expanding Universe, effectively introducing the scale factor $R(t)$, though he did not present his model in that way. De Sitter's model included flat space ($k = 0$), and a cosmological constant but no matter, so there was nothing to oppose a continuously accelerating expansion of space. In 1922, Alexander Friedmann, a mathematician from St Petersburg, published a general analysis of cosmological models with $k = +1$ and $k = 0$, showing that the models of Einstein and de Sitter were special cases of a broad family of models. He published a similar analysis of $k = -1$ models in 1924. Together, these two publications introduced all the basic features of the Robertson–Walker spacetime but they were based on some specific assumptions that detracted from their appeal. In 1927 Lemaître introduced a model that was supported by Eddington, in which expansion could start from a pre-existing Einstein model. Lemaître later (1933) proposed a model that would be categorized nowadays as a variant of Big Bang theory and he became interested in models that started from $R = 0$.

By 1936 Robertson and Walker had completed their essentially mathematical investigations of homogeneous relativistic spacetimes, giving Friedmann's ideas a more rigorous basis and associating their names with the metric. This set the scene for the naming of the Friedmann–Robertson–Walker models. (Sometimes they are referred to as Lemaître–Friedmann–Robertson–Walker models.)

Acknowledgements

Grateful acknowledgement is made to the following sources:

Figures

Cover image courtesy of NASA/CXC/Wisconsin/D.Pooley and CfA/A.Zezas and NASA/CXC/M.Weiss; Figure 1: Science Photo Library; Figure 1.1: Emilio Segrè Visual Archives/American Institute of Physics/Science Photo Library; Figure 1.4: Science Photo Library; Figures 2.2 and 2.5: Copyright © CERN; Figure 2.6: US Department of Energy/Science Photo Library; Figure 2.8: © Corbis; Figure 4.1: Sheila Terry/Science Photo Library; Figure 5.1: Emilio Segrè Visual Archives/American Institute of Physics/Science Photo Library; Figure 6.1: Physics Today Collection/American Institute of Physics/Science Photo Library; Figure 6.2: Los Alamos National Laboratory/Science Photo Library; Figure 6.3: Emilio Segrè Visual Archives/American Institute of Physics/Science Photo Library; Figure 6.18: Time and Life Pictures/Getty Images; Figure 6.21: Corbin O'Grady Studio/Science Photo Library; Figure 7.10: NASA Marshall Space Flight Center (NASA-MSFC); Figure 7.12: ESA, NASA and Felix Mirabel (French Atomic Energy Commission and institute for Astronomy and Space Physics/Conicet of Argentina); Figure 7.13: http://www.star.le.ac.uk; Figure 7.17: NASA Johnson Space Center Collection; Figure 7.18: ESA, NASA, Richard Ellis (Caltech, USA) and Jean-Paul Kneib (Observatoire Midi-Pyrenees); Figure 7.19: Weisberg, J. M. and Taylor, J. H. (2004) 'Relativistic binary pulsar B1913+16: thirty years of observations and analysis', *Binary Radio Pulsars*, Vol TBD, 2004. ASP Conference Series; Figure 7.20: Volker Steger/Science Photo Library; Figure 8.1: M. Collness, Mount Stromlo Observatory; Figure 8.2: Robert Smith, The 2dF Galaxy Redshift Survey; Figure 8.3: Bennett, C. L. et al. 'First Year Wilkinson Microwave Anistropy Probe (WMAP) Observations: Preliminary Maps and Basic Results', *Astrophysical Journal* (submitted); Figure 8.6: NASA Goddard Space Center; Figure 8.11: Ria Novosti/Science Photo Library; Figure 8.13: Adapted from Landsberg, P. T. and Evans, D. A. (1977) *Mathematical Cosmology*, Oxford University Press; Figure 8.15: Sky Publishing Corporation; Figures 8.17 and 8.18: Mark Whittle, University of Virginia from Carr, B. and Ellis, G. (2008) 'Universe or multiverse?', *Astronomy and Geophysics*, Vol. 49, April 2008.

Index

Items that appear in the Glossary have page numbers in **bold type**. Ordinary index items have page numbers in Roman type.